Implementing
Preventive Maintenance
for Industries The Right Way

(5th Discipline on World Class Maintenance Management)
Series 11

By Rolly Angeles

Implementing Preventive Maintenance for Industries the Right Way
(5th Discipline on World Class Maintenance Management)

10 9 8 7 6 5 4 3 2 1

First Edition: Printed in the Philippines by **Central Books**
Head Office: Phoenix Bldg. 927 Quezon Avenue, Quezon City, Philippines 1101

centralbooks

The National Library of Philippine Catalogue

Kindle Asin Amazon	B09Y114J4Z
Paperback Amazon	979-8802320440
Hardcover Amazon	979-8802369319
Paperback Ingramspark	979-8885260053
Hardcover Ingramspark	979-8885260060

Published by:

Published by Amazon KDP for Kindle Version 2022
Published by Ingram Spark for Hardcover for International Edition 2022

This book was designed and produced by:

RSA Reliability and Maintenance Consultancy Firm
Sta. Rosa, Laguna, Philippines 4026
Website: http://www.rsareliability.com
Email: rollyangeles@rsareliability.com

First Printing: April 2022

Original Concept of World Class Maintenance Management
The 12 Disciplines (Series 1)

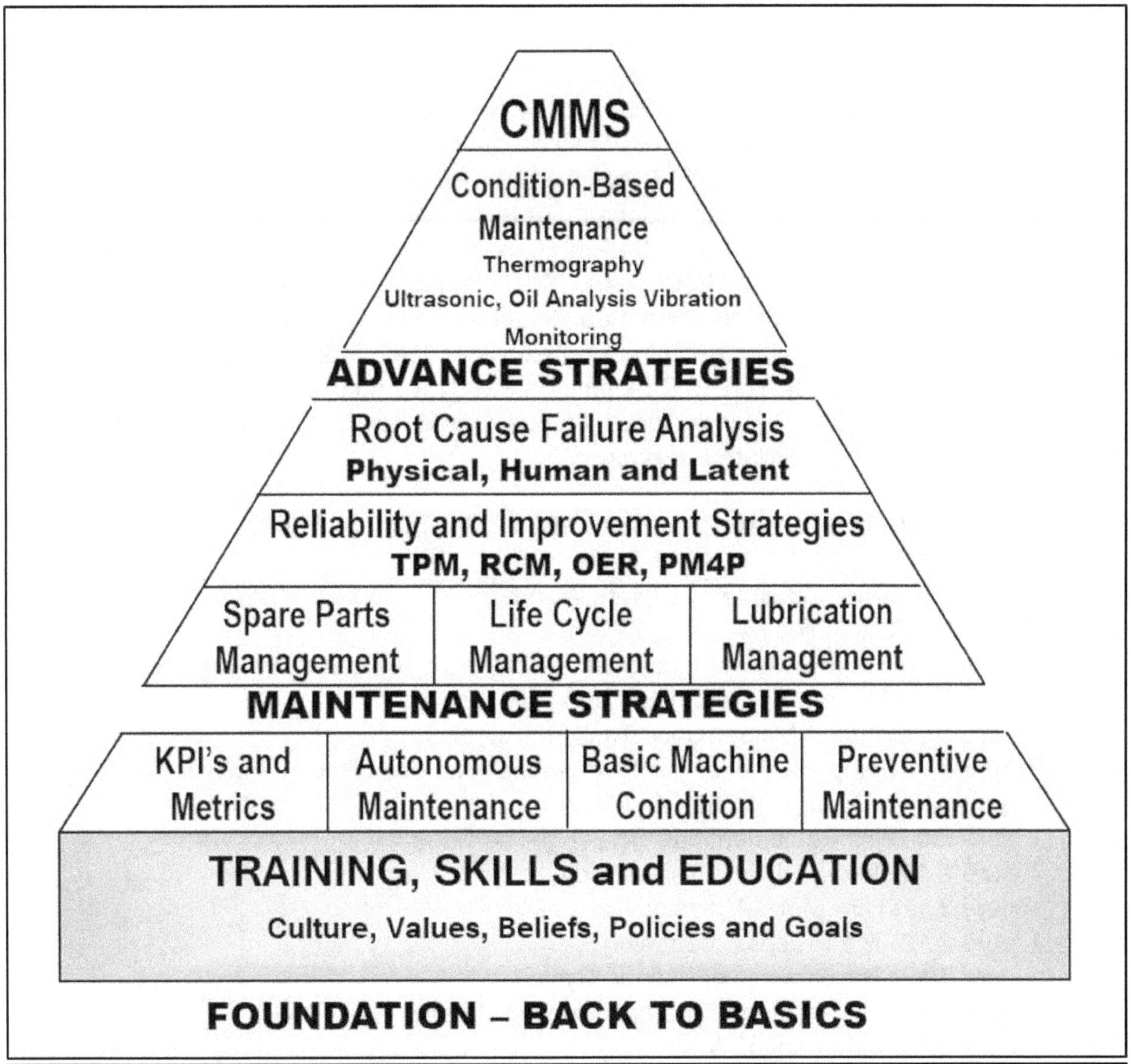

Figure A: Original Concept of World Class Maintenance

By: Rolly Angeles

Figures

Note: For simplification purposes, all pictures, graphs, charts, tables, and drawings in this book will be referred to as figures, which are consecutively numbered below.

Figures

Figures

Figures

Figures

Table of Contents

Table of Contents

Table of Contents

Table of Contents

Acknowledgment

First, I would like to thank all my students who have attended my training in the past. I always believe with all my heart that training is a two-way process. I have to admit that I also learned a great deal from the people I taught in the past. I hope that the knowledge gained from reading this book enlightens the reader in doing Preventive Maintenance the right way and that industries benefit from their PM activities.

Special thanks to my wonderful family especially my three kids: Marie Vic, Kathleen Kay, and Christian Joseph, my wife, Marites, and my granddaughter Kalie.

I would like to mention my mom Cecilia Angeles who recently passed away this July 2021 for the love, care, values, moral support, and wisdom that this wonderful woman has provided me throughout my life. To me, she is the greatest mom I could have and I could not ask for more. She also edited my first three books on World Class Maintenance Management, The 12 Disciplines, Maintenance – Roadmap to Reliability and Reliability – A Shared Responsibility for Both Operators and Maintenance.

Lastly, I would like to thank God Almighty for allowing me to complete this book. May he grant me the time and wisdom to complete the remainder of the books that still need to be written which I hope to serve as my small way of contributing to the knowledge and wisdom to all maintenance mankind in industries.

Trust in the Lord with all your heart, and lean not on your own understanding, in all your ways, acknowledge Him, and He will make your path straight – Proverbs 3.5

About the Author

Rolly Angeles is a seasoned technical and international reliability and maintenance trainer and book author. His portfolio of reliability and maintenance training includes maintenance management and reliability courses on Total Productive Maintenance (TPM), Planned Maintenance, Autonomous Maintenance, Lubrication Strategy, Tribology, Oil Contamination Control, Condition-Based (CBM), Predictive Maintenance, Reliability-Centered Maintenance (RCM), Root Cause Failure Analysis (RCFA), Planned Maintenance, World Class Maintenance Management (WCM), Meaningful Measures of Equipment Performance, and more.

Rolly is a graduate of Mechanical Engineer from Mapua Institute of Technology in the Philippines, batch 1985, and passed the Licensure Board Examination the following year in 1986. With more than 30 years of solid experience, he had worked in various industries from shipping, woodworking, foundry, cast-iron machining, assembly lines, semiconductor manufacturing, and the mining industry. From 1994 to 2002, Rolly worked as a TPM senior engineer at Amkor Technology Philippines, a Multi-National company engaged in the manufacture of integrated circuit products, and spearheaded Amkor's Planned Maintenance Organization, composed of maintenance managers and engineers. He was responsible for the dramatic reduction of their machine's unplanned breakdowns in their TPM journey as well as RCM implementation on their Facilities Air Handling Units (AHU) and their sub-station equipment. Here is where he had gained hands-on experience and understanding of both TPM and RCM, respectively. His last corporate employment was in 2002, where he worked as a technical training specialist at Lepanto Consolidated Mining Industry. In 2005, Rolly retired early from the industry and decided to establish his own consulting business, **RSA Reliability and Maintenance Consultancy Firm**, where he dedicates his time and passion to working as an independent reliability and maintenance consultant. His email is rollyangeles@rsareliability.com and his website is https://www.rsareliability.com. Rolly has written the following books in a series that is all about his passion for Reliability and Maintenance.

- Volume 1: World Class Maintenance Management – The 12 Disciplines
- Volume 2: Maintenance – Roadmap to Reliability
- Volume 3: Reliability – A Shared Responsibility for Both Operators and Maintenance
- Volume 4: Cutting–Edge Maintenance Management Strategies
- Volume 5: Problems and Solutions on MRO Spare Parts and Storeroom
- Volume 6: Lubrication Tactics for Industries Made Simple
- Volume 7: Decoding Reliability-Centered Maintenance Process for Manufacturing Industries
- Volume 8: RSA Reliability and Maintenance Newsletter Vault Collection, Subscribers Edition
- Volume 9: Investigating Equipment Failures through Root Cause Failure Analysis
- Volume 10: Maintenance Indices – Meaningful Measures of Equipment Performance
- Volume 11: Implementing Preventive Maintenance for Industries the Right Way

Take Quiz on Preventive Maintenance Part 1

1. In RCM the activities of performing inspections whether through the use of human senses or gauges will default to
   a) Preventive Maintenance Tasks
   b) Predictive or On-Condition Tasks
   c) Scheduled Maintenance Tasks
   d) Failure Finding Tasks

2. The Basic Equipment Condition includes cleaning the equipment, proper lubrication, addressing leaks and
   a) Parts replacement
   b) Overhauling the machine
   c) Tightening of Bolts
   d) Extending parts' lifespan

3. Other terms to denote Condition-Based Maintenance are also known as
   a) Reliability-Based Maintenance
   b) Predictive Maintenance
   c) On-Condition Tasks
   d) Non-Destructive Tests

4. PM will only apply to patterns (____) of the six failure pattern
   a) Pattern D
   b) Patterns A, B, and C
   c) Patterns D, E, and F
   d) Only Pattern F

5. According to the author, the best people suited for the role of maintenance planners would be the
   a) OEM or Vendor
   b) Reliability People
   c) Engineering
   d) The Rolling Stones (People with White Hair)

6. In RCM, Run to Fail Tasks are not recommended for failures consequences which will result to
   a) Operational Consequences
   b) Hidden Consequences
   c) Safety and Environmental Consequences
   d) Non-Operational Consequences

7. The best approach for determining the interval for replacement and overhauls will be the use of
   a) Age Exploration Method
   b) Using MTBF
   c) Using On-Condition Tasks
   b) Using Guestimates

Take Quiz on Preventive Maintenance

8. In RCM, Preventive Maintenance Task will include Schedule Restoration and
   a) Scheduled Discard Tasks
   b) Calibration
   c) Inspection for Hidden Devices
   d) Inspection through the Use of Human Senses

9. If 80% of 100 impellers or more will reach a lifespan of 2 years, the best maintenance tasks to perform will be
   a) Preventive Maintenance
   b) Predictive Maintenance
   c) Failure Finding Tasks
   d) On-Condition Tasks

10. Preventive Maintenance is designed to capture
    a) Infant Mortality Failures
    b) Random Failures
    c) Hidden Failures
    d) Age-Related Failures

11. The most feasible maintenance tasks for failure modes that will provide signs or symptoms that it is on the verge of failing will be
    a) Predictive Maintenance Tasks
    b) Corrective Maintenance Tasks
    c) Scheduled Overhaul Tasks
    d) Scheduled Restoration Tasks

12. These maintenance tasks can be integrated with Preventive Maintenance to minimize the chances of Infant Mortality Failures
    a) On-Condition Tasks
    b) Precision Maintenance
    c) Corrective Maintenance
    d) Redesign or Modification

13. In determining the interval for conducting Failure Finding Tasks or Functionality Inspection, the following are needed, Availability and (_____).
    a) MTTR
    b) MTBA
    c) MTBF
    d) OEE

14. What we are looking for in conducting Predictive Maintenance tasks is to establish
    a) Root Cause of Failure
    b) Structure of Failure
    c) P-F Interval
    d) MTBF and Failure Rate

15. In RCM, the term No Scheduled Maintenance refers to
    a) Run to Fail

b) Breakdown Maintenance
c) Corrective Maintenance
d) Reactive Maintenance

16. One unique feature of implementing an operator's Autonomous Maintenance is the application of;
a) MTBF Analysis
b) Redesign and Modification
c) Poke-Yoke
d) Visual Controls

17. A Preventive Maintenance Indicator that will indicate the percentage of completed tasks divided by the overall Scheduled Planned tasks is;
a) PM Compliance
b) PM Effectiveness
c) Replacement Asset Value (RAV)
d) PM versus Breakdown Ratio

18. According to the survey on Preventive Maintenance, the top problem on Preventive Maintenance is;
a) PM is waived by operations
b) Lack of training
c) Infant Mortality Failures
d) Ageing Workforce

19. One technique to reduce infant mortality failure is to adapt;
a) Preventive Maintenance
b) Precision Maintenance
c) Corrective Maintenance
d) Proactive Maintenance

20. According to the author of this book, which Preventive Maintenance tasks are prone to human errors?
a) Calibration Tasks
b) Replacement Tasks
c) Overhauling Tasks
d) Routine Inspection Tasks

21. An Oil Analysis test to determine that a part or item is starting to wear out is by monitoring the oil thru;
a) Spectroanalysis and Ferrography
b) TAN/TBN
c) Particle Counter
d) Absolute Viscosity Test

22. The main cause of Infant Mortality Failures is,
a) Operator Error
b) Human Error

c) Improper Tools Used
d) Delayed Delivery of Spare Parts

23. MRO Analysis which tracks the movement of Spare Parts is
a) VED Analysis
b) SDE Analysis
c) ABC Analysis
d) FSNO Analysis

24. The secret to the success of Autonomous Maintenance will be the
a) Maintenance Planner
b) JIPM Consultant
c) Planned Maintenance
d) Quality Maintenance

25. This Oil Analysis instrument will indicate the rate of wear for mechanical parts inside the equipment
a) Particle Counter
b) NOACK Volatility Test
c) Karl Fisher Titration Test
d) Ferrography and Spectroanalysis

26. For industries implementing Autonomous Maintenance, the longest step will be
a) Step 1: Initial Cleaning
b) Step 2: Address Sources of Contamination
c) Step 3: Establish Cleaning and Lubrication Standards
d) Step 4: Develop Equipment Inspection Standard and Training
e) Step 5: Autonomous Maintenance Standards
f) Step 6: Systematic AM and Manage the Workplace
g) Step 7: Practice Full-Self Management

27. When implementing Autonomous Maintenance, consolidation of maintenance tasks for operators and maintenance will happen on Step;
a) Step 1: Initial Cleaning
b) Step 2: Address Sources of Contamination
c) Step 3: Establish Cleaning and Lubrication Standards
d) Step 4: Develop Equipment Inspection Standard and Training
e) Step 5: Autonomous Maintenance Standards
f) Step 6: Systematic AM and Manage the Workplace
g) Step 7: Practice Full-Self Management

28. What makes RCM implementation unique and more precise is that it derived the maintenance task based on.
a) OEM recommendations
b) Operating Context
c) Production Needs
d) Government Recommendations

29. According to the Circadian Rhythm most human errors occur during the following;
   a) Normal time
   b) Day Shift
   c) Morning Shift
   d) Night Shift

30. For components with redundancies or standby, the recommended switching interval between the duty and standby should be;
   a) Weekly
   b) Monthly
   c) Duty to run for 6 months and Standby to run for 1 month
   d) Run the duty until it fails and switch to standby

Take Quiz on Preventive Maintenance Part 2

1. It is recommended that the Maintenance Planner should just be part-time so that they can perform other tasks on maintenance
   a) True
   b) False

2. PM is designed to address both random and infant mortality failures.
   a) True
   b) False

3. A good PM program will eliminate all breakdowns on the equipment.
   a) True
   b) False

4. All parts after consistent use will reach a point of wear and tear.
   a) True
   b) False

5. A bearing that fails randomly at any given period is likely a candidate for Preventive Maintenance scheduled replacement.
   a) True
   b) False

6. To be more effective, both Precision and Predictive Maintenance should be integrated into the Preventive Maintenance Structure.
   a) True
   b) False

7. All critical parts should be stocked in the storeroom.
   a) True
   b) False

8. The best indicator to measure the effectiveness of Preventive Maintenance will be PM Compliance.

a) True
b) False

9. All identical equipment with the same model, maker, and vendor will receive the same maintenance tasks whether these are weekly, monthly, quarterly or yearly tasks. This means that the PM task needs to be identical since the equipment is the same.
   a) True
   b) False

10. One of the reasons why Preventive Maintenance is costly is that it uses the concept of JIC or Just In Case and may replace parts that are still in working condition.
    a) True
    b) False

11. According to the Author of this book, maintenance is a profit center?
    a) True
    b) False

12. According to the book on TPM World Congress, the number 1 reason most TPM initiatives fail is due to a lack of management support and commitment.
    a) True
    b) False

13. According to the author of this book, the decision on whether to stock or not to stock parts in the storeroom will be based on the criticality of the part or item.
    a) True
    b) False

14. According to the author of this book, the most important KPI or indicator for the storeroom is the inventory accuracy which should be 98% and above.
    a) True
    b) False

15. According to a survey by James Reasons and Alan Hobbs, most human errors in Preventive Maintenance occur during the reassembly process of overhauls.
    a) True
    b) False

16. Planners should focus on future works and not on reactive works.
    a) True
    b) False

17. In RCM, inspection using the human senses is considered to be part of the On-Condition tasks.
    a) True
    b) False

18. The best course of action in deriving the Preventive Maintenance tasks of our equipment and assets is to follow OEM recommendations.

a) True
b) False

19. OEE is a good measurement to indicate the effectiveness of executing PM Tasks.
    a) True
    b) False

20. A CMMS software is much bigger in scope than an EAM or Enterprise Asset Management software.
    a) True
    b) False

Preface: Those Were the Times

When I graduated in 1985, finding work was difficult and scarce during that time. A single position of cadet engineer will have at least a couple of hundred applicants or more. Just imagine your chance of being selected for the job. I still remember typing my resume and sending postal mail. The majority of the time, the industry will reply via postal mail which I will receive in a couple of weeks indicating that they currently have no opening. There was no internet during that time and every Sunday I need to buy a newspaper called Manila Bulletin since this is where most industries will post their job vacancies.

John, a friend of mine in high school told me if I wanted to work as an apprentice in their ship in the engine room as his Dad owns a big shipping business. We both came from the same high school at Don Bosco Makati. His father interviewed me and wanted me to act as a Preventive Maintenance Specialist (although in real life, I worked as an Oiler). I do not know what it meant but it was something nice to hear from anyone's ear and I accepted the offer. We work in shifts and there were 6 shifts in a day where I worked from 4 am to 8 am and from 4 pm to 8 pm again in the evening. The other oilers were experienced people and I was new and it was my first taste of work. They always advised me to look dirty so that the Chief Engineer (the highest paid person from the engine room) would think that I am working my sweat out. I followed their advice and place some black diesel oil in my overhauls. When we need the Chief Engineer in the engine room, I make sure to pour some old black diesel oil in my hands before knocking on his room making sure the oil stains will mark his door. Those were the times. CMMS, EAM, or these smart sensors today or any form of automation does not exist and we do everything manually. When I enter the engine room, we need to perform a walk-around inspection of the generator making sure it contains fuel, check the main fuel tank of the ship, inspect the bulbs to ensure they are working, and all those kinds of stuff. There were 2 Mitsubishi DAIYA Diesel Engines as the ship has a twin-propeller. We experienced problems a few times where all engineers and oilers needed to be in the engine room since only one engine is running which means that only one propeller is functioning cutting the speed of the ship in half. When this happens, the ship's Captain will go down to the engine room asking how much time is expected for the other engine to run. It was always a heated argument between the Captain and the Chief Engineer. After a year of working on the ship, I decided that I would be better off working on land than on the sea. This was where I first experience myself working in maintenance.

After working on the ship, I was finally hired in a plant that manufactured sewing machines and was assigned to the woodworking station, where we have all sorts of cutting and sanding machines for knock-down types of furniture. The same advice has been given to me by those pioneer supervisors that my clothes need to be dirty at all times so that the Operations Manager would think that I was working my sweat out once again.

Sometimes if I come to reminisce about those times, we were always reactive. We have no standards and our only goal in life on the ship is to bring the ship from one port to another at all means. I remember one oiler placing an engine oil in the gearbox in which his reason is that the machine is better off with oil than none. Time had passed and with all the technology, information, internet, and books on maintenance (my books included), it seems that little or nothing has ever changed and the majority of industries are still reactive. I hope that it would not be the case in the distant future for industries. This is the main reason why I decided to write this book about Preventive Maintenance.

This book refers to Discipline Number 5 based on the original book I wrote in 2009 on World Class Maintenance – The 12 Disciplines. Sharing my thoughts about this book, I think that I should have written this book before my other books since this is one of the basic disciplines. Writing this book is quite challenging as I need to reminisce about my experiences, close my eyes and think of what is really happening right now in industries. Despite so many consultants (Myself included), experts, training, webinars, information, automation, and publications about maintenance, the truth is that the majority of industries are still reactive. Almost all industries have their Preventive Maintenance, but still, many problems emerged and the question is can we do something about it. Many maintenance people from industries are not satisfied with their current Preventive Maintenance even if they are complying with every single maintenance task listed. My sole intention in writing this book is to provide a guide and direction to help maintenance in industries get the results in doing the correct Preventive Maintenance on their equipment and assets. In today's digitalization era, and with all these advanced CMMS and EAM software, apps, technology, cloud, smart sensors, Industrial Internet of Things, and automation, the majority of industries are jumping on the bandwagon thinking that these technologies will optimize their asset. I have nothing against technology, but what I am stating is that everything will boil down to one thing and that is addressing the basics first. The goal and objective of having Preventive Maintenance is to prevent or anticipate a failure or breakdown from happening first. The keyword in this case is being first. This means that we need to perform an activity or task before failure happens. But the thing is we need to be precise in what failures can be addressed by PM since not all failures can be prevented in the first place. This book includes 15 chapters and is summarized accordingly.

Chapter 1: Changing The Way We Think About PM: Explains the traditional belief of most maintenance people in industries about PM. This is one of the main reasons why industries still remain reactive. Another important topic in this chapter is an updated count of the top 10 problems on PM. This survey was consistently being done religiously during my training classes where the delegates came from different industries, countries, cultures, and races where they select the top 3 problems that currently exist in their plant. Learn why Preventive Maintenance is costly in most industries and what can we do about it.

Chapter 2: Understanding The Concept of Preventive Maintenance: Decipher what Preventive Maintenance is all about. This chapter also unravels the process of wear, how it occurs, and the most common types of wear that can occur on mechanical components. A thing of interest is that there is a wide range of terms and nomenclatures industries used for the different maintenance tasks. A good example of this is the term Outage where which is a common term for power plants. This chapter also explains the role of different people in the maintenance organization and how they are all connected to serve a common goal. This means that a failure of one will be a failure of the entire maintenance organization. Just like a chain, its strength lies in its weakest link.

Chapter 3: Building a Solid Preventive Maintenance Structure in the Plant: Clarifies the three main elements of a Preventive Maintenance Structure, which include Preparation, Execution, and Feedback. The preparation is a list of activities that need to be done before executing Preventive Maintenance. The execution involves the different maintenance tasks performed on Preventive Maintenance. The feedback is the activities done after the PM has been executed so that we can further improve the way we execute PM for future works.

Chapter 4: Planning and Scheduling for PM: Clarifies the concept of planning and its role in PM. While most planners in industries are overwhelmed and drowned by the number of repair works that must be done, what is important is that planners should focus more on future works and not on reactive works. Planning is not always perfect, it is a continuous process that can be improved through feedback from the people who execute the PM tasks. What is important is that the planner should communicate and collaborate with the PM crew continuously since these are the people who will carry out the task. If there are adjustments that need to be made to the plan, then this must be looped back to the planner.

Chapter 5: Integrating Precision Maintenance in PM: This chapter deals with the subject of Precision Maintenance and how it can be integrated into conducting PM. It also discussed the main requirements needed to adopt Precision Maintenance effectively in the plant. Training is very important because this is where we acquire the knowledge to build the skills and develop mastery over the subject. Mastery is achieved once we can transfer our knowledge to other maintenance people in the organization. Precision Maintenance is not actually a maintenance task such as PM or PdM. Still, it should be included and integrated as part of Preventive or any other tasks on maintenance, including repair and what benefit it can provide if Precision Maintenance is implemented correctly in the plant. The author also believes that implementing this maintenance strategy will definitely reduce infant mortality as well as random failures. Precision Maintenance simply means that whoever is performing the task whether the person is the most or least experienced in the craft should provide the same outcome and results.

Chapter 6: PM Interval, Are We Doing PM Too Late or Too Soon? While many industries derived their maintenance interval from guestimates or OEM recommendations, this chapter provides a summary on how to determine the correct interval for the different tasks on maintenance such as routine Preventive Maintenance replacement, scheduled overhauls, greasing bearings, an interval for failure finding tasks, or functionality inspections, frequency of switching interval for redundant components, and interval for performing Predictive Maintenance tasks on the equipment and assets. This chapter also explains why MTBF cannot be used to determine the interval for PM overhauling or replacements. Lastly, we also explain how we can further refine several PM tasks.

Chapter 7: The Role of MRO Spare Parts on PM: MRO Spare parts and Storeroom will play an important role in Preventive Maintenance. This means that if a part is not available for PM the PM will stall. The most important thing to consider in the Storeroom is the lead time to deliver the parts needed by the user. This chapter also covers several factors before stocking them in the storeroom. An MRO Algorithm or Decision diagram is provided with sample case studies to decide if parts need to be stocked or not inside the storeroom. Practical tips to shorten the travel time to the storeroom are provided in this chapter.

Chapter 8: Involving Operators in Maintenance: Explains the importance of having operators that are knowledgeable about the equipment. This chapter also discusses the basic maintenance tasks and activities the operator should be doing, such as routine lubrication, cleaning, and having equipment with no missing or loose bolts and no leaks. Three standards will be performed by the operator which will include cleaning, lubrication, and inspection standards. Once these standards are finalized, they will be consolidated with the PM task so that further duplication can be avoided. This Chapter also explains the key secret to implementing Autonomous Maintenance successfully and the transformation of operators from just operating to those who know their equipment intimately. Autonomous Maintenance will take time to implement, but if correctly implemented, then it can transform operators from those who are merely routine switch flickers to those who are empowered.

Chapter 9: The Role of Predictive Maintenance on PM: This chapter explains the concept of Predictive Maintenance tasks. How it differs from Preventive Maintenance. Why is the P-F interval important to any of these Predictive Maintenance users? This chapter also explains a slight difference between Condition-Based Maintenance and Predictive Maintenance. Other topics covered in this chapter include vibration monitoring, things to consider when buying infrared thermography, ultrasonics, and the benefits of conducting an oil analysis program. Finally, we end this chapter with the crucial role Predictive Maintenance will play to make Preventive Maintenance more effective.

Chapter 10: Deriving the Maintenance Tasks: RCM and other streamlined versions can be used to develop and derive the PM tasks but the starting point is to understand how the equipment will be operated in the plant. This is called the operating context or the condition in which the equipment will be operated. Another important point to consider is that all maintenance tasks should address a particular failure mode and where to source these failure modes. This chapter also details what the PM task should include as well as the different functions that will be involved in executing Preventive Maintenance

Chapter 11: Measuring PM Effectiveness and Performance: This chapter examines the different maintenance indicators that can be used to measure our efforts in Preventive Maintenance such as PM Compliance, PM Effectiveness, Ratio of Preventive Maintenance versus Breakdown Maintenance, Maintenance Cost, Percentage of Maintenance Cost to RAV (Replacement Asset Value), Maintenance Backlog, Wrench Time, MTBF and Reliability. Also detailed in this chapter are some weaknesses of these measurements. Some useful measurements and indicators for Predictive Maintenance are likewise explained.

Chapter 12: Automating the PM Tasks and Performance: Explains that although CMMS or EAM software can benefit maintenance, most industries that have this software are often underutilized. What is important is to understand what is it that we want to automate and if the system is capable. This maintenance software such as CMMS is only as good as what we populate in them. This chapter also illustrates what are the important things and information that should be automated for Preventive Maintenance besides the Work Orders generated. Also included are the important things to automate in the storeroom that is needed for Preventive Maintenance.

Chapter 13: FAQs and Tips on PM is a collection of Frequently Asked Questions and answers on Preventive Maintenance. Some of these questions have been raised during my training in the past. This chapter also provides some helpful tips that industries can apply to improve their existing Preventive Maintenance activities.

Chapter 14: Improving Existing Preventive Maintenance Program: This chapter provides detailed guidelines for improving the current and existing Preventive Maintenance for industries which includes inspection activities, routine, and major Preventive Maintenance shutdown, routine greasing, and procedures for conducting a functionality inspection for protective devices.

Chapter 15: The Conclusion: Finally, this chapter concludes that maintenance should not be treated as a cost or even as a profit center because maintenance is both. Rather it is better if we treat maintenance as a business since in any business there will always be an investment but what is important is that whatever we invest should have a return or payback since this is where we generate the profit. This chapter conceals solutions to the top ten problems on PM and lastly, it states that PM is not only for the planner and those who will execute them but several functions of the organization also have their role and responsibility that they need to understand.

Chapter 1

Changing The Way We Think About PM

Our traditional belief leaves maintenance people to think that all parts have a life and that replacing or overhauling these parts before they fail on an assumed period will eventually restore these parts back to its original condition. If you think that this is right, then you are so wrong. This thinking is the main reason why most Preventive Maintenance activities result to being reactive itself.

1.1: The Traditional Belief in Preventive Maintenance

The traditional belief of most maintenance people is that if a certain part is used consistently, it will definitely reach a point where the part will eventually wear out; therefore, overhauling or replacing the part before it fails on an assumed schedule will ensure the equipment's reliability. By doing this, the equipment will continue to run and operate as expected. The truth is that not all parts will eventually wear out. Just like people, especially during this **covid 19** pandemic days, not everyone will reach their natural life expectancy. Meaning not everyone will become a grandfather or grandmother. This means that not all human beings from the United States or other countries will reach 79 years; others will die young, while others will even die before they will be born. Japan can be considered one of the most advanced countries in the world, but its rate of people committing suicide is also high. This means that these people will not reach their natural lifespan. Just like in industries, the problem is simple, but people make it complicated. In maintenance, we assume that all parts will eventually wear out; that is why our primary defense on our equipment is to perform intrusive Preventive Maintenance to prevent failures and this had been the culture of the industry since the beginning of time. The more breakdowns, the more repairs are experienced and the people's time is eaten up by troubleshooting the equipment, which has been their day-to-day task in the plant.

In figure 1.1, let us assume that the x-axis represents the time or the period, while the y-axis represents the rate of wear or rate of deterioration. The traditional belief on maintenance is that if the equipment is running continuously, 7 days a week, 30 days a month, 365 days a year. Maintenance will assume that definitely, something will fail. Maintenance will assume that the failure will happen at Point 4 and because of this thinking, maintenance will create a contingency plan and specify several maintenance tasks that should be done on the equipment. They will execute these activities either on Point 2 or Point 3. What is important is that these tasks must be done before the said failure. These activities will include replacement and overhauls, where the equipment will be stopped completely for these tasks to commence. What maintenance is thinking is that these parts should be replaced before they fail so that production will not be interrupted. However, when these tasks are finally completed and the machine is returned back to the operators, it seemed that the operators are having a hard time running their equipment. What's on the mind of the operator is that, if maintenance did not touch this I bet this will be running smoothly without any problems. Technically, the reason why we performed these tasks is for the equipment to run smoothly, but the problem is that the opposite happens. So how can we explain this phenomenon? In fact, if the reader is from maintenance, perhaps you have experienced this event yourself. If yes, the truth is you are not alone.

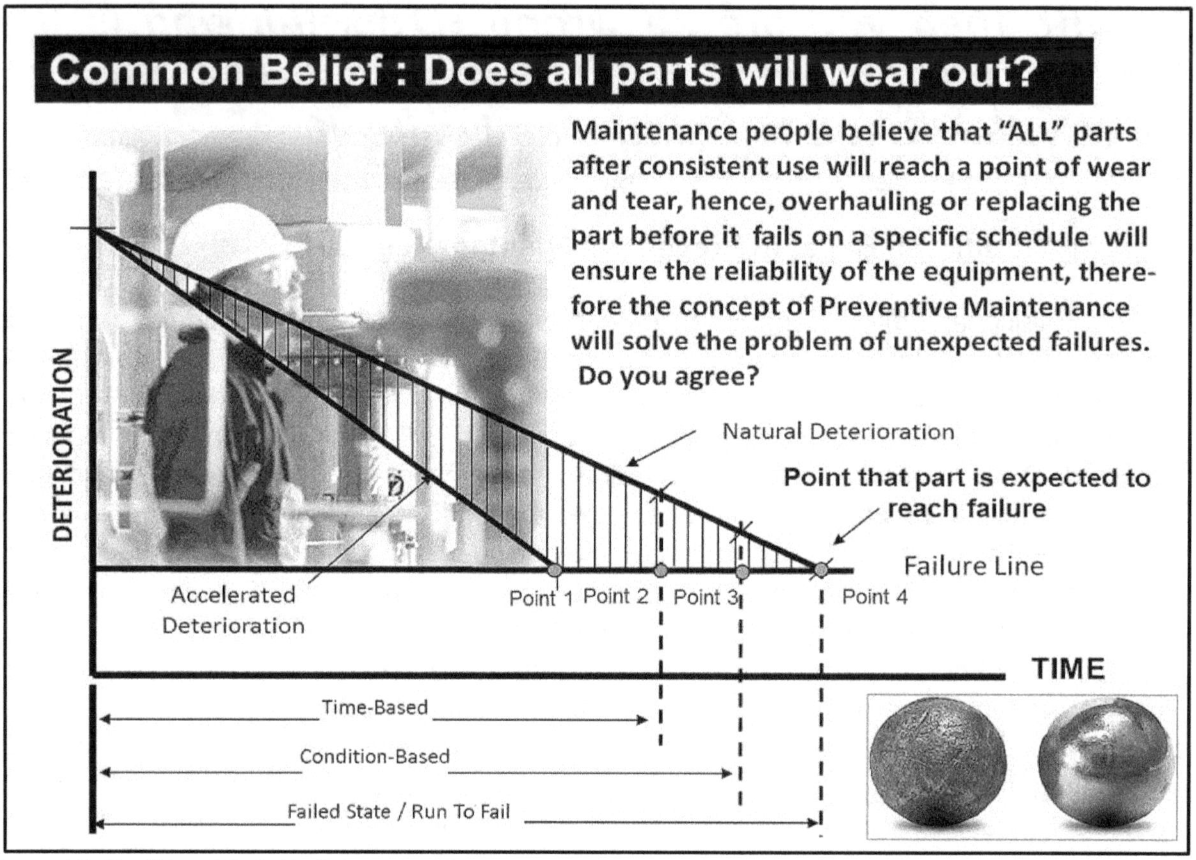

Figure 1.1: Traditional Belief in Preventive Maintenance -1

Everything can be traced back to one thing, and that is our traditional belief in maintenance. According to the book of John Moubray on RCM II page 143, he quotes that [1]it is also borne out by the machine operator who said that every time maintenance works on it over the weekend, it takes up to Wednesday to get it going again. If we analyze this case, the scheduled maintenance tasks will be done on a weekend, which is a Saturday, and Sunday,

[1] Moubrey, John, ***Reliability-Centred Maintenance II***, Butterworth-Heinemann Page 143

but the machine will take up to Wednesday to run smoothly. This means that the operators are having a difficult time running the equipment on Monday and Tuesday. In reality, many operators experienced that after performing scheduled Preventive Maintenance on their equipment, they are having a hard time operating their equipment and machines. But isn't it supposed to be that it should be the other way around? If we performed a scheduled Preventive Maintenance in our equipment, it should be running smoothly, but the thing is right after conducting Preventive Maintenance tasks, especially when replacement and overhauling are present, operators complain that they find it difficult operating the equipment. These early failures are called infant mortality failures. Many factors contribute to infant mortality failures, but the main contributor to Infant Mortality Failure is human error.

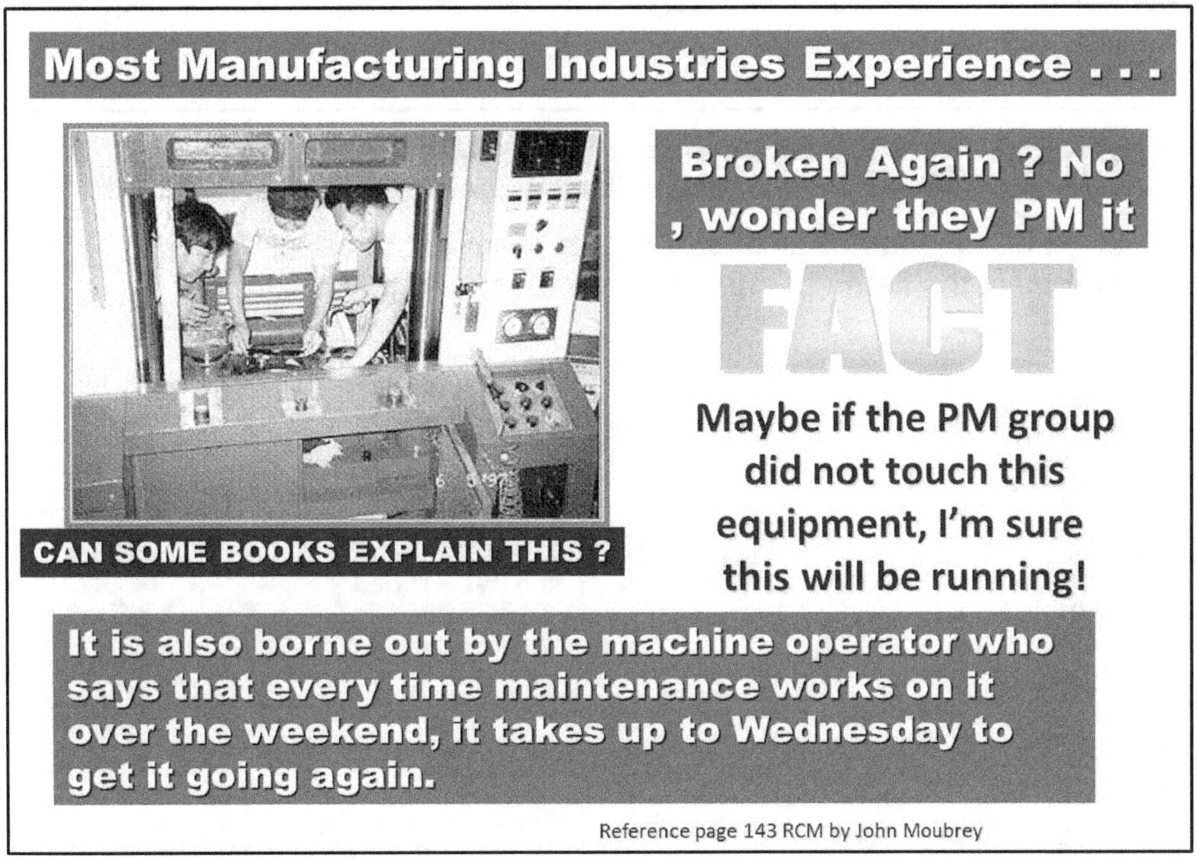

Figure 1.2: Traditional Belief in Preventive Maintenance -2

There are two common types of human errors that can happen in this case. It can either be a slip or a lapse. A slip occurs when somebody does something incorrectly; for example, an electrician wires a motor incorrectly that it runs backward. These are human actions that take place which was not intended. The occurrence of a lack of one's train of thought is derived from unconscious behavior. On the other end, lapses occur when someone misses a key step in a sequence of events or activities. For example, a mechanic leaves a tool behind after working in a machine or simply forgets to fit a key component while reassembling it. Some say it has something to do with the process of aging as we grow old. Both slips and lapses happen because the people were distracted, preoccupied, thinking of something else, or the person was simply absent-minded. Putting this in another perspective, a slip occurs when you need to perform step 1, step 2, step 3, step 4, and step 5, but what you did was step 1, step 2, step 3, step 5, and step 4. A lapse occurs when you need to perform step 1, step 2, step 3, step 4, and step 5, but what you did was step 1, step 2, step 3, and step 5, where you missed out on step 4. According to James Reasons and Alan Hobbs on Managing Maintenance Error, 64.5% of human errors associated with maintenance tasks involved the omission of necessary maintenance tasks. Their findings were from the airline industry. In a series of interviews with the maintenance aircraft crew, they conducted throughout the years. The largest single category was omission, which accounted for around 48% of all human errors. The scary thing about human errors in the airline industry is that most of these human errors can lead to fatality. In fact, one of the key indicators for FAA (Federal Aviation Administration) is the Accidents per Million Take-offs, which means that for every 1,000,000 planes that will take off on the runway how many will go down and fail.

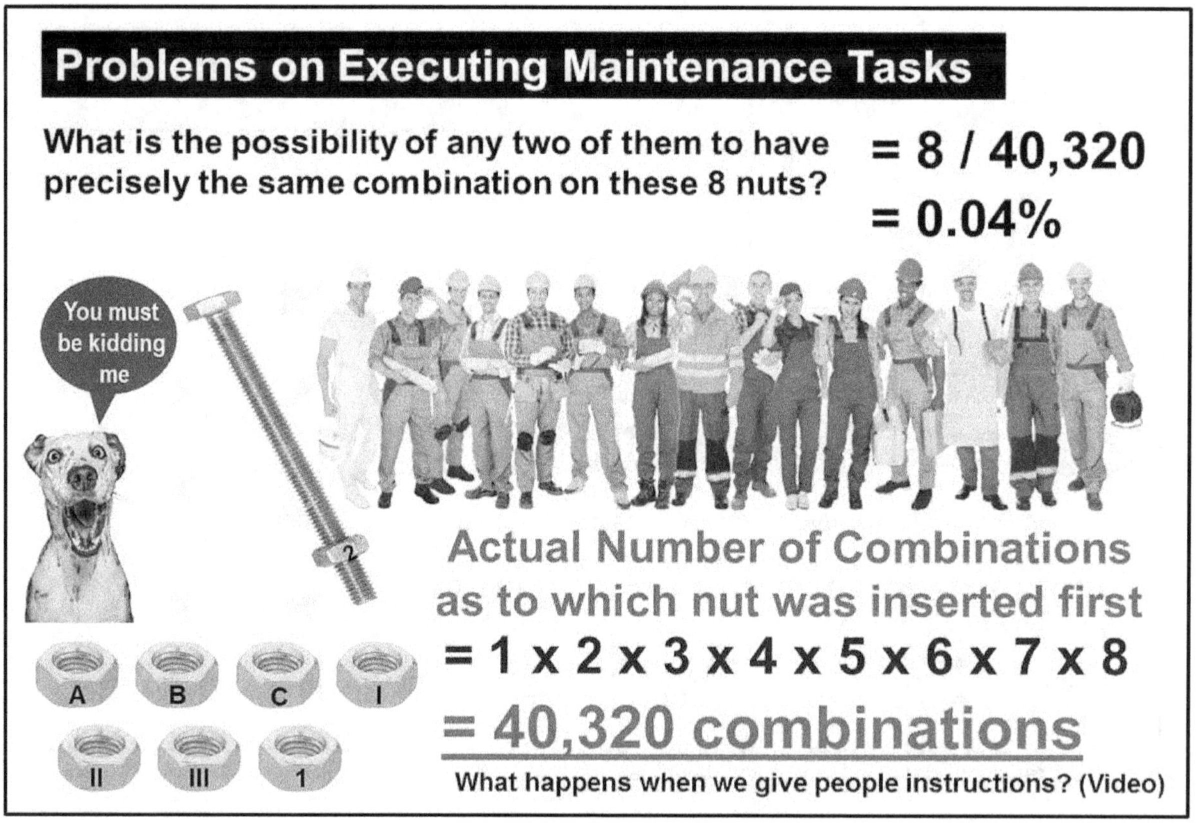

Figure 1.3: How Many Combinations there is for One Bolt and Eight Nuts

If I may explain this in the simplest way I can, let us say that I have one long bolt with eight nuts. I have marked each nut with a code. I have fifteen people with me, and I asked the first person to place all the eight nuts on the bolt. I asked them to write their sequence on which bolt came first and so on. After the person completes assembling and inserting the 8 nuts on the bolt my instruction is to disassemble the nuts once again and give them to the next person with the same instructions. Once the next person completes, he passed on to the next person until all the fifteen people complete the assembly and disassembly process. At the end of this simple exercise, my conclusion is that it is unlikely that any two people or more will have the same precise combination of what was the first nut they placed until the last nut. Why? The reason is simple, if we speak about one bolt and eight nuts, there will be at least around 40,320 combinations in this case and we are only speaking about one bolt and eight nuts, what more if we speak about entirely overhauling the equipment.

When we overhaul the equipment, there will be two main activities. First is dismantling the equipment and finally reassembling it once again. The chances of human error occurring will be negligible during the dismantling process, but when maintenance will attempt to reassemble the equipment the way it was originally, this is where human error usually happens. You do not have to be a rocket scientist to figure this out; all you need is a modicum of common sense.

Most maintenance people believe that the more often an item is overhauled, the less likely it is to fail. In reality, this is just the opposite. The truth of the matter is that these scheduled overhauls and replacements can actually increase the chances of early failures by introducing what we called infant mortality into otherwise stable system. This means that if maintenance is not equipped with the correct knowledge, tools, and skills to disassemble and reassemble the equipment or have some doubts about returning it back to its original condition, then think twice or thrice before dismantling. Perhaps the best thing to do is leave the equipment alone. Do not disturb stable systems unless we are equipped with the confidence to bring the equipment back to its original condition. Performing Preventive Maintenance for the mere fact of complying with the scheduled maintenance can create the possibility that we might be doing more harm than good to the equipment.

Again, if I may explain in the simplest way I can if you stay in a 5-star hotel, there are many rooms on each floor along the aisle. Almost all the rooms are exactly identical except for the bed. All the rooms for a specific floor, let us say the 3rd floor will have its own TV, toilet, bed, pillow, chair, desks, mirror, window, sofa, ac, and whatnot, but there will be one thing that will be different inside in every room, and that is the person occupying that room. This means that the person inside the room can either place a sign outside the door, saying, **Do Not Disturb, or Clean My Room**. If the sign says do not disturb, the hotel staff must respect the sign and avoid any means of disturbing the person inside that room, as the person may come from another country with a different time zone and is adjusting to the time. What I am saying is, that the same thing goes for our equipment and assets. There may be sensitive portions in our equipment that need not be disturbed, especially electronics or instances where the clearance of two mechanical parts should be precise since if we disturb them, we might not be able to put them back in their original condition, which can cause the equipment to fail or malfunction.

Another important point we need to consider in performing Preventive Maintenance is what we called the operating context of the equipment. The pioneers of RCM, Stanley Nowlan and Howard Heap discovered two important points that change the thinking on Preventive Maintenance worldwide. First, resulting schedules are used for all similar assets again, without considering that different consequences apply in different operating contexts. Second, there are many items for which there is no effective form of scheduled maintenance. What this simply means is that when we have identical assets in the plant, the same model, maker, OEM, and type, we assume that the maintenance requirements will be exactly and precisely identical, which may not be the case based on its actual operating context or how the equipment is being operated. Let us consider these three cases.

Case 1: Three Identical Pumps: Let us take a sample of three identical pumps in figure 1.4. One is a stand-alone pump. Let us call this pump A, while the other pump has a standby pump connected in parallel with the duty pump. Let us call these pumps B and C. These three pumps are exactly identical with the same flow rate, model, maker, and specs. The question is, will these three pumps receive the same identical maintenance tasks? We can say that Pump A will be the critical one of these three pumps since it is a stand-alone pump. If Pump A fails, then the operation stops, while for Pump B and C, we can tolerate the failure to happen, or we can have a switching interval for Pump B since if it fails, operations will not be affected since it has redundancy or standby pump. For Pump C, which is the standby pump, we can occasionally run it to check if it is still running and rotate it weekly to avoid the chances of false brinelling which could happen on the raceway of the bearing. This simply explains that the conditions in which the equipment is being operated should be taken into account in determining the most feasible maintenance tasks and that similar equipment and assets might actually require different maintenance tasks even if they are exactly identical, just like the person inside the hotel room.

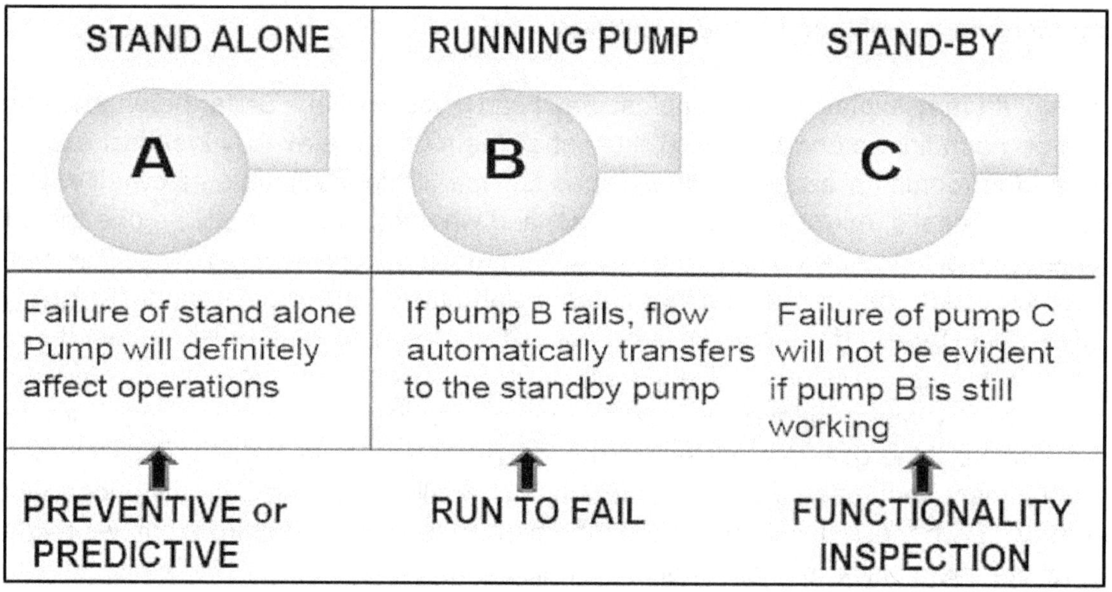

Figure 1.4: The Case of Three Identical Pumps

Case 2: Three Identical Cars: In figure 1.5, we have 3 identical cars. Let us assume that the three cars are brand new. It states that Car A will be traveling with a distance of 60 kilometers daily, Car B at 20 kilometers daily with bumper-to-bumper traffic, and Car C at 600 kilometers weekly, averaging around 86 kilometers daily. Therefore if we answer the following questions, what would you think will be the correct answer?

- Does their degree of maintenance requirements be the same?
- Does the scheduled maintenance frequency be the same?
- Which car will require the most amount of maintenance?

If we speak about the maintenance tasks and the frequency of conducting the tasks for the three identical cars, it would not be the same. When we ask which car will require the most amount of maintenance, we might be tempted to select Car C since it required the longest distance that the car will be traveling, but if you live in my country, the Philippines, we have the worst traffic conditions in the world. In fact, with Car B, expect bumper-to-bumper traffic when you travel, even for a short distance of 20 km/day. This means that you will be pressing the brake pads more frequently, so the brake pads and drum will wear out faster. If you ask me which car will require the most amount of maintenance? To be honest, I do not know the answer to this question. Why? Because we are only looking at one side of the story. To determine precisely which car will require the most amount of maintenance tasks, we need to understand other factors besides the distance. This will include the following:

Figure 1.5: Which Car Will Require the Most Maintenance Tasks?

- How far will the car travel? (Given)
- How much time will it consume?
- What is the traffic condition?
- Is the car loaded or carrying something when traveling?
- What sort of road or terrain will the car travel?
- What is the average speed of the vehicle?

- How about the skills of the driver?
- Does the driver have good eyesight?
- Does the driver perform routine maintenance in his car?
- Will the vehicle be traveling day or night or both?
- What is the average temperature in the city?
- How many passengers will the car carry?

Understanding how the car will be operated, and knowing the answer to these questions then only can we decide which car will require the most amount of maintenance tasks. It is quite premature to assume that Car B or Car C will require the most amount of maintenance. Another problem experienced mostly in Preventive Maintenance is replacing perfectly good parts with new parts since maintenance assumed that it would fail after reaching a specific period. In most cases, the assumption is based on guestimates. This is what makes Preventive Maintenance more costly for industries because PM adheres to the concept of JIC. Meaning if the boss asks us why we replace the part, our response is JIC or Just in Case, it might fail. Doing Preventive Maintenance does not guarantee that the parts that will be replaced really need to be replaced. The thing is, if the equipment is in good running condition, then it should run and continue operating. Basic maintenance, such as keeping the equipment clean, routine lubrication, and inspection, will be the best course of action to sustain the equipment.

Case 3: The Bulb in your House: I used to ask the students in my class if they have a bulb in their house, which they use regularly, perhaps around 10 watts, 20 watts, 30 watts, and whatnot. All of them would respond and say yes. When I ask them if they perform any form of daily inspection on the bulb to check if it is working or still capable of providing illumination, then the delegates in the class will laugh and say, of course not, or others might be thinking, is this guy speaking stupid or something? Some will even think that it is the dumbest thing to do. The logic is simple, if the bulb in your house is busted, we would replace it with a brand new one. In industries, we termed this as reactive maintenance, run to fail, run to destruct, breakdown maintenance, unplanned maintenance, or whatever your industry termed it to be. When I told the people in my class that in 1986, I too have a bulb in my home, and we used to inspect these bulbs not once, not twice, but 6 times a day. When I asked the class if I am stupid or something for inspecting the bulb 6 times a day, their reply was a resounding yes! Perhaps some delegates will not only think that I'm stupid but my brain is also busted. I tell my class that I have my reasons for doing it, and I will prove to them beyond a reasonable doubt that I am not stupid, nor crazy in any way. So let me explain.

I tasted my first work in 1986 as an Oiler (although that company gave me a good title as Preventive Maintenance Specialist) in a shipping company. It was a cargo ship named MV Sea Raider. That was my home for a year and my first taste of experience in maintenance. We have twenty-two crew, and they were my family during that time. The Captain of the ship served as our father since he was the highest-ranking officer on the ship. I belong to the engine room, and my duty was from 4 am to 8 am in the morning. It will resume from 4 pm to 8 pm again in the afternoon till evening. There were six shifts in a day where we were required to work two four-hour shifts a day, totaling eight hours.

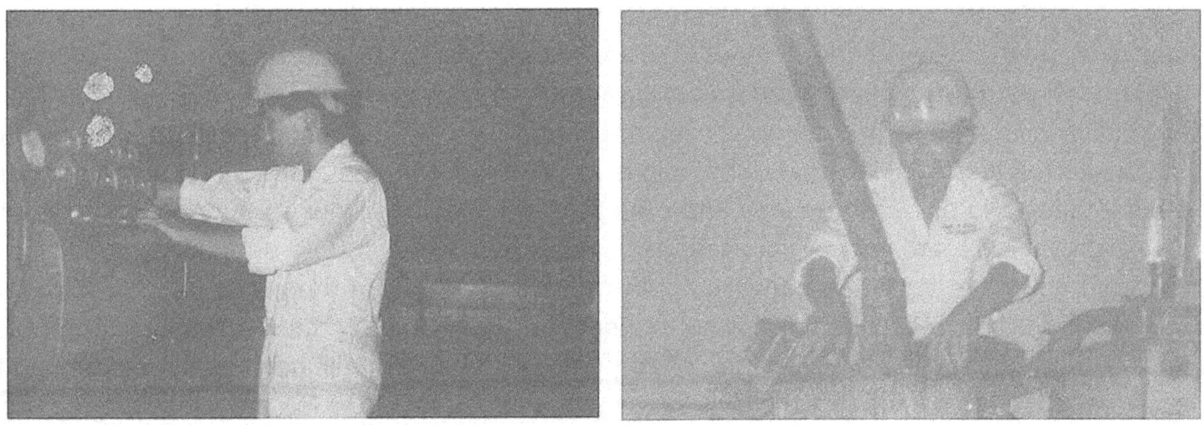

Figure 1.6: Myself Working at MV Sea Raider as an Oiler in 1986

Inside the engine room were two generators and two 750-horsepower diesel engines. It means that this cargo ship has two propellers. The back was a wall made of steel sheets with six bulbs; three bulbs were located on the port side, while the other three bulbs were located on the wall's starboard side. In the shipping industry, port means left and starboard means right. Each of these bulbs has a switch where we can turn it on and off. This is exactly the same bulb that you have in your house. The only difference is that we paint the two bulbs green, the other two bulbs red, and the remaining two bulbs yellow, just like in a traffic light. When we're on a voyage, I wake up at 3:30 am, take my morning coffee and breakfast, then go down to the engine room five minutes before 4:00 am for my duty. When I go down to the engine room, we made a walk-around to check everything based on the checklists, monitor the engines, check the gages, fuel of the generator, and so on and forth whatnot. One activity that we perform religiously during the start of the shift was to inspect each of the six bulbs on the wall one by one if they are still working or not and log them on the record book. We always keep an abundance supply of these bulbs, especially if we were on the voyage.

Perhaps in your house, these bulbs do not need to be inspected, but not in my case. Why? Because we also use the PM Concept of Just in Case. This means that if the ship is sailing, both the port and starboard green bulb is lit on, then the ship is in motion and moving forward at its designated speed. Still, if the green bulb blinks continuously, then the captain or the officers on deck want us to increase the rpm to speed up the ship, and we comply. If the green bulb lit off, and the yellow bulb lit up automatically, then this means that eyes were on the bulb. If the yellow bulb blinks, it means that we need to be alert, our eyes wide open, and we cannot even blink for a single millisecond. Whoever was on duty in that engine room should be closely watching the bulbs. When both the port and starboard indicate that the red bulb is lit up, then we need to stop both port and starboard engines and place them in idle motion where the engines were still running. If the red bulb blinks continuously, then that will be the scariest command you can ever imagine. It means that you need to have both engines in full reverse immediately, without wasting a single second. There is only one thing in your mind if you see the red bulb blinking, it means that the ship will hit something forward, which can cause the ship to sink, and the lives of all the 22 crew on board, myself included, will be in imminent danger. We will be completely in danger in which we can lose the ship and the crew, myself included. Remember that ships have no breaks like a car, and the only way to stop a ship is to reverse the engines. After finishing my story, the people in the class were silent; I asked them again if I am stupid or something for inspecting the bulb six times a day and fulfilling my duties and responsibilities. The answer was a resounding no, why because the consequence of failure on the red bulb greatly differs from the bulb you have in your house even if these two bulbs are exactly and precisely identical. If you were the one in my situation way back in 1986 and part of your responsibility is to check the bulb during your shift and log it, would you have done the same? Again, we are speaking of the same bulb here, just like the one in your house. Imagine failing to inspect the red bulb, and it was busted, and no one knows, what do you think will be the consequences? Well, I'm still here today and writing this book because we did inspect the bulb religiously 6 times a day. This explains what operating context is all about.

We may have the same bulb in our house and in the ship where I previously worked, but the difference has something to do about its operating context or how the bulb is being operated, and what will be the consequences if the bulb fails on its own. In this case, the failure of the red bulb or any of these bulbs was considered hidden, and the only time we will know if it is hidden is if we struck something which can damage the ship and put the lives of the crew in danger. It means that the failure of any of the bulbs, most especially the red bulb, will put the ship and the crew in a very dangerous situation. If we neglect to inspect the bulb, then we have no way of knowing if these bulbs were still working or not. These bulbs in the ship I previously worked for served as a protective device. If we speak about the equipment in industries, this case holds true for equipment's alarm, sensors, LED (light-emitting diode), emergency stop buttons, lighting arrester, and the like. Suppose we do not perform any functionality inspection or failure finding tasks on these protective devices; then, in that case, there is no way of knowing if they are still working on your equipment or not.

Industries may have identical equipment with the same type, make, and model. Before concluding that the tasks will be the same since we have similar equipment types, we need to factor out if their operating context is also identical. This means that if we have 10 identical machine models with the same type and machine 1 requires 20 maintenance tasks to be done for a semi-annual PM, it does not necessarily mean that the remaining 9 machines will also have exactly 20 activities. They may have fewer or more activities depending upon their operating context or how the equipment or machine is operated in the plant as in figure 1.7. Ask yourself the following;

- Are all these 10 machines running at the same time?
- Are all these 10 machines operating for 24 hours continuously?
- Are any of these 10 machines serving as a backup or standby?
- Are the environmental conditions to which these 10 pieces of equipment are exposed identically?
- Are all operators equipped with the same skills?
- Are all settings and parameters on the equipment identical?
- Are all machines producing identical products with the same running hours?
- Are all the operators' twins? And even if they do, are their DNA exactly identical?

PERFORMING EQUIPMENT ON IDENTICAL EQUIPMENT

DETERIORATIONS UNCOVERED	1	2	3	4	5	6
• Air leak in the system	●	●	X	X	○	○
• Excessive belt tension	●	●	●	○	●	○
• Excessive vibration on the fan wheel	●	●	●	○	X	○
• Soft foot on the base foundation	●	●	●	X	X	○
• Worn out bearing	●	X	●	X	●	○
• Worn out drive coupling	●	X	○	○	●	○
• Excessive grease in ball bearings	●	○	●	●	○	○
• Excessive misalignment of shaft	●	○	●	○	●	○
• Abnormal end thrust load	●	X	X	X	X	X
• Incorrect direction of rotation	●	X	X	X	X	X

○ ⟶ Restoration still ongoing ● ⟶ Restoration completed X ⟶ Deterioration not present

Figure 1.7: Consider the Operating Context When Performing PM on Similar Machines

1.2: Top Ten Problems on Preventive Maintenance

Have you ever read a book, or watched a movie where the beginning of the story starts with the end. Others refer to this as a framed story or perhaps a flashback. Just like this book, instead of starting with an introduction to Preventive Maintenance, I started off with the problems of Preventive Maintenance. The reason is that everyone is familiar with these tasks. Almost all industries are doing PM, but if we talked about if industries are currently happy and satisfied with their PM, or when we talked about the outcome of performing PM task 80% of the respondents or more will that it has a negative outcome, or they are not satisfied with their current Preventive Maintenance program. This means that industries still experience many problems and breakdowns right after implementing Preventive Maintenance tasks on their equipment and assets. Although everyone agrees that Preventive Maintenance is important, yet many industries also admit to the fact that many are still in reactive mode and are trapped in the never-ending cycle of stop the bleeding syndrome mode.

I started this survey in May of 2009 when I first wrote my first book, in which I have included this as part of my training activities on World Class Maintenance and other training on reliability and maintenance. I believe that it is about time to conclude this survey about the top three problems in Preventive Maintenance. When we did this survey in my class, my instructions were to select at least three problems from the top 10 lists they currently experience on their plant's Preventive Maintenance activities. This survey was conducted during my public, in-house, and recently during my online training. The delegates in my training classes come from different industries, countries, races, and cultures. When I published my book on Cutting-Edge Maintenance Management, the number 1 problem with PM was the lack of training. As of March 2022, lack of training still seems to be the number 1 problem, followed by PM being waived by operations or other departments. For those who have not read my other books, let me briefly explain each of the problems experienced in Preventive Maintenance.

SURVEY ON TOP TEN PROBLEMS ON PREVENTIVE MAINTENANCE FINALLY CONCLUDED (March 2022)

No	Problem on Preventive Maintenance	WCM Book		MRR Book		OAM Book		CEMMS Book		RCM Book	
		As of May 2009	Rank	As of Oct. 2016	Rank	As of June 2018	Rank	As of Mar 2020	Rank	As of March 2022	Rank
1	Add on PM Checklist Syndrome	35	5th	159	6th	334	5th	357	6th	381	6th
2	Introduction to Infant Mortality Failures	39	3rd	190	4th	343	4th	391	4th	409	5th
3	Replacement of good parts to conform with PM	28	7th	141	7/8th	262	7th	293	7th	310	7th
4	The case of Random Failures	33	6th	141	7/8th	259	8th	284	8th	301	8th
5	Ageing Workforce, maintenance nearing retirement	22	8th	74	10th	138	10th	159	10th	175	10th
6	Lack of training for the maintenance function	37	4th	244	2nd	504	2nd	**584**	**1st**	**608**	**1st**
7	Still reactive even with a sound PM program in place	41	2nd	177	5th	326	6th	369	5th	412	4th
8	Frequent reorganization on PM	17	10th	83	9th	156	9th	186	9th	194	9th
9	Lack or poor documentation on PM	43	1st	204	3rd	394	3rd	434	**3rd**	**469**	**3rd**
10	PM is waived by operations or other departments	20	9th	273	1st	537	1st	**575**	**2nd**	**607**	**2nd**

Note: These numbers represent the number of people who voted

Figure 1.8: Survey on Top 10 Problems of Preventive Maintenance Concluded 2021

1) Add on PM Checklists Syndrome: When the asset was new, and the OEM had completed commissioning the plant's equipment many years ago, they left us with a PM manual, which includes a list of scheduled maintenance activities performed regularly on the equipment. These activities include inspection, parts replacement, scheduled overhauls, routine lubrication, and other tasks that need to be done in a scheduled and timely fashion. As time passed by, breakdowns and failures were experienced on the equipment where maintenance would add up a Preventive Maintenance task due to their corrective action to address the failure. This now becomes a habit of the plant where their Preventive Maintenance lists on the equipment activities seem to grow. The problem experienced by industries would be that, even if we complied religiously with all these activities, unexpected and unplanned breakdowns still happen on the equipment all the time. When the boss' attention was finally caught down by these breakdowns, his only question was, what was the root cause? The easiest excuse you can give is wear and tear. The boss' will make a follow-up question and ask you if there are any PM activities included to address this problem? If you say no, then the boss will insist that you include a PM activity as your corrective action. This sort of episode had been going on for many years. In short, when the machine was newly commissioned, there were just a few activities to be done on Preventive Maintenance. Nevertheless, as time passed by, more and more activities have been added up. The PM list of activities that need to be accomplished never stops increasing minute-by-minute, day-by-day. This is what the industry has been doing since the reign of the Japanese empire during WW2.

2) Introduction of Infant Mortality Failures: As a result of too many activities done on Preventive Maintenance, the chances of infant mortality grow. These infant mortality failures occur right after a major maintenance intervention; Preventive Maintenance replacement or overhauls on the equipment is performed. Too much activity on Preventive Maintenance will often result in Infant Mortality Failures. Others refer to this as commissioning failures, or start-up failures. This problem exists right after a major Preventive Maintenance overhauling, or replacement is performed on the equipment. The PM is done, and when it is endorsed back to the operators, they have a hard time running the equipment. Sometimes the operator will whisper that if this equipment was not disturbed by the Preventive Maintenance crew, I am sure that this will be running smoothly. There are also cases where infant mortality failure can result from a poor changeover or set-up performed on the equipment that will eventually lead to breakdowns. The majority of the causes of infant mortality failures are human errors during the reassembly process of overhauls.

3) Replacement of Parts Just to conform to the PM Specs: Preventive Maintenance uses the JIC concept. In the Toyota Production System, JIT means just in time, but with Preventive Maintenance, JIC simply means "**Just in Case**." This was the advocacy they adhered to when you asked them why this and that part needs to be replaced or overhauled. Their response is simply "Just in Case" it might fail. Most sentiments on maintenance people believed that when a machine is scheduled for a Preventive Maintenance shutdown, this would be the only time they have left in this world to replace these parts and components that they think are on the verge of failing and needs to be replaced. Their thinking is that if this part fails and the boss finds out that you did not replace this during the shutdown period, then you are doomed and most likely will be blamed. You will be the culprit as to why the machine failed. Maintenance assumes that the parts they replace are nearing their rupture. However, in most cases, the parts that were replaced are still in working and in good running condition. This is partly what makes Preventive Maintenance expensive because most of the assumptions made are incorrect. When a part is still in working condition, then it should remain in service. The only exemption will be those regulated by the government. It is as simple as that. When we replace parts that are still working, then we are throwing the company's money away for good.

4) The Case of Random Failures: Random failures occur at any given period. They are also called "chance failures." It simply means that the probability that an item or part will fail in any period is the same as it is in any other given period. In short, the part or item has no life and can fail at any given time. For human beings, consider them just like Kobe Bryant, Michael Jackson, or Prince, who died young and did not reach their lifespan. One characteristic of random failure is that wear-out age is not known or not identifiable. When failures that occur are random in nature, then this is where Preventive Maintenance will be at its weakest point. In simple terms, it is not a recommended option. Other tasks to use will either be a run to fail. Still, this option will only be feasible when the failure's consequences are low, or some form of redundancy is in place in which the system will not be affected. If a run to fail is not valid, then Condition-Based Maintenance or modification takes place. Sample of random failures includes electronic boards, bulbs, ball bearings, seals, hydraulics, etc. Most industries think that Preventive Maintenance can address random failures by doing scheduled maintenance activities on them, such as scheduled replacement and overhauls.

Inspections can help but not always. Doing these activities will induce more infant mortality failures in the equipment, especially when you dismantle the equipment. This is when maintenance people got it all wrong. One mistake industries make is that if the failure is random, Preventive Maintenance such as overhauling or replacement will be done on the equipment. Doing this will only increase the chances of infant mortality failures from occurring.

5) Ageing Workforce and Nearing Retirement: When good old maintenance people from the University of Hard Knocks, just like the Rolling Stones, retire for good since they reached their working-age limit, their experience goes with them to the grave. Industries will now hire new fresh graduates with totally no experience in the asset because they will be paid much less in salary and benefits. And because these people's experience was not captured nor documented when they were still employed, these new people will tend to experiment with how to repair the failure, and the cycle goes on for generations. These older people have experienced most of these failures on the equipment since they have stayed with the plant long enough to witness them. Most of the time, industries do little or nothing to document how they perform their maintenance tasks or repair failures. This is one of the reasons why these white-haired maintenance people should be included in the RCM team. When the time comes for them to retire, their experience had been captured so that whatever experiences they shared can be taught to the newly hired maintenance crew so they will no longer experiment with how to do the tasks.

6) Lack of Training in the Maintenance Function: Many say that people are the company's greatest asset. I totally disagree with that statement because the correct way of stating this is that ***the right people are the company's greatest asset***. The wrong people are simply called liabilities. Besides, we can only have the right people if they are equipped with the right skills and knowledge to perform their jobs correctly right the first time around. Training is where we acquire knowledge, but sad to say that when cost-cutting is the name of the game that management plays, the number one department or organizational function that will be affected will be training. I often tell people that maintenance is not only a verb (action word), but maintenance is also a noun. This means that maintenance are also people. They are also human beings. When we invest in the people, these people will do the right things to do their jobs. However, sad to say that most time consumed on maintenance is about repairing and troubleshooting equipment failures and breakdowns. I am also saddened to hear that when I conduct in-house training in industries, many said that this was the first time they have attended a training of this kind. These people have stayed with the plant for more than 10 years or even longer. I even have experienced more than a dozen times in my training where white-haired maintenance people approached me during break time and told me how they wish they met me 15 years ago. When I ask them why did they say that? They told me that they were scheduled to retire next month. These good old people have been deprived of even the very basic knowledge of maintenance. The problem is that we cannot bring back the time. What a shame, and it breaks my heart. ☹

7) Still Reactive and still many Failures Even with a Sound PM Program: When I asked good maintenance people if they are satisfied with the results of their Preventive Maintenance activities, the majority will say no. Despite their most noble efforts on maintenance, failures, and breakdowns still happen and are still rampant. Industries have their own way of doing Preventive Maintenance. Yet, most of them are not satisfied with the outcome of their PM activities, even if they have high PM compliance. Industries must understand that doing Preventive Maintenance includes risks because there is always an assumption and temptation that everything dismantled will be put back together in one piece just like before. Overhauls are prone to human errors. Hence, despite the very best Preventive Maintenance efforts, failures will still happen and will not be captured solely by doing Preventive Maintenance. Human errors occurring during disassembly can frequently lead to infant mortality failures. Zeroing out all breakdowns is like catching a bolt of lightning with a Polaroid camera, and why the heck use a Polaroid camera if you can capture the lightning with a video camera? If we ask, why doesn't Preventive Maintenance capture all these failures? Typically, only around 20 to 30% of equipment failures will wear out or are directly related to the equipment's age. This means that around 70 to 80% of all failures will fit the random and infant mortality failures. When the failure is random, there is no amount of Preventive Maintenance that can address this failure. Increasing the amount of PM will only increase the chances of infant mortality failures in the equipment.

8) Frequent Reorganization in the Plant: When an organization is dynamic, so are the people. People are transferred from one department to another because management wants them to become a jack-of-all-trades. That is why these industries keep on reorganizing people. I have seen many improvements, projects, and modifications abandoned because the person in charge had been transferred. When a new boss heads the department, a new system occurs, and the culture also changes. However, the validity of that system only lasts when the boss is still around the corner. When Elvis has been replaced, a new system occurs once again for the people to comply because that is what the new boss wants. If you want to stay long in the industry, just dance to the music. If the boss wants us to dance to the music of cha-cha, then everyone does the cha-cha. If the new boss wants to boogie, then everyone dances the boogie. The maintenance system now becomes a flavor of the month while the maintenance people just wait and see.

9) Lack or Poor Documentation on PM: Sometimes, documentation on PM is not only lacking but at times, it is also being faked or done half-baked. If you have 10 people from maintenance in your department handling more than 100 machines and have a lot of fire-fighting and troubleshooting going on around, do not expect these PM documents to be filled up completely because maintenance has no time to document every single event that is happening around him. What is important in doing maintenance is documenting only what is relevant and important. Maintenance makes decisions based on these documents and if these documents are missing, flawed, half-baked, or faked, then the maintenance decision will be jeopardized, and at times there will be repercussions. Documentation is important; hence let us do this not because we will be audited by our boss, but let us do this for the right reasons. Ignoring these activities may backfire on us someday.

10) PM is waived by Operations: When the equipment was subjected to a scheduled maintenance intervention or scheduled Preventive Maintenance, operations or management frequently deferred or waived the equipment because the output is always the top priority. It supersedes everything. When the demand for the product is high, or there were months when production is at its peak, the maintenance crew complains that they cannot get the equipment for a scheduled maintenance. At this stage, maintenance will find it very difficult or impossible to get the equipment for a scheduled Preventive Maintenance activity. As a result, more unexpected failures occur, and the feud between operations and maintenance gets worst over time. Maintenance ends up always on the bargaining end because the operation's decision prevails. Operations will give the equipment to maintenance only when the demand is low.

1.3: Mistaken Belief about Preventive Maintenance

Before setting up a Preventive Maintenance strategy in the plant, everyone in the organization must understand the real mission of what maintenance is all about and what it can and cannot do. Most PM efforts fail because past sins that unleash mayhem still continue to occur despite undergoing Preventive Maintenance on the equipment. Even after running a few days after completing PM, there are still so many rampant failures occurring that it seems we cannot make a headway start. The true maintenance mission is to provide reliable physical assets and excellent support for operations by allowing smooth operations after executing PM activities. PM is only a way of trying to determine what parts will eventually wear out so that we can replace them before the failure occurs. To gauge the performance of any PM activities, compare the number of hours spent during Preventive Maintenance to the number of emergency or breakdown hours after completing PM activities in the plant. This means that if the breakdown is still high despite having high compliance in doing PM, then we need to review the tasks that are done in PM. Something is wrong here. PM is not a cure-all medicine for all failures experienced in the plant. Here are some mistaken beliefs about Preventive Maintenance that you need to know.

PM is Solely for the Maintenance People: Operators must be involved in maintenance. They can do some light PM tasks such as cleaning, lubricating, tagging leaks, and checking the bolts for any form of looseness, which is the essence of establishing Basic Equipment Conditions. This is possible if the industry understands that reliability is always a shared responsibility for both operators and maintenance. If operators are involved, then maintenance can focus more on improving the design flaws of the equipment.

[2]**PM are All the Same:** We can just copy and paste the maintenance tasks from the PM manual or follow OEM recommendations, and everything will work out just fine. Equipment may be identical, but how it is being operated might not be exactly identical. Before designing PM activities for the equipment, it is important to determine the conditions in which the equipment is operating. RCM called this the operating context. PM tasks derived from the manual by OEM are done generically and may not be exactly and precisely the tasks that will be done on the equipment. If PM is done correctly, equipment uptime increases, and this reduces reactive work and emergency response to failures and breakdowns. Remember that

[2] Levitt Joel, **Complete Guide to Preventive and Predictive Maintenance**, Industrial Press Inc. 2003, Page 30

if the cost of performing PM is higher compared to the cost of the consequences of failure it is meant to prevent, then PM is not necessary.

PM will Address All Failures and Breakdowns: PM is a series of activities and tasks applied at specific intervals and should address all failures and breakdowns in the equipment. This statement is not true. PM can only address age-related failures that will wear out due to its operating age. PM cannot make a 5 kW motor do the work of a 10 kW motor. Even with the most structured and detailed PM program, there will still be breakdowns from abuse, misapplication, human errors, and some failures such as electronic failures that will not be solely captured by any PM tasks. PM is not a silver bullet solution for every single failure occurring in the equipment. It is important to understand that PM has its own limitations and that knowing when to use PM will derive the most benefit from its application. Predictive Maintenance must be part of a good maintenance strategy and must not be an isolated case. Both people involved in Preventive and Predictive must work hand in hand in proper harmony and should communicate, coordinate and collaborate regularly.

In Performing Lubrication, The More, The Better: People involved in lubrication must understand that their equipment must receive only the right amount of lubrication. Over lubrication or under lubrication will have a detrimental effect on the equipment. The correct amount of lubricant should be applied with great care to the equipment. Adding too much grease builds pressure, pushing the rolling elements through the fluid film and against the outer raceway of the bearing. The increase in friction and pressure will increase the temperature of the bearing. Likewise, inadequate lubrication will cause friction and increase the temperature of the equipment.

If PM Compliance is High, Then PM is Effective: The truth of the matter is that PM Compliance is the weakest and most abused metric or KPI in maintenance. To gauge the effectiveness of PM is to compare the results of PM Compliance versus the rate of breakdowns and failures occurring on the equipment. If PM compliance is high and breakdowns are also high, then there is something wrong with the tasks performed during the Preventive Maintenance execution. It should be the other way around.

All Parts Replacement During PM Should Be Executed: If parts or spares are replaced during PM, they should be done religiously without questions. The truth is PM uses the concept of JIC or Just in Case, and in most cases, PM replacement will be done even if the part is still in working condition. If a part is still working, then use it. Replacing parts, which are still working is one of the main reasons why maintenance is extremely expensive to perform.

PM will Address Random Failures: Random failures are failures that can happen at any given time. This is like when you are driving, a small rock hit your windshield, the question is, when will the 2nd rock once again hit your windshield. Insisting on a Preventive Maintenance Task for a random failure will only create more problems as this can induce more chances of infant mortality failures. When a failure is random, this will require a different strategy such as

Predictive Maintenance if the random failure will provide symptoms, or run to fail if the consequences of failure are minimal. If both Predictive Maintenance and run to fail are not feasible, then we can modify or redesign the item. If the failure is random, this is beyond the scope of Preventive Maintenance.

1.4: Why Preventive Maintenance is Costly in Majority of Industries

As the name implies Preventive Maintenance will schedule a piece of equipment for several maintenance tasks at a defined interval or frequency. These tasks may include cleaning, routine lubrication, inspections, calibrations, part replacement, overhauls, and other activities based on the Work Order generated for a specific interval which may be in running hours, calendar days, number of volumes produced, number of cycles, or any other means depending on the industry's way of scheduling their equipment.

[3]According to a good friend of mine, R. Keith Mobley in his book An Introduction to Predictive Maintenance, quote that recent surveys of maintenance management effectiveness indicate that one-third or 33 cents out of every dollar of all maintenance costs is wasted as the result of unnecessary or improperly carried out maintenance. When you consider that US industries spend more than 200 billion dollars every year on the maintenance of plant equipment and facilities, the impact on productivity and profit that is represented by the maintenance operations becomes clear. The result of ineffective maintenance management represents a loss of more than 60 billion dollars each year. According to John Moubray, the author of RCM II, a common problem with mature maintenance programs is that if not correctly designed, then between 40 to 60% of the PM tasks serve very little purpose, and therefore, evaluating our current Preventive Maintenance System should lead us to the following:

- Many tasks duplicate other tasks.
- Some tasks are done too often while others are not enough.
- Some tasks serve no purpose whatsoever.
- Many tasks will be forced or intrusive and overhauled based whereas they should be condition-based.
- Some tasks cost more to do than the failure it is meant to prevent.
- Maintenance is costly by replacing perfectly good parts since we are basing replacement on time-based.

Preventive Maintenance overhauls and replacements should be based only on the useful life and not on the average life of a spare or item. This means that replacements and overhauls should only be done if there is an age-related pattern of these parts in our equipment and assets. Rolly Angeles

All failures have a pattern that can either be infant mortality, random or age-related. Infant mortality failures are failures that occur at the beginning of its life. We can experience infant

[3] Mobley, R. Keith, ***An Introduction to Predictive Maintenance***, Butterworth-Heineman Elsevier, 2002, Page1

mortality failures right after commissioning new equipment or right after a major Preventive Maintenance shutdown where the operators cannot operate their equipment smoothly. This usually happens when Preventive Maintenance includes any form of contact with the equipment such as replacing parts and overhauling. The major cause of infant mortality failures is due to human errors committed during the actual execution of PM. Others also called them start-up failures or commissioning failures that are likely to occur after a major overhaul or Preventive Maintenance had been initiated. Random failures are those failures that occur at any given period. Others call these chance failures. There are many reasons for random failure such as dirt, materials-related problem, human errors, environment, lack of lubrication, premature fatigue, or the part having an inherently short lifespan. Finally, age-related failures are those that will continue to run until their dictated life span consistently. Age-Related Failures simply mean that the failure is directly related to its specified age and there is a clear wear-out zone. Age specified may be in the form of running hours, time, number of strokes, revolutions, number of stress applied, or any other form. The best maintenance strategy to use for this type of failure will be to identify when the majority of the parts will fail and to apply Preventive Maintenance.

What industries must realize before anything else is that they need to accept that these patterns exist. In fact, these patterns can be reflected in the study of Matteson, Mentzer, Nowlan, and Heap of United Airlines 60 years ago which led them to the development of the MSG document, which is the basis of the concept of Reliability-Centered Maintenance. What these failure patterns meant is that for every failure that we experience from the equipment whether this is a bearing seizure, an electronic failure, electrical or mechanical failure, there is always a pattern. The pattern is always given, we just do not realize it.

Yet even before RCM was discovered by Stanley Nowlan, Howard Heap, Thomas Matteson, and Bill Mentzer, all from United Airlines, Conrad Hall Waddington also experienced Preventive Maintenance problems that contributed to infant mortality failures way back during World War II. Waddington observed that British bomber squadrons whose mission was to sink German U-Boats consisted of around 40 bomber planes, but during the mission, only half of them were available while the rest were either in a failed state only after a short 50 hour flying time. The reasons for the unavailability varied from one bomber to another, but a trend was clear that after a scheduled Preventive Maintenance performed on the RAF-Bomber planes, most of them were unavailable. Waddington tried to study the data and plotted all these downtime events concerning their last Preventive Maintenance. What he discovered was that unscheduled repair and downtime increased immediately after performing maintenance. He observed that the scheduled Preventive Maintenance performed on these bombers provided them with more harm than good because it was done too short at 50 running hours of flying time.

RATIO OF AGE-RELATED vs NON-AGE RELATED FAILURES

Failure Rate by type	United Airlines 1968	Bromberg 1973	US NAVY 1982
Pattern A	4 %	3 %	3 %
Pattern B	2 %	1 %	17 %
Pattern C	5 %	4 %	3 %
Pattern D	7 %	11 %	6 %
Pattern E	14 %	15 %	42 %
Pattern F	68 %	66 %	29 %

Majority of the failure modes realized by these industries are random and Infant and cannot be addressed by performing Preventive Maintenance replacement and overhauls

Patterns A to C	11 %	8 %	28 %	→ Age Related
Patterns D to F	89 %	92 %	77 %	→ Non-Age Related

From page 60 RCM – Gateway to World Class Maintenance by Anthony Smith

Figure 1.9: Study of United Airlines, Bromberg, and US Navy on the 6 Failure Patterns

Waddington's recommendation was to increase the time interval and duration of scheduled Preventive Maintenance on these RAF Coastal Command Bomber planes. He removed all intrusive and unnecessary maintenance performed, and the results were an astounding increase of up to 60% in the availability of these RAF Bomber planes. This was known as the Waddington Effect, and the principle behind it is the very core principle of Reliability-Centered Maintenance. Therefore, the less invasive the Preventive Maintenance is, the better the outcome, according to the Waddington effect. As the Waddington effect shows, more is not always better from a maintenance point of view. The truth is that the more PM done on the equipment, then the more problem it can experience and the less PM then the better the equipment will perform.

[4]Figure 1.9 represents a study by United Airlines, Bromberg, and the US Navy regarding the percentage of failures that belong to the 6 failure patterns. This figure means that if we can see the summary of the percentage of failures reflected from patterns A, B, and C, United Airlines has a total of 11%, Bromberg 8%, and the US Navy at 28%. On the other end, the percentage of failures that fall on patterns D, E, and F will be United Airlines at 89%, Bromberg at 92%, and the US Navy at 77%. This means that these 6 patterns can be grouped into 2 which are age-related for patterns A, B, and C and non-age-related for patterns D, E, and F. What this figure means is that Preventive Maintenance will only be feasible if the

[4] Anthony Smith, ***Reliability-Centered Maintenance, Gateway to World Class***,(Mc Graw Hill Inc., 1993), Page 60

pattern of failure conforms to either A, B, or C, while patterns D, E, and F will require a different task as indicated in figure 1.10.

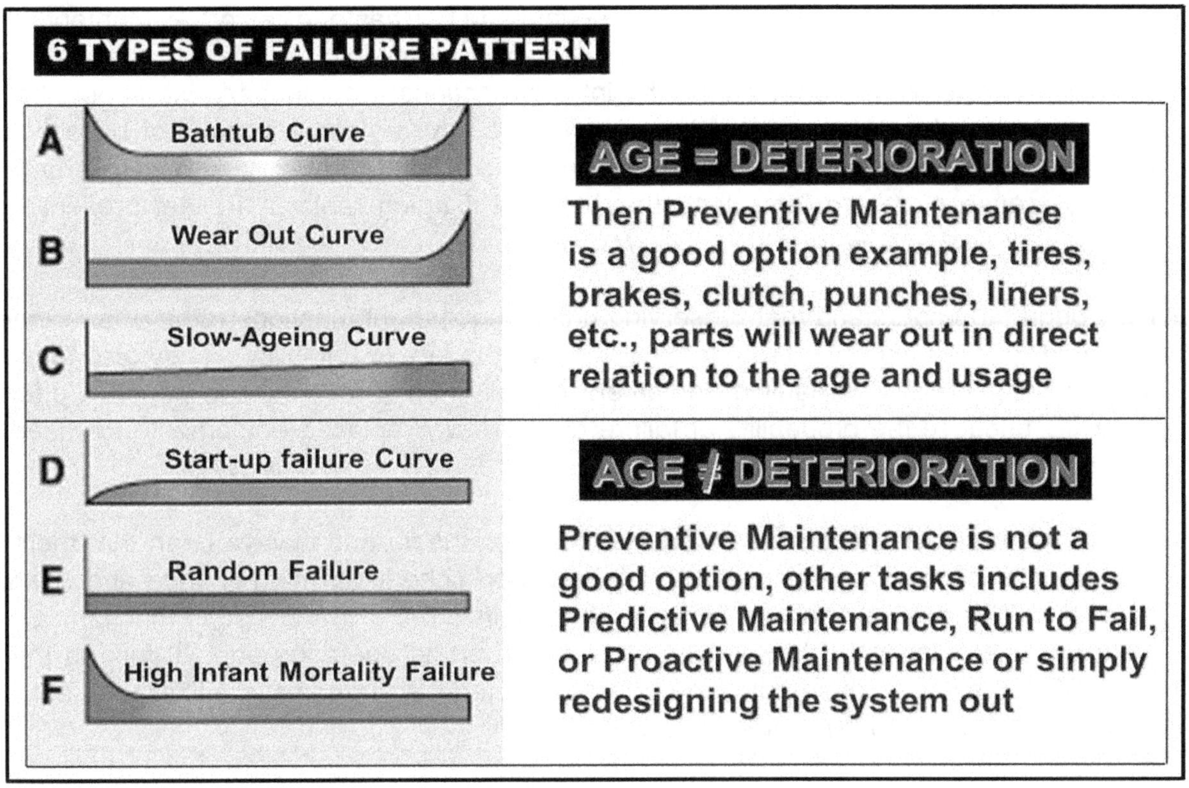

Figure 1.10.: The 6 Failure Patterns

Failure to accept the existence of the 6 Failure Patterns will just result in having more expensive Preventive Maintenance in the plant. Why? It is because traditional Preventive Maintenance will replace parts regardless of their condition, and maintenance will assume that replacing good parts with new ones will inevitably restore the equipment back to its original condition. That's why in every essence, overhauls and replacement is costly and may not always be the right decision to pursue. Most Preventive Maintenance today in industries adhere to the concept of JIC which means Just in Case. A maintenance planner will dictate the Preventive Maintenance activities for scheduled maintenance on the equipment which will include replacements and overhauls. Parts and items will be replaced even if they are still working and do not exhibit any wear-out mode. Equipment will be overhauled intrusively and human errors occur when reassembling the equipment once again to its original state. Quality or Safety people will conduct some audits if the Preventive Maintenance crew complies with all the activities written on the plan. If a task is not performed by the PM group, a non-compliance ticket will be issued to them which will become a backlog for the PM to comply. After the Preventive Maintenance tasks are completed, the PM crew endorsed the equipment back to operations, and the operators once again will be a victim of infant mortality failures. As the equipment had finally been fixed, new and existing failures and breakdowns emerged from the equipment, and once again for every failure that evolved on the equipment, the default corrective action is to place a Preventive Maintenance task on the never growing lists of PM tasks to be done ending in more failures and the cycle of doom continues leaving maintenance in a reactive mode and in a merry-go-round of never-ending problems. Here are the following quotes from several books about Preventive Maintenance.

- Overhaul actions may actually be harmful because in our haste to restore equipment to new pristine conditions, we may have advertently pushed it back into the infant mortality region of the curve due to human error during the intrusive actions. In summary, we should be very careful about selecting overhaul PM tasks because our equipment may not have an age-reliability pattern that justifies such tasks. In addition, due to human errors, overhauls are likely to cause more problems than they prevent if aging regions are not present. *From Anthony Smith on RCM Gateway to World Class Maintenance*

- They began to understand that preserving critical equipment functions rather than randomly and arbitrarily tearing an entire aircraft apart was the key to reliability. They also found out that indiscriminately overhauling equipment actually had a reverse and negative effect on reliability because the probability of failure of the newly replaced equipment increased due to premature failures and infant mortality. *From Neil Bloom on RCM Made Simple*

- One of the underlying assumptions of maintenance theory has always been that there is a fundamental cause-and-effect relationship between scheduled maintenance and operating reliability. This assumption was based on the intuitive belief that because mechanical parts wear out, the reliability of the equipment is based on its operating age. It follows that the more often the item is overhauled the better protected it was against the likelihood of failure. *From Nowlan and Heap on Reliability-Centered Maintenance*

[5]An example of some of the benefits achieved from this learning on infant mortality failures by the civil aircraft industry was based on a dramatic reduction in their scheduled overhauls in their DC-8 aircraft from 339 items subject for an overhaul to only 7 items in their DC-10 which is a bigger aircraft. One of the items that were no longer subject to overhaul was their turbine engines. Likewise, on the initial program developed for BOEING 747, it took United Airlines 66,000 man-hours on major structural inspections to reach an inspection interval of 20,000 hours compared to their traditional 4 million man-hours inspection on smaller and less complex aircraft such as DC-8. Truly, it had been a remarkable achievement and feat which saved them millions of dollars on maintenance without compromising its safety. But perhaps the greatest benefit of learning from Infant Mortality Failures is that it saved lives.

It is important to understand that Preventive Maintenance has its own limitations, and understanding when to use and not to use it will derive the most benefit from its application. When applied correctly, PM will increase the equipment's uptime. However, if we abuse, misuse, or overuse its use, it will cause us more harm by experiencing more failures instead of preventing them. Preventive Maintenance is feasible if the part or component will wear out directly concerning its operating age, and that its life will definitely survive this defined age. I think we need to resound the learning's from Nowlan, Heap, Mentzer, and Matteson, which they discovered in the 1960s since it will create a lasting impact if we really absorb and apply

---

[5] Stanley Nowlan and Howard Heap, ***Reliability-Centered Maintenance***, (Springfield : US Department of Commerce, National Technical Information Service, 1978) Section 1-1, Page 5

these learning's that the lessons they provided to us today are of great importance for every person whose responsibility is to maintain their equipment and assets in industries. There is no substitute for training since this is where we acquire knowledge, and this knowledge is where we develop our skills, and whatever learning we gain should be shared among our people. 60 years had pasts, yet most industries are still trapped in a reactive culture of maintenance. What Stanley Nowlan, Howard Heap, Bill Mentzer, and Thomas Matteson have discovered is not about some rocket science or silver bullet solution to all maintenance problems, but rather understanding that Preventive Maintenance has its own limitations too. If I can put this in the simplest words possible, more is not actually better. The truth of the matter is, with PM, the more you touch, the more it fails, and the less you touch, the less it fails. Preventive Maintenance will always have a place in the overall maintenance strategy, just don't abuse it for it to work properly.

Hence, to answer the question, why doesn't Preventive Maintenance reduce the amount of reactive maintenance being performed in the plant? Preventive Maintenance is a good strategy, but most industries simply abuse, misuse, or overuse its concept making PM more reactive. My experience tells me that activities included in the PM checklists continue to grow daily due to deviations from audits, customer issues, quality and safety issues, day-to-day breakdowns, and so on. We must understand that the lesser the amount of PM performed on the equipment, especially those that involved coming in contact, the better the equipment will run smoothly. Intrusive or forced maintenance will only create more problems in operations. Our people must understand and realize the lessons from Nowlan, Heap, Mentzer, and Matteson that there are not one, not two, but six types of failure patterns in the real world of doing maintenance. This means that every single part or spare we have in our equipment may actually fail according to any of these six failure patterns. Not all parts of our equipment will wear out just like people. The majority of failures we experienced in the past in the equipment are random failures and premature wear. If industries are experiencing many random failures, Preventive Maintenance is not the solution. Remember that Preventive Maintenance was designed around the theory that equipment failures are directly related to the equipment's age. In reality, only around 20 to 30% of equipment failures will fit this pattern, which means that 70 to 80% of equipment failures will not be managed effectively by doing time-based or scheduled Preventive Maintenance activities.

1.5: Grave Concern about Government Laws on Outages

I asked this question to a couple of my friends on LinkedIn. Here is my question. Would you find it contradictory for the Department of Energy (DOE) or other government agencies to provide a mandatory requirement for power plants and power distribution or other similar industries to comply with a Preventive Maintenance Shutdown or Scheduled Outage? In this case, the equipment will be shut down for a specific amount of time to do PM, where we all know that not all failures will be captured solely by PM. Hence, what happens then is that even if Predictive Maintenance activities are in place in those plants, those under the watch of Predictive Maintenance will be replaced during shutdown or scheduled outage periods just to comply with government regulations and policies. What are your thoughts on this?

- In general, I would like to see organizations do the right thing for their equipment (whatever that is). Still, I am always suspicious when a government gets involved. As you mentioned, you could have a program better than required and still have to do the mandated activity. On the other hand, companies under maintaining assets to "save money." It is no dishonor to run to failure, but if the consequences of a run to fail are not acceptable, then it is not the right thing to do. *By Joel Levitt, Maintenance and Reliability Book Author*

- Rolly, yes, it is contradicting, but regulatory agencies are slow to change. As the joke goes, it takes an act of Congress to make changes literally. The logic behind electric power generating plants doing a 5-year turnaround, where they take an extended outage, tear everything down, inspect, repair, and then return to service was the only safe choice decades ago. Condition monitoring did not exist and based on history, this was the best approach to avoid catastrophic failures. Are there better ways now? Of course. Will the regulations change anytime soon? Probably not. Some of the questions regarding these regulations are illogical. They are not good business practices, but they are regulations and must be followed until they can be changed. I have no argument with your comments but consider this. Many people, including hospitals, depend upon the utility to provide dependable power. Should a failure occur, it would take the utility weeks or months to repair unscheduled failure. Is it better to spend more every five years to replace questionable parts or risk a catastrophic failure? I think you will agree that none of the known technologies can predict 3 to 5 years into the future. A change from the current turnaround philosophy means that instead, the utility would take multiple, shorter outages based on their predictive data. Assuming the utility has enough reserve power to remain fully online during the shorter outages, the deciding factor is how much do these outages cost? As you know, repair of power plant equipment is time-consuming and costly. Add to that the potential loss in revenue (assumes they have to buy power to keep the grid fully serviced). It might be more cost-effective to continue with the turnarounds. All I am saying, is you have to look at the bigger picture. You cannot just look at asset reliability. If you do, you could drive the client into bankruptcy. Hope all is well with you. *By R. Keith Mobley, reliability and maintenance book Author*

My Deep Concern about Power Plants in the Philippines: I have taught many power plants in the Philippines ever since I started teaching, but they find it difficult to use the learning in their actual situation. Why? Because my teachings contradict government procedures and compliance such as DOE (Department of Energy) that a plant needs to have a shutdown to avoid forced outages. What happens is that when the actual shutdown comes, maintenance will not only comply but will tend to replace parts even if the part is still in working condition or even a Vibration Monitoring indicates that it is still on its normal frequency thinking that it might fail before their next shutdown thereby these maintenance people from Power Plants perform intrusive maintenance. According to a study conducted by TA Cook and Solomon Associates in 2019

- 50% or more of all turnaround shutdowns experience significant delays which cause a loss of revenue for the industry due to infant mortality failures or failure to operate after the plant's turnaround activities.

- Study shows that 50% of accidents in manufacturing industries occur during a maintenance outage.
- 82% of all turnarounds do not satisfy performance requirements.
- 80% of all turnarounds are over the initial budget by more than 10%.

The truth of the matter is that a Shutdown, Scheduled Outage is not a guarantee of a continuous operation of equipment, as said many times Preventive Maintenance alone cannot capture all failures, and performing too much PM or Outage will just induce more infant mortality failures and that these PM outages, Turn-Around or whatever you call it is only design for age-related failures and for industries it is not age-related but random failures that you face on a day-to-day basis. For crying out loud Preventive Maintenance and shutdown will not and never address random failures. For as long as the government procedures and policies on maintenance remain unchanged, then nothing will ever change in maintenance. Perhaps if I am in politics then things will change.

1.6: Understanding When to Use Preventive Maintenance

PM Mindset: Preventive Maintenance is an important strategy for industries, however, many maintenance either abuse, misuse, or overuse this strategy which is why many industries are not satisfied with their current PM program. Preventive Maintenance must stand firm in their decision on understanding what PM can do and what it cannot do in their equipment. PM will yield good results if it is correctly and properly applied, but havoc and chaos will reign if people abuse its use. PM is not designed to eliminate all equipment failures. This is an impossible feat to accomplish. Maintenance must understand that all activities performed on Preventive Maintenance are geared towards anticipating, preventing, prolonging, or delaying the process of failure to occur. PM replacements and overhauls should only be performed on parts and components that have worn out or have an age-related pattern. All random cases of failures should be removed from the PM list.

Correct Frequency: Determining the correct frequency of replacements and overhauls for Preventive Maintenance activities should be based on the useful life and not on the average life of the part, which inhibits an age-related pattern. Parts that do not have a useful life should not be replaced on a time-based dominated frequency because their pattern is different. Only parts that will age due to stress-induced upon the part should be subject to parts replacement and overhaul in the PM activities. Aging is the process where certain parts of the equipment deteriorate due to stress-induced upon a part over a given period. Any part that conforms to this process will be the subject of any Preventive Maintenance intervention.

Economics: The feasibility of performing Preventive Maintenance must be looked upon from the economic point of view and cost of performing Preventive Maintenance. Preventive Maintenance tasks must be worth doing. One measure of worth is that doing the tasks furthers the business goals of the organization. The economic question is very critical, and most of the time, it does not exist in the PM program. Hence, spending $1,000.00 to maintain an asset worth $500.00 is usually a waste of time and resources. Understanding the economic point of view on PM will allow us to understand that PM is not just a cost center. Still, if this is done correctly, then it will yield revenue and profit.

Engineering: The tasks have to be the right tasks, done with the right technique at the right time and frequency. If Preventive Maintenance tasks are performed, and the amount of breakdown is still high, it means that either the wrong tasks or the wrong frequency are performed on the equipment. Preventive Maintenance tasks performed on the equipment should address a particular failure mode. Preventive Maintenance is not about eliminating all failures in the equipment. This is an impossible feat to accomplish. However, a good Preventive Maintenance program will see to it that the consequences of failure are either prevented, eliminated, reduced, or mitigated. To do this, all activities performed on Preventive Maintenance must be geared toward anticipating and planning for the failure itself.

Safety on Maintenance: Preventive Maintenance must only be done when the equipment is completely stopped. Safety signs on the equipment must be in place before conducting any PM activities such as the slippery floor, tags, danger signs, high voltage, live cable wires, flammable fluid, gas signs, and others. Protective clothing and apparel such as a hardhat, gloves, safety clothes, and shoes must be completed as a precautionary measure before conducting any Preventive Maintenance activities on the equipment.

Psychological Factor: People involved in conducting Preventive Maintenance must not only be well motivated but must also be knowledgeable so that they can perform their tasks correctly. A good PM inspector can work little or almost with no supervision if the person is well equipped with the tools, knowledge, and skills to perform the tasks confidently and correctly. The person involved should be interested and trained in advanced Predictive Maintenance technology and knows how to review the history of equipment undergoing Preventive Maintenance. People doing PM should be capable of deciding and acting on predictions rather than acting in a situation where failure is already imminent.

Built into the System: Preventive Maintenance has to be built into the system. Detailed documents and procedures must be in place and must not be intrusive. As Edward Deming, a renowned quality guru, once said that quality must be built into the system of production and not in the individual effort. Whoever is performing Preventive Maintenance must be consistent in performing these tasks. Information on PM has to be integrated into the flow of business information. Precision Maintenance must be applied and integrated when conducting PM, mostly during overhauls and replacement.

Chapter **2**

Understanding the Concept of Preventive Maintenance

Preventive Maintenance is a good strategy to adapt. However, many industries, abuse, misuse, or overuse this maintenance strategy providing them more harm than good on their equipment and assets, and many are not satisfied with the outcome of their Preventive Maintenance. Remember we can only benefit from this strategy when we understand both its feasibility as well as its limitations.

2.1: The Objective of Maintenance

Figure 2.1: Why Do We Maintain?

All of us humans are maintenance in every way. We might not realize it, but even if we have a different degree such as an accountant, doctor, carpenter, lawyer, pilot, or whatnot, just like each one of us, we maintain things that are important to us regularly. We brush our teeth twice or thrice a day. We go to a doctor for a regular check-up. We wash our car or change its oil after reaching a certain number of miles or kilometers, bathe our pets weekly, and so on. But for industries, the question is raised, why do we maintain our equipment and assets. From an operations point of view, we maintain our equipment so that operations can push through with production. But from a maintenance point of view, we maintain so that we can preserve and take care of our equipment. Although both points are reasonable, there is still one more reason why we need to maintain our equipment and assets. By far, what I believe is that this is the most important reason for them all. The most important reason we maintain our equipment and assets is that there are times when the consequences of failure are far more important than the failure themselves.

Just like humans, failures are not created equal. Every failure we experience on the equipment has its own degree of consequences. Some failures are acceptable since we have a form of redundancy or standby which will not affect the entire operations, but there are also failures in which the consequences are simply unacceptable.

- Some failures can halt our operations for a few hours.
- Some failures can halt our operations for several days
- Some failures can injure or kill our employees
- Some failures can totally shut down our industry permanently
- Some failures can totally damage the environment or breach any government laws
- Some failures can totally damage life itself

This means that the degree and amount of maintenance we need to carry out on our equipment must be based not only on the failure modes but on the degree of consequences of each failure itself. Three kilometers from where I'm writing this book is a place called the Enchanted Kingdom. The Enchanted Kingdom is the largest and most popular theme park destination in the Philippines where families and friends go for a myriad of fun, adrenaline-pumping ride adventures and a whole lot more family time together. The 17-hectare sprawling theme park is located in Sta. Rosa Laguna in the Philipiines. They got all the rides such as a large ferries wheel, bump cars, and a roller coaster which is an 11 story they called the Space Shuttle Max. This is one of the most popular rides in which the queue is always long just like when getting your ticket at an international airport. People patiently wait for their turn as it is a thrilling experience that is not recommended for people with a weak heart. Meaning if you take the ride, you cannot request for it to stop and board down. One unique feature of this roller coaster is that it has a 360-degree loop. A decade ago, it was in July 2013, a malfunction happened on the coaster. When it entered the 360-degree loop, the roller coaster completely stopped and the people were upside down. The management of Enchanted Kingdom called the police, and the firemen to conduct a rescue operation to relieve the people trapped in their seats. Soon one by one, the media people are covering the news live on TV. Regular TV programs were interrupted so that the public can see the live update rescue going on at the Enchanted Kingdom. The last person to be rescued took more than a couple of hours since the seat's lock was jammed. The following day, Enchanted Kingdom was locked down. The government suspended the Enchanted Kingdom activities for more than a year just for a single malfunction on the roller coaster. What I am saying is that it can only take one single failure for your industry to stop operating permanently.

Figure 2.2: Major Industrial Disasters Worldwide

Figure 2.2 are just a few of the major industrial disasters that occurred in the past. The majority of these industrial disasters have something to do with maintenance and human error. Not mentioned in the lists are accidents in commercial aviation and airline industries. The daily news we watch daily in which truck's brake malfunctioned, as well as the newspaper we read where an explosion occurred killing several people. Unless we do something about the way we do maintenance in our industry, these disasters will continue to happen in the future. These disasters do not select what type of industries we work in but expect them to happen in industries, especially where their Top Management does not take maintenance seriously. The question is just a matter of when.

The objective of performing maintenance on our equipment and assets is to take care of the equipment's Total Life Cycle at the most reasonable cost under optimal conditions with compliance to safety, quality, and the environment. To sustain, conserve and preserve our equipment and assets, people involved in maintenance should be equipped with the correct knowledge to perform their jobs right the first time around. Knowledge is the very basic foundation where we derive and build our skills, in which the end result is mastery where one is capable of mentoring others on what they know on how to do the job correctly whether we are executing a Preventive Maintenance or performing a repair on the equipment. The sad thing is that this is the Number 1 problem in Preventive Maintenance as mentioned in Chapter 1 regarding the Top 10 Problems on PM. Once maintenance has the knowledge and skills, then we can sustain the equipment. For parts with design flaws, we need to improve them by challenging how to lengthen their lifespan. Doing these will reduce the cost of doing maintenance of our equipment. Improving reliability and reducing costs are not one and the same. If industries want to reduce their maintenance and operating cost, the best way to do that is to improve the reliability of their equipment because if reliability improves, then cost will automatically go down and this process cannot be reversed.

Figure 2.3: The Objective of Maintenance

I have been often asked in my class what is the best way to reduce the cost of our maintenance, or how do we reduce our cost of doing maintenance? My response is simple, the only way I know to reduce the cost of doing maintenance is not to focus on it at all. Our role is to preserve the equipment, then look for design flaws and weaknesses and improve them so that reliability will be improved. By doing this, the cost will be reduced automatically whether we like it or not. There will be times when focusing on reducing costs the wrong way will be detrimental to the reliability of the equipment. This is one lesson top management, and leaders must take note of and understand. This is done by adapting the following;

- Improving the knowledge and skills of all our maintenance craftspeople
- Performing maintenance tasks to sustain and preserve our equipment and assets
- Challenge maintenance to improve the design weaknesses of our equipment and assets so that the cost of doing maintenance will be reduced.

Preventive Maintenance is the most basic and commonly used task by industries, yet the irony behind this is that majority of industries are not satisfied with the aftermath of what happens after implementing Preventive Maintenance tasks on their equipment and assets. The end result is that maintenance gets the blame for everything. The thing is, how did we derive the Preventive Maintenance task on our equipment? Was it dictated by the OEM or vendor? Were the PM lists generated by the people who previously worked in the plant a couple of decades ago? Were the PM tasks generated by the complaints of our clients and customers? And so on. What is important to consider is if the PM tasks we are currently performing on our equipment and assets are still valid and worth doing?

2.2: Different Terminologies on Maintenance Tasks

Before we discuss Preventive Maintenance, there are far many types of industries from all around the world and spending around 30 years working on maintenance previously as an employee, and serving them today, what I observed is that industries have their own terminologies, vocabulary, nomenclature, or even jargon used for these different maintenance tasks. In the Philippines, the word run to fail also means the MacGyver Move (from the original TV series 1985 on MacGyver). Although countries from the west are familiar with the term corrective maintenance as repairing a piece of equipment when failure happens, this definition means differently for industries implementing TPM (Total Productive Maintenance). The term corrective maintenance for these industries means improving the equipment, while the term corrective maintenance for power plants means repairing or troubleshooting the equipment or system after a failure occurred. Still, I know of one water utility plant in which the term corrective maintenance means that the failure has not yet occurred on the equipment or system but is likely to occur. Hence they need to do a maintenance activity before the failure happens, right after a Predictive Maintenance detected a potential failure, and the functional failure is nearing its toll. In a way, for the water utility plant, corrective maintenance is proactive. So a maintenance person who recently resigned from work on a power plant and started to work in a manufacturing plant implementing TPM might be confused if he will be asked to perform Corrective Maintenance on the equipment since the two industries have different meanings for the term. Figure 2.4 is a table I made to categorize and group these different maintenance tasks and terminologies.

It really does not matter what name or terminology your plant uses as long as they have the same definition and meaning for all maintenance in that plant. For example, many will be familiar with the term Predictive Maintenance or Condition-Based Maintenance, but RCM prefers to use the term On-Condition Tasks. But the term On-Condition tasks also include inspection using the human senses, which the majority of us will include as a routine Preventive Maintenance task. The term, On-Condition means any tasks that entail checking the actual condition of the equipment either using state-of-the-art non-destructive instruments or through the use of human senses. Performing inspection through the use of gages or even the use of SPC (Statistical Process Control) will default to On-Condition Tasks for those industries implementing RCM. In RCM, Preventive Maintenance is only known as those activities that include scheduled overhauls and replacements. Performing PM Overhaul is known as scheduled restoration tasks and replacing parts during PM is known as scheduled discard tasks.

My advice is to stick to what you feel is comfortable. If you are comfortable using your own terminologies for the different maintenance tasks, that would be fine as long as someone interprets the meaning of these RCM tasks for your people. What is important is that we need to be consistent with the terms that people use to identify the different maintenance tasks. Another example I would like to discuss is the inspection or routine lubrication tasks process for both operators and the equipment's maintenance. Typically, this will be included as part of our Preventive Maintenance tasks, but in RCM, these inspection tasks will default once again to On-Condition Tasks and not PM.

MAINTENANCE TASKS TERMINOLOGIES USED BY INDUSTRIES

	Reactive Maintenance	Preventive Maintenance	Predictive Maintenance	Proactive Maintenance
	Fix it when it fails	Maintenance at Scheduled Intervals	Maintenance is Base on Condition	Equipment Improvement
TPM	• Breakdown Maintenance • Unplanned Maintenance	• Preventive Maintenance • Periodic Maintenance	• Predictive Maintenance	• Corrective Maintenance • Maintenance Prevention
RCM	• No Scheduled Maintenance	• Scheduled Restoration • Scheduled Discard	• On-Condition Tasks	• Redesign • Modification
OTHERS	• Run to Fail • Run to Destruct • Reactive Maintenance • Band-Aid Maintenance • Corrective Maintenance • Forced Outage • Unplanned Outage	• Time-Based Maintenance - Running Hours - Stroke-Based - Volume-Based • Scheduled Outage • Shutdown • Planned Outage • Calendar-Based Maintenance • Turn-around Maintenance	• Condition-Based Maintenance • Condition Monitoring • Equipment Diagnostic Technique • Reliability-Based Maintenance • Non-Destructive Testing • In-line monitoring	• Proactive Maintenance

Figure 2.4: Different Maintenance Tasks Terminologies Used by Industries

2.3: Different Maintenance Tasks Explained

Being familiar with both the east and the west concept of maintenance, if we talked about the west, Preventive and Predictive Maintenance are separate tasks, but if we speak about the east, Predictive Maintenance tasks are under the scope of Planned Maintenance. Let us first define the different maintenance tasks as follows.

Reactive Maintenance is a maintenance task that allows the equipment to run until it fails. This means that the failure will happen first, then maintenance reacts by repairing the failure on the equipment. This is an activity that tells us that when a machine fails, then it is time for maintenance to fix it. The limitation of doing reactive maintenance is that it will cause unplanned downtime and production delays resulting in revenue loss since the equipment will be stopped. Another disadvantage of this type of maintenance task is that repairing the equipment after it fails usually creates the possibility of secondary or even tertiary damage, which can increase maintenance costs. Allowing failures to occur on the equipment can be justified if the consequences of failure and repair costs will be minimal and acceptable to both the user and the maintenance. This maintenance task can also be applied if the equipment has some redundancy in place. Some failures have no significant consequences when they occur; in this case, a run-to-fail or breakdown strategy is valid. Note that it is not feasible to apply Reactive Maintenance if the consequences of failure will have safety or environmental consequences. Other terms to denote reactive maintenance includes, run to fail, corrective maintenance, breakdown maintenance, unplanned maintenance, or RCM refers to this as No-Scheduled Maintenance.

Preventive Maintenance is a maintenance task performed on a scheduled basis. The schedule varies widely from one industry to another. Scheduling Preventive Maintenance can be in the form of time, such as daily, weekly, monthly, semi-annually, and so on. The schedule can also be in the form of running hours, number of cycles, number of strokes, or amount of volume produced. There will be a list of activities performed on the equipment, such as inspection, routine cleaning, lubrication, parts replacement, calibration, and overhauling activities. A scheduled plan of activities will be provided where equipment will be scheduled for a Preventive Maintenance shutdown. The equipment will stop at this point for maintenance to perform these activities. Scheduled replacement and overhauls will be feasible only for age-related failures or those parts that have a wear-out pattern. This means that an item, spare, or component has reached its remaining useful life. ISO definition for Preventive Maintenance is maintenance performed according to a fixed schedule, or according to a prescribed criterion, that detects or prevents degradation of a functional structure, system, or component from sustaining or extending its remaining useful life.

Predictive Maintenance is a maintenance task that takes into consideration the actual condition of the equipment when deciding whether maintenance is necessary to be done or not. In Preventive Maintenance, the tasks are usually based on a fixed or regular schedule. In contrast, Predictive Maintenance will be based on the actual condition of the equipment. This requires regular monitoring of critical parts of the equipment. The instrument used will predict problems and faults before they occur. Condition Monitoring or Condition Directed Tasks are the activities of monitoring these tasks on the equipment with the aid of instruments. Condition monitoring provides data that will be useful in determining what the problem is and the possible causes of the problem. This allows us to schedule repairs of specific problems before the equipment actually fails. There are two ways of doing Predictive Maintenance that can be done either in-line (others prefer to use the term online) or offline. In-line monitoring is done by placing sensors on the equipment, which is monitored 24 hours a day in a control room, much like in oil, gas, and power plants. Offline monitoring will include the user, and the instrument, extracting data from the equipment and trending it to compare data from previous measurements.

Condition-Based Maintenance Although they are almost the same both will be based on maintenance of the actual condition of the equipment. Predictive maintenance tends to be equipment-oriented, while CBM is system-oriented. CBM deals with the entire system as an entity. Besides these instruments, Predictive Maintenance will include monitoring techniques based on variations in product quality such as Statistical Process Control Charts, process parameters monitoring such as recording and monitoring charts recorders, pressure, temperature, and also the use of inspection using the human senses. What is important for both Predictive and Condition-Based Maintenance is to determine the P-F interval or P-F curve where P is the potential failure and F is the functional failure. A functional failure is when the equipment is already declared in a failed state. This means that when the potential failure is determined, the question that needs to be answered is how long before the functional failure will happen so that a maintenance task can be performed before reaching the functional failure. The only difference between using the human senses and these instruments is that the P-F curve of humans will be much shorter compared to the Predictive Maintenance instruments.

Difference Between Condition Based and Predictive Maintenance

Condition-Based Maintenance	Predictive Maintenance
• Condition techniques which involved the use of specialized Predictive Maintenance instrument to monitor the condition of equipment and assets • Techniques based on variations in product quality (Ex. SPC Charts) • Primary effects monitoring techniques which entails the intelligent use of existing gauges • Process parameters monitoring equipment chart recorder etc., • Visual Inspection techniques based on the human senses	• Vibration Monitoring • Oil Analysis • Ultrasonic Monitoring • Infrared thermography Monitoring • Temperature Monitoring • Wall Thickness Decrement • Corrosion Monitoring • Motor Analyzer • Radiographic Testing using x-rays • Eddy Current Testing • Can be online or offline monitoring

Figure 2.5: Different Between CBM and PdM

On-Condition Tasks: Although similar to Predictive Maintenance which will use instruments to predict the failure based on the condition of the equipment. The main difference is this will be on a much wider scope which may also include, condition monitoring techniques that involved the use of specialized instruments to monitor the condition of the equipment, techniques based on variations in product quality such as Statistical Process Control Charts, primary effects monitoring techniques which entails the intelligent use of existing gauges, process monitoring instruments, smart sensors, chart recorder, and inspection techniques based on the human senses. On-Condition tasks are any tasks that monitor the actual condition of the equipment by any means.

Detective Maintenance is also termed as Failure Finding Tasks or Functionality Inspection. Detective Maintenance is the activity of inspecting protective devices and redundant components to avoid the chances of any hidden failure. The reason for inspecting hidden failures is to reduce the chances of multiple failures. Multiple failures are when the protected function fails and is not detected because the protective device is also in a failed state. Examples of these protective devices include alarms, led, sprinkler systems, emergency stop buttons, lightning arresters, and overcurrent protection.

Proactive Maintenance: According to the Merriam-Webster Dictionary, Proactive Maintenance means acting in anticipating a future problem. It can also mean controlling a situation by causing something to happen rather than waiting to respond to it after it happens or acting in advance to deal with an expected difficulty. Proactive maintenance is trying to anticipate something before a functional failure occurs. Both Preventive and Predictive Maintenance when done correctly, is already a form of Proactive Maintenance. This is also about understanding the root cause of why a part keeps on failing and performing corrective measures to prevent the recurrence of failures. In Proactive Maintenance, we are one, two, or more steps ahead of the failure. It involves activities about analyzing why failures occur so that their recurrence can be avoided, thereby extending the life of the part, component, or equipment. Proactive Maintenance activities include performing root causes to understand why failures do happen and performing corrective action such as redesign or modification to avoid the failure from recurring.

Precision Maintenance: According to the dictionary, precision is the state or quality of being precise or being exact. Precision Maintenance is the act of maintaining equipment in a consistently reproducible manner. This means that the maintenance performed is done exactly the same way and delivers the same results and outcome regardless of who performs the activity. It is a method of performing maintenance in a consistent, precise, and accurate way. If Precision Maintenance is properly implemented, maintenance should yield exactly the same results no matter who performs the tasks. This can be integrated into the Preventive Maintenance tasks to minimize the chances of human errors and infant mortality failures.

Maintenance Prevention: [6]According to the Japan Institute of Plant Maintenance (JIPM), Maintenance Prevention is defined as the use of the latest maintenance data and technology when planning or building new equipment to promote greater reliability, maintainability, economics, operability, and safety while minimizing maintenance costs and other problems. This means that equipment is designed from the start with easy maintenance and trouble-free operations. When developing new equipment, the design, fabrication, and installation may proceed smoothly with a few problems. Still, the test run and commissioning stages may take some time due to debugging problems during the test period. The key to preventing these problems is to ensure that all processing and operational requirements for the new equipment must be fully incorporated into the equipment requirements used by the OEM. It is better to develop equipment in-house in the plant so that the designers of the equipment understand the operating context and environmental conditions of the plant as well as provide much quicker communication to the users and maintenance of the asset.

Autonomous Maintenance: Refers to one of the major pillars of TPM (Total Productive Maintenance) that involves the operators. This is a detailed approach that is done through the 7 steps of Autonomous Maintenance to empower operators to perform basic and routine maintenance which includes detailed cleaning, routine lubrication, and inspection standards on their own equipment. One unique feature of Autonomous Maintenance is the operator's ability to use Visual Control to spot abnormalities and deviations from the norm.

---

[6] Fumio Gotoh, ***Equipment Planning for TPM, Maintenance Prevention Design***,(Portland, Oregon, 1991), Page 72-73

Planned Maintenance: This is also considered one of the major pillars of TPM whose activity is to deliberately improve the current maintenance system. Planned Maintenance aims to optimize reliability and achieve both a Predictive and Proactive stage. This is done in the generic 4 Phase approach, excluding Phase 0 which is the Preparatory Phase. Planned Maintenance will also act as the coach and mentor of Autonomous Maintenance and should be done in parallel with Autonomous Maintenance activities.

DIFFERENT MAINTENANCE TASKS TO ADOPT

The correct maintenance tasks is not about limiting to a single tasks but using all the available maintenance tasks with the aid of an algorithm or Decision Diagram

CM – Corrective Maintenance
CBM – Condition-Based Maintenance
RED – Redesign or Modification
FFT – Failure Finding Tasks
RTF – Run to Fail
PM – Preventive Maintenace

Figure 2.6: Different Maintenance Tasks to Adopt

Turnaround is a scheduled event where an entire system or process of an industrial plant such as an oil and gas refinery, petrochemical plant, power plant, pulp and paper mill, and similar plants will be shut down for an extended period for major rehabilitation and restoration. Unlike ordinary turnaround. This will be done after several years, for example, 5 years depending on the plant's scheduled turnaround activities. Turnaround activities are a common occurrence in many asset-heavy industries. The common thinking is that equipment and assets reach a turning point and will be assumed to deteriorate or wear out, hence the need to shut it down is required to perform upgrades, repair, restoration, inspections, and other maintenance tasks. This is done to comply with government regulatory compliance.

Even when the design, fabrication, and installation of new plant and equipment have gone smoothly, problems still often emerged during the equipment's test operation and commissioning phases. Production people and maintenance engineers struggle to get the plant working properly. They achieve normal operation only after repeated modifications have been made. After the plant has begun operating normally, checking, lubrication, and cleaning to prevent deterioration and failure may be difficult to carry out, as may also be the case for set-up, adjustment, and repair. When equipment is not designed to ease operating and maintenance, both operators and maintenance people will find it difficult to operate, maintain, or even repair the equipment. Some people claim that numerous problems at the initial operation stage are inevitable because of rapid technological advancement and increased speed, size, and equipment automation. Never try to justify the problem like this. Maintenance and equipment engineers must incorporate new processes and operating conditions into the equipment's initial design conditions. To ensure that equipment is highly reliable, maintainable, operable, and safe to operate, avoid relying excessively on outside purchasing. Make full use of in-house technology that your own production, design, and maintenance engineers have accumulated from problems they experience in the past. The thoroughness of the investigations performed at the design stage largely determines the amount of maintenance a plant requires after installation.

These different maintenance tasks must be used to address all the possible failure modes that can occur in the equipment or asset. There will be cases where a combination of tasks should be deployed to address the failure. The role of maintenance is to prevent, predict, prolong, control, manage, or anticipate the failure. This can be made possible by having an algorithm or decision diagram to understand when each of these maintenance tasks is feasible to use or not. Figure 2.6 represents a band, my favorite band, that is the Rolling Stones. They compose of different musicians playing different instruments, with Mick Jagger being the lead vocalist of the group. A band contains a vocalist and several musicians playing different instruments, just like in maintenance, we should not confine ourselves to a single maintenance task, but utilizing all these available tasks will provide a better outcome in addressing the majority of possible failure modes that are likely to affect the equipment or asset.

2.4: Preventive Maintenance Explained

[7]Borrowing from the concept of my dear friend Joel Levitt's book Complete Guide to Preventive and Predictive Maintenance, Preventive Maintenance is a set of activities or maintenance tasks that is performed on the equipment and assets on a scheduled basis. The main goal of performing tasks on a scheduled basis is to extend the equipment's life and to assure its capacity in support of the industry's goals and targets. In Preventive Maintenance, the basic law to consider is that the cost of performing PM must always have to be lower than the cost of the consequences of failure it is meant to prevent. This means that if the maintenance planner estimates that the Bill of Materials (BOM) for conducting a quarterly scheduled Preventive Maintenance is $ 50,000.00 and the cost of the consequences of failure it is meant to prevent is $ 15,000.00, then performing Preventive Maintenance is not feasible in this sense. It should be the other way around. The main reason for performing Preventive Maintenance on our equipment and assets is to be ahead of the failure and prevent them from happening. There will be several activities that will be done on the equipment at different intervals.

[7] Levitt Joel, ***Complete Guide to Preventive and Predictive Maintenance***, Industrial Press Inc. 2003, Page 6

Figure 2.7: Complete Preventive Maintenance Activities Should Include

Preventive Maintenance is also considered as scheduled maintenance of the plant's assets designed to improve equipment lifespan and avoid any unplanned activity. PM tasks include routine cleaning, lubrication, adjustments, calibration, minor component replacement, and overhauls. The reason for performing these tasks is to extend the life of equipment and facilities. Its purpose is to minimize breakdowns and excessive depreciation. In its simplest form, Preventive Maintenance can be compared to the service schedule of a car. The amount of Preventive Maintenance needed at a facility varies greatly from one industry to another. It can range from walk-through inspection of facilities and noting equipment deficiencies for later correction up to shutting down the equipment after a certain number of hours or after a certain number of units had been produced. The primary goal of PM is to anticipate the failure of equipment before it actually occurs. It is designed to preserve and enhance equipment's reliability by replacing worn parts and components before they actually fail in operation. Almost all industries have their regular and routine Preventive Maintenance that is being done in a timely and scheduled fashion. The interval for conducting PM activities varies from one industry to another. While most industries will base their scheduled activities on calendar time, others would be using running hours, the number of volumes produced, the number of cycles, and there is even one industry I know which is a sugar milling industry in which they perform their scheduled Preventive Maintenance when all the sugar cane have already been harvested by the farmers.

In figure 2.7, developing a Preventive Maintenance strategy in any organization should begin with having an inventory of all equipment and assets in the plant. Once the inventory has been completed, the equipment is ranked accordingly based on its criticality. We can categorize all Rank A to be the worse, Rank B is in between the worse and good, and Rank C will be considered good machines. The planner in charge can use this machine ranking to prioritize which equipment and assets will be on the top list for the Preventive Maintenance schedule. The planner will write the detailed tasks, estimate the labor, bill of materials needed, spares, consumables, and the interval of performing the tasks. To make the PM more effective, we can coordinate initially with the Predictive Maintenance people to conduct a reading before initiating the actual Preventive Maintenance activities and we can call this a Pre-Monitoring. The CBM or PdM team can meet with the PM and planners to discuss their findings on what parts, modules to focus on the equipment. After the meeting, the PM tasks will be reviewed for any addition or deletion of tasks. The maintenance responsible for executing the tasks will be withdrawing in the storeroom the needed parts, consumables, and tools needed to execute their PM. Once the needed items had been withdrawn, then the PM tasks will be executed. Once the execution of the Preventive Maintenance tasks had been completed, they can call the CBM or PdM team once again to conduct a Post PM to verify if the readings improved. This Pre and Post PdM is also important to minimize the chances of Infant Mortality or start-up failures. Once these PM tasks are completed, then the PM crew will close the report and a feedback session with the planner will be done if there will be further adjustments in their next PM activities. The planners make the necessary adjustments based on the feedback and the next machine will then be scheduled for the next PM and the cycle continues.

While Preventive Maintenance is generally considered worthwhile, it is important to note that there are risks involved in the process. One of them is human errors. Preventive Maintenance is conducted to keep equipment working to extend the life of the equipment to support our plant goals and targets. Corrective maintenance, sometimes called repair is conducted to get equipment working once again. The primary goal of PM is to anticipate failures in the equipment before it actually occurs. It is designed to preserve and sustain equipment's reliability by replacing worn-out parts and components before they actually fail or break down in operation. In addition to this, operators detect early signs of problems and inform maintenance before it causes a system failure.

2.5: Understanding the Process of Wear

It is very common to hear our maintenance people say they have replaced the part they assumed to have worn out in their equipment. If the boss asks why the part failed? The most common and saturated root cause the maintenance will provide is wear and tear. First, let us define the process of what wear is all about. Wear is defined as damage to any solid surface caused by the removal or displacement of material caused by some means of mechanical action of either a contacting solid, liquid, or gas, causing significant surface damage. The damage is usually thought of as gradual deterioration. It means that it can be a solid to solid mechanical item that will cause wear or a continuous bombarded of either a gas or fluid to a solid object that will cause the object to directly wear out for as long as the bombardment is continuous, just like in the case of a continuous bombardment of a fluid on a pump impeller. If a fluid is constantly bombarding the impeller, after some time the impeller will erode which will cause a decrease in the flow rate. Wear may also mean the undesired removal of material from any contacting surfaces through some means of movement or mechanical action. This means that several parts and spares have movements. These mechanical parts can either slide up and down, left and right, or rotate inside our equipment. Any part, item, or spare that has movements inside our equipment, machines, or assets will be subject to wear and degradation after a definite amount of period.

Although TPM (Total Productive Maintenance) books claim that there are two types of wear, including accelerated deterioration and natural deterioration. Accelerated deterioration means that the part, item, or spare had failed and has not reached its lifespan. We can state that we have a case of premature wear. The latter means that the part has reached its lifespan and eventually and gradually wears out. Suppose we purchase a bearing from a given vendor; in that case, the manufacturer has a projected life associated with the bearing, which is given as L_{10}. Let us assume the L_{10} life of the bearing is 5 years. This means that the reliability of the bearing is given at 90% under a given load and speed with a 10% chance that the bearing will fail under a normal load. However, this L_{10} life of the bearing is theoretical in which it is assumed that everything is in perfect condition. This means that if the load, RPM, speed, lubrication, and environmental condition satisfy the requirements of the bearing, then the bearing will only fail on one condition, which will be a factor of normal fatigue. This means that we are speaking about an ideal condition of the bearing. The bearing's life is determined by the number of hours it will take for the metal to fatigue and that is a function of the load on the bearing, the number of rotations, and the amount of lubricant a bearing receives. Still, in reality, that is not the case in industries. So the end result is that the bearing will fail prematurely before reaching its dictated life because there will be more than a dozen ways that a bearing can actually fail in operations.

To explain the process of wear, every movement that a particular item, part, or spare will cause stress on the part after some time. If I explain this wear process in the simplest way possible, I can speak about a boxing match where the opponent is 1 foot taller and his reach is much 15 inches longer than his opponent. A tactic that will be used by the taller boxer will be to stick and jab his opponent. As the round progresses from the middle rounds 5, 6, 7, and so on, the opponent's face will begin to puff and become swollen since the jabs continuously hit the face. If we talk about mechanical parts inside our equipment and assets, the number of jabs that hit the face represents the number of cycles or movements. Each of these cycles or movements contributes to a small incremental amount of stress. There can only be an (n) amount of stress that the mechanical part can tolerate before the stress finally wins and exceeds the strength of the material. In any failure analysis involving metal fracture, the subject of both metal strength and metal stress must be considered because they cannot be separated or divorced from one another (I think the Philippines is the only country in Asia that does not allow divorce). This means that as long as there is stress, wear will occur until the material's strength has finally been exceeded. Once this happens, the process of wear has been completed. By definition, stress is the force per unit area, often considered as the force acting through a small area within a plane. Strength in the metallurgical sense is the metal part's property to resist the stress imposed upon the part.

Therefore, all fractures will be caused by stress, a version of the weakest link theory, which applies that a fracture will occur if the local stress finally exceeds the strength of

the materials.

I have several books on the subject of metal fracture and metallurgy in my home. Each time I intend to absorb this message, it tries to confuse me more than I can comprehend regarding the subject of wear. I seek help from metallurgical experts on the subject matter, yet their explanation was so technical and profound that it was simply too complicated for me to comprehend. I almost gave up on this subject of how the wear process took place until I visited my old-time friend and sensei Mang Tibo. I asked him his thoughts about the wear process. I have so many questions to ask. He told me to calm down and gave me a glass of water. I asked him how the wear process occurred in the equipment parts and components and what this message means that all fractures are caused by stress which is a version of the weakest link theory. If the local stress exceeds the material's strength, then wear takes place. Mang Tibo is like a father to me. He was my earliest mentor on maintenance. He explains things in a very simple way, that even a small kid can comprehend. Mang Tibo lives in the slums and does not possess any certification in maintenance, but oh boy, he knows his stuff.

Mang Tibo told me that he has 8 children and all of them have already grown up. All his children were already married and have their own families to live with. He once told me that there were many times in his life that his children convinced him to move up with them so that he could live a better life and live the comforts of life for the remainder of his days, but the old man was simply too stubborn and refused them all the time. He preferred to stay in their small home, which he and his family once stayed in for more than 50 years. He had set aside his ambition for a good life, tried to live with his principles, and would rather settle down in their old home with his loving wife. Mang Tibo invited me inside his house. His house had no furniture and just some wooden stool that he made a long time ago. His house had no ceiling, and you can see so many holes in the roof when you look up. I can just imagine his life and be toasted mostly during summertime in our country, the Philippines. Mang Tibo never finished school because when his parents died when he was 15 years old, he was left responsible for taking care of his seven brothers, so he started working early. He was the bread earner during that time. He invited me to sit down (just like in the days of the La Cosa Nostra and the Mafia Bosses) and gave me some biscuits as we discussed his humble version of the wear process. As we sat down, he told me that he had stayed in this house for more than 50 years, as far as he can reminisce. He told me to look at the floor and asked me what I saw. I said that I could see a lot of small dents and fractures on the floor. The floor in his house was just cemented with no tiles. He told me that it used to contain no fractures when their house was new and asked me what I thought would have caused these tiny bits of fracture on the floor. As I tried to think of an answer, I looked up to think, which is a common gesture when people think. They look up in the sky, not to look for birds or superman but just to think. As I looked up, I saw these tiny holes in the roof since his house had no ceiling at all, and each of the holes in the roof was located perpendicular to where the fracture was located on the floor, and it gave me an idea as to what caused the damage on the floor. It was definitely the leak caused by the rain throughout the years. You see, in the Philippines, there are only two weather seasons: the summer months and the rainy season. I told him it was the leak caused by the rain that damaged the floor. He smiled and said that I was correct. He said that when this floor was new, cemented, and shiny fifty years ago, he did not place tiles on the floor. The floor has its own strength of materials. Each drop of rain that fell on the floor represents one stress, and each stress continues to drop drip by drip into the floor for fifty years during the rainy seasons. A few drops definitely will do nothing and will not damage the floor. Then as the days passed by, a hundred drops, a thousand drops, and as the year passed by, a million drops, a billion drops, he said that although he never counted the drip, which was understandably foolish. All he knew was that the drop of rainwater continued until the stress had ultimately exceeded the floor's strength eventually causing the fracture and damage on the floor. That was when I finally understood how the wear process occurs. Mang Tibo does not have any Master's, Ph.D. degrees, or maintenance certifications. Neither is he a metallurgist nor did he finish his schooling, but his experience tells it all, and that is all I have to know about the process of wear. You don't need to be a rocket scientist to know that; all you need is just a modicum of common sense.

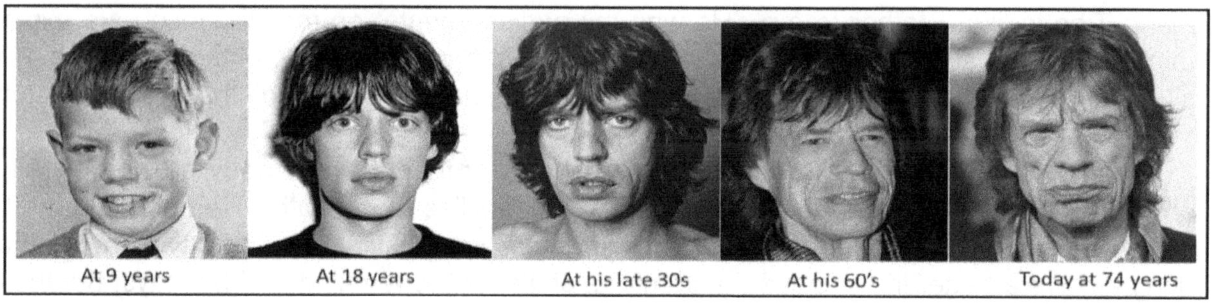

Figure 2.8: Mick Jagger, Then and Now

Our equipment is not a plug-and-play appliance or instrument, just like a TV set or mobile phone that when you plug it in, it will run even when you fall asleep while watching your favorite Netflix movie or sitcom show without any sort of problem. Inside our equipment are many parts composed of electrical parts, mechanical parts, electronic parts, and electromechanical parts. Some of these parts, especially mechanical parts, have movements, and every single movement that these parts make is equivalent to one stress, which is also one complete cycle. A piston, for example, in a diesel generator when set in motion, will move in an upward and downward movement. This up and down movement will stress the piston out after (n) several cycles. This simply means that as the number of cycles increases, there will be tiny bits of metal fragments that will chip off on the metal. It will not last forever. It can last longer once we increase the strength of the material to a stronger one, but again, a new problem will arise as the stronger the material's strength, the more the material is prone to brittleness. A piston that moves from an upward and downward motion would be subject to stress after an (n) number of movements. When the equipment is not loaded or is at rest, the piston remains on the centerline. When we start the equipment, the piston will move from the centerline to an upward motion, back to the centerline, then from a downward motion, and once again back to the centerline, which is equivalent to one stress or one cycle. In vibration analysis, this simply represents one sign wave. Once a small metal debris fractures from the surface of the metal some of this metal debris can be trapped between clearances thereby abrading the surface which generates a new batch of metal debris.

One movement is called one cycle or one stress. After hundreds of thousands of upward and downward movements, tiny metal debris is removed from the piston and the wall, and the engine's blowby increases. This is now the start of the wear-out process. Once the wear process has reached its limit, the equipment can no longer continue its operation where the maintenance people intervened and finally replaced the part that had actually worn out. This is the completion of the wear-out process. The process of wear does not happen overnight or even in a week. It is a gradual process, just like when we were kids, up to this point in time where we have grown up and get wrinkles on our faces.

Wear is a big problem for mankind; on the other hand, wear is also good because this has provided employment to billions and billions of people worldwide. If we try to speculate for a while and think. What if wear does not exist at all? What if God comes down to earth and asks us if we want to remove the process of wear on everything? What will be your answer? I think there will be chaos and disorder, and many people will stop breathing and die. If wear no longer takes place, we don't need to buy any pair of shoes because our shoes will last a lifetime, and the entire shoe industry in the world will close down as a result, and its entire people will be unemployed. Because wear will no longer take place, then your hair will no longer grow, and your barber will no longer have work. You will no longer suffer any tooth decay, and that goes for your dentist too. They will close down their clinic. These people will lose their work and result in other means just to survive. All factories that provide replacements such as tires, clothing, brakes, clutches, bricks, food, drugs, and much more will close down, and billions and billions of people from all around the globe will be laid out of work and become unemployed, not because of this **covid 19** pandemic, but because the wear process was eliminated. Therefore, let us thank God that there is wear in this world and so to conclude, always remember that wherever you are or wherever you go, wear will always be with you

Many experts say that everything that is man-made will eventually wear out; well, I tend to disagree with that. There is one thing that will never wear out, and that is the music of the Rolling Stones. Just would like to mention, Rest in Peace, Charlie Watts (Rolling Stones drummer) who passed away last August 24, 2021, at the age of 80. Oh Yeah. ☺.

2.6: Most Common Types of Mechanical Wear Explained

It is important for every maintenance in industries to associate themselves and be familiar with the different types of wear and how each of these wear actually looks like so that they can have some idea on how the wear process is actually taking place and what is really happening inside our assets, equipment, and machines.

According to Dr. Ernest Rabinowicz, equipment may lose its usefulness because of three factors. First, the equipment becomes obsolete, where either the equipment or the product it is producing will no longer be used, and the equipment will therefore be retired or decommissioned. This consists of around 15% of why equipment can no longer be used. The second factor is accidents where the equipment encounters a major breakdown or failure. The cost of rebuilding or repairing the equipment will almost equal to purchasing new equipment, which constitutes around 15% of why equipment loses its usefulness. But the bigger percentage of why equipment loses its usefulness is due to surface degradation, which constitutes around 70% of why equipment is no longer used. And under surface degradation, 20% will be due to corrosion, and about 50% will be caused by mechanical wear. In this regard, mechanical wear can either be caused by abrasion, adhesion, erosion, or fatigue. Dr. E. Rabinowicz states that surface degradation cannot be eliminated and is inevitable, but the good news is that the process of surface degradation can be controlled, or it can be prolonged. What is important for our dear maintenance people is distinguishing between the different types of wear, how they looked on mechanical parts, their markings, and, most importantly, what caused them to occur. With this kind of knowledge and information, maintenance can understand how the wear process happens, how to prolong it, and what should be done to control it. When we can identify the type of wear, then we have already solved 50% of the process, the next question is determining the reason why and how it happened. The following are the most common types of wear.

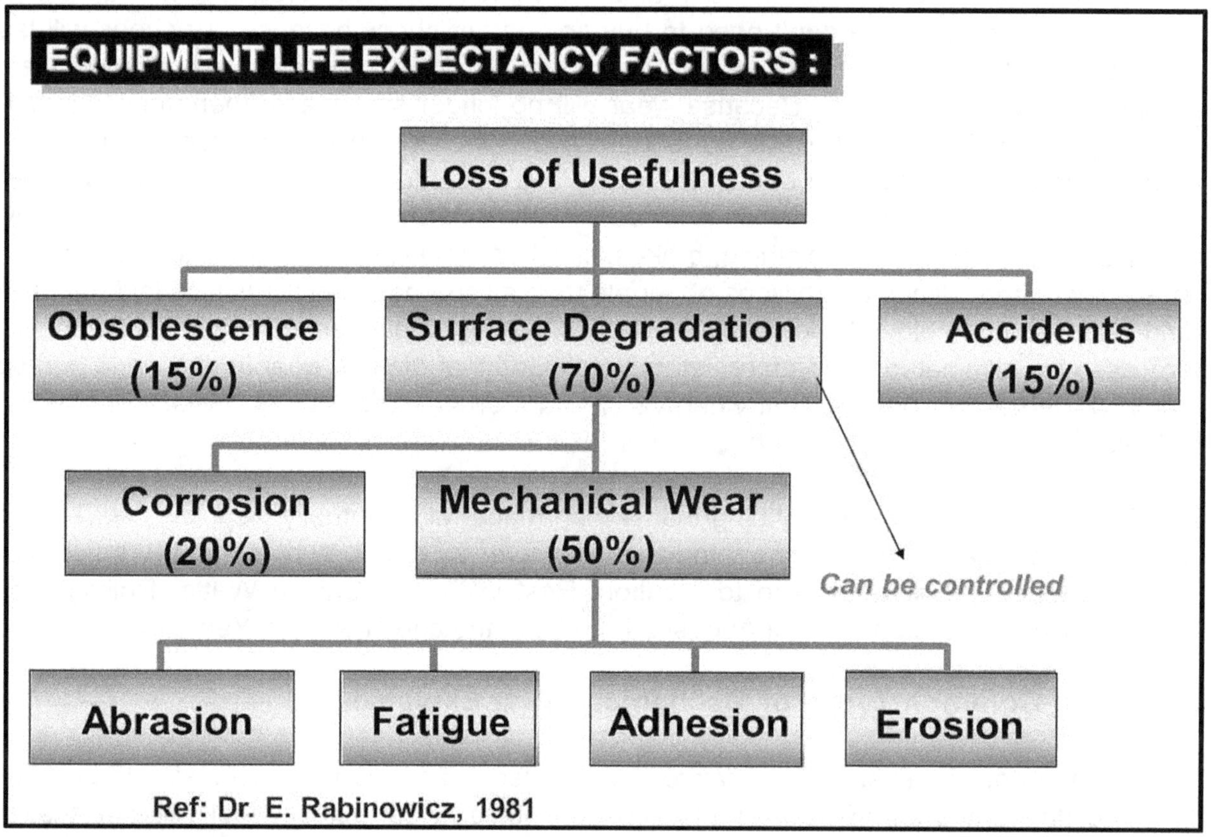

Figure 2.9: Why Equipment Can No Longer Be Used

Abrasive Wear: This type of wear can be categorized by a single keyword, which is cutting. This type of wear occurs when hard particles or contaminants are suspended in a fluid or projections from one surface roll or slide under pressure against another surface, thereby cutting the opposite surface. This wear mechanism is basically the same as machining, grinding, polishing, or lapping for shaping materials. When you go to a machine shop, and a lathe machine is in operation, cutting a shaft produces long metal debris as the cutting tool comes in contact with the metal; this is actually how abrasion occurs. If we get a mechanical part and placed it on a Scanning Electron Microscope (SEM), and enhance the magnification to 1000 times. We noticed that the surface of the mechanical part is not actually perfectly smooth. Meaning that there are micro or very tiny peaks and valleys that surround the mechanical parts. These peaks and valleys are usually called asperities. When the peaks and valleys of two opposite mechanical parts moving in the opposite direction come in contact with one another, they will fracture or tear off. One might be thinking that this metal debris will settle at the bottom which is not the case. Most lubricating oil contains an additive called dispersant whose function is to suspend these solid metal fragments so that they can flow together with the oil so that it can be captured by the oil filter. The problem is that these small metal fragments will just pass upstream of the filter and return back once more to the system. These small fragments of metal removed will flow together with the lubricating oil once again and might end up in tight clearances and start the process of abrasion. In layman's terms, if you are involved in Preventive Maintenance overhauls and replacements, you see several spares or mechanical parts and items with long hairline scratches, then they are caused by abrasion.

There are two types of abrasion. A two-body abrasion occurs when the lower portion of the mechanical component's peak or asperity comes in direct contact with the upper component's peak or asperity since they move in opposite directions. The weaker of the two asperities will break off, creating a third party, which later on can become the other type of abrasion, which is a three-body abrasion. In a three-body type of abrasion, a solid contaminant or particle usually as hard or harder than the two mechanical components gets trapped in the two mechanical components' clearance. Suppose the contaminant gets trapped in the upper portion of a mechanical part; it will cut or rub the lower portion of the mechanical part. Signs or distinguishing marks from abrasion will be some hairline scratches on the component or part. There are two ways to control abrasion. The first is to harden the mechanical part since the principle for abrasion to occur is that the contaminant or particle trapped should be as hard or harder than the mechanical part itself. But making the material harder will create a secondary problem, which will make the material more prone to a brittle fracture. A brittle fracture is usually a fracture with little or no permanent deformation. Another way of controlling abrasion is to remove the contaminants themselves through the oil by applying absolute high efficiency and beta rating type of oil filtration in which the filter can guarantee to remove more than 99% of the solid contaminants present in the oil. The higher the beta rating of filters in lubricating hydraulic or engine systems, the more solid contaminants it can remove. The cleaner the oil, the longer the wear process can occur in mechanical parts.

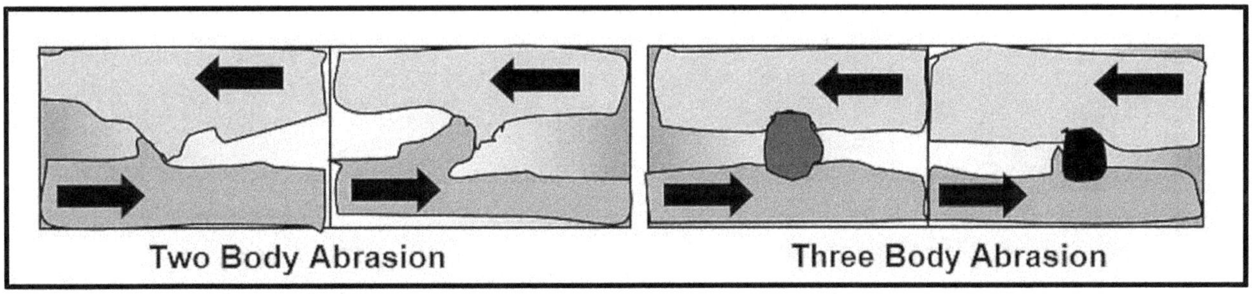

Figure 2.10: Two-Body and Three-Body Abrasive Wear

Adhesive Wear: This type of wear occurs when a peak or unevenness from one surface comes in contact with the other surface's peak or asperity. Adhesive wear has been commonly identified by the terms galling or seizing. There may be instantaneous micro-welding due to the friction or heat involved due to the loss of the film on the lubricant. For adhesive wear to occur, there must be a metal-to-metal contact. When the metal comes in contact with each other, friction occurs, and heat is generated. There will be some form of discoloration that occurs in the process. Micro-welding occurs, and the final stage will be a metal breakage or fracture on the mechanical component. Adhesive wear is produced by the formation and subsequent shearing of welded junctions between two sliding surfaces. For adhesive wear to occur, surfaces must be in intimate contact with each other. One way of controlling adhesive wear is by applying the oil's correct viscosity to the mechanical component. Surfaces held apart by very thin lubricating and oxide films will reduce the tendency for adhesion to occur. Adhesive wear starts out on a small scale but rapidly escalates as the two sides alternate weld and tear metal from each other's surfaces. That is what adhesive wear is all about, microscopic welding. The heat produced at the contact interface is very high which is near the two metals' melting points touching each other. Although it started as adhesive wear, these small metal fragments that break off can also become abrasive wear once these metal fragments join the flow of lubricating oil and get trapped in tight clearances between mechanical parts.

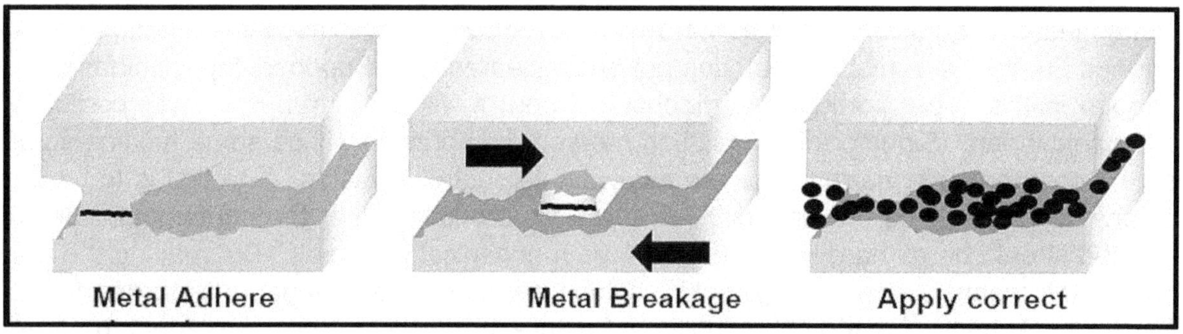

Figure 2.11: How Metal Adhesion Takes Place

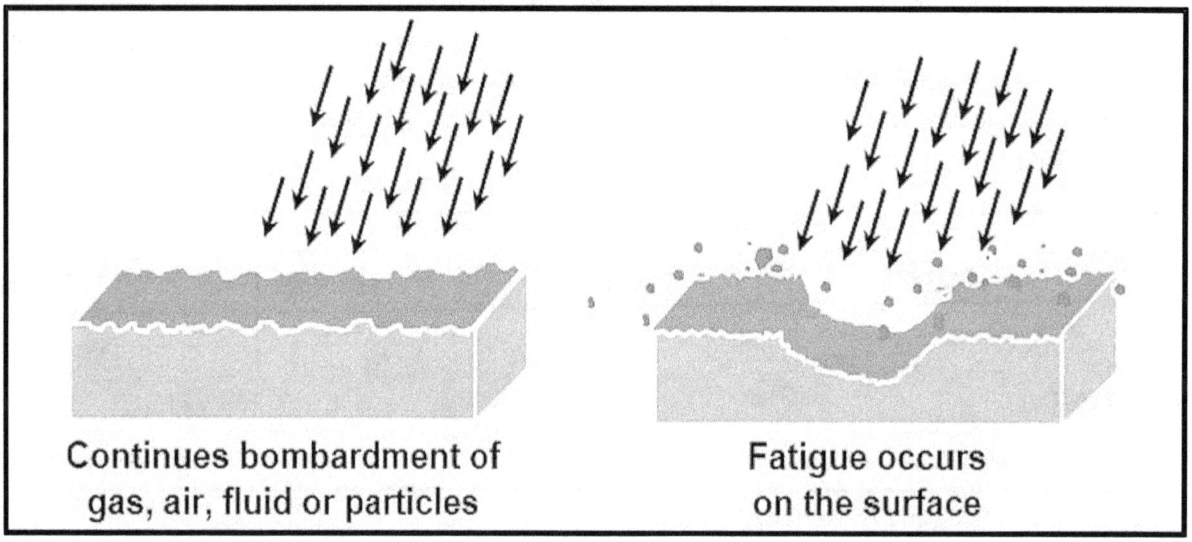

Figure 2.12: The Process of Erosion

Erosive Wear: This type of wear happens due to mechanical interaction between a surface and a fluid, a multi-component fluid, or impinging liquid or solid particles. Erosive wear occurs by the impact of particles upon a surface with a resultant material loss due to fracture. This can occur both on metals and ceramics. The continuous bombardment of particles, either solid, fluid, or air, contacting the surface cuts a tiny particle from the component over time. This tiny particle that is removed is insignificant at the moment but, over some time, will have its impact. Erosive wear can be expected to occur in metal parts and assemblies where the above conditions are present. Common problem areas found are pump impellers, fans, steam lines, nozzles, inside of sharp bends in pipes and tubes, and sand shot blasting equipment. Sudden sharp curves or bends cause more erosion problems than gentle curves in tubes and pipes. One way to control erosion is to strengthen the material's hardness, which can prolong the erosion process, but again can subject the metal more prone to a brittle fracture. When a fluid continuously bombards the pump's impeller, the pump's flow rate will be reduced after a given period. This is because the impeller's tip will start to erode, which results in an increase in the clearance size of the impeller and the volute casing inside the pump. When this happens, internal leakage occurs where not all of the fluid is pumped out by the impeller but rather returns to the pump. The end result is that the flow rate will drop. When the reduction in the flow rate is no longer acceptable to production, maintenance will schedule to replace the impeller. Again erosive wear can also turn out to become a source of abrasive wear once the metal fragments get trapped in tight clearances.

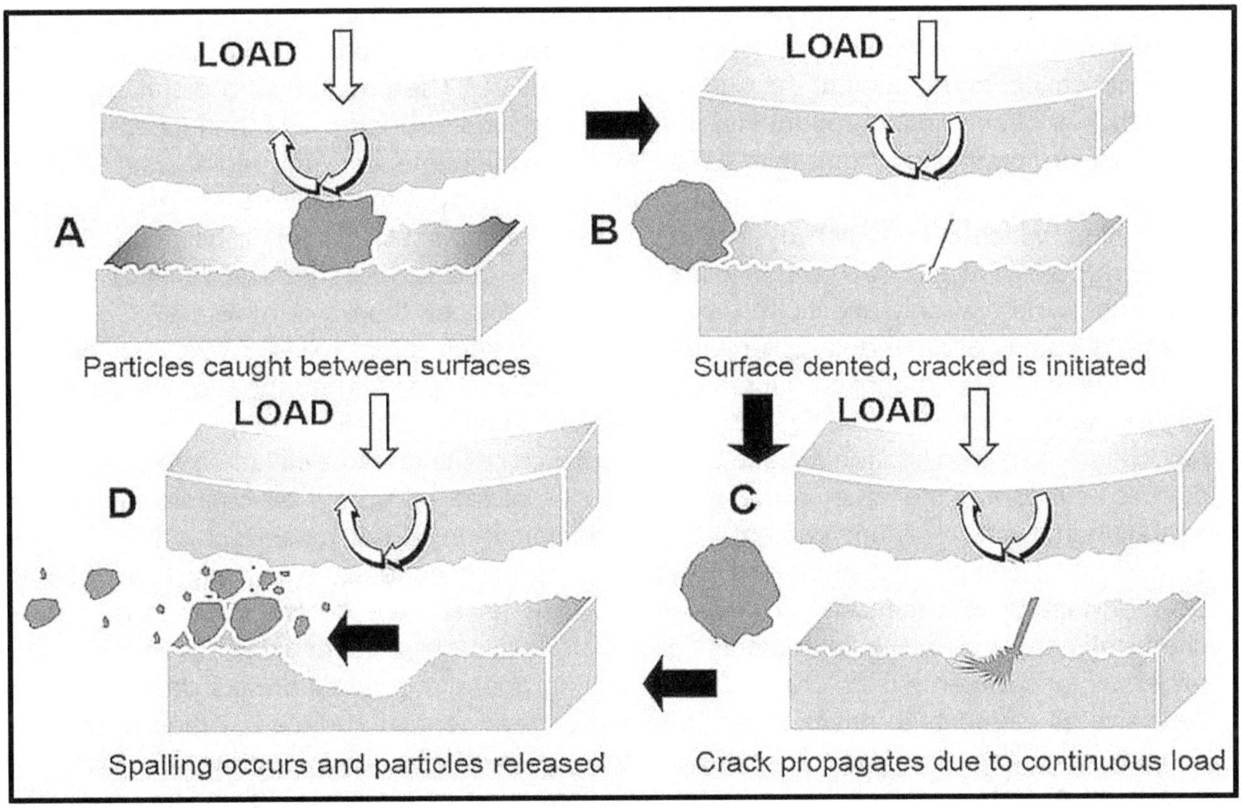

Figure 2.13: The Process of Fatigue Cycle

Fatigue Wear: This type of wear results from a continuous cyclic slip under repetitive load applications for thousands or even millions of load cycles. These fragments removed can cause abrasive wear and other damage when carried by the lubricant to other parts of the mechanism. Some fractures will start as microscopic cavities and may stay microscopic if the equipment is no longer loaded. Still, it will start off as a microscopic fracture most of the time but gradually become larger after continuous load. The last stage is a fracture or crack. Fatigue wear is also a phenomenon leading to fracture under repeated or fluctuating stress having a maximum value less than the material's tensile strength. Fatigue fractures are progressive, which can start as a micro-crack and propagate and grow into larger ones that cause the destruction of mechanical components. For fatigue wear to happen, a contaminant or particle is caught between clearance, the surfaces are dented, and a crack will be initiated. After (n) number of cycles, the particle trapped is released, and the crack starts to propagate. The last stage will be a fracture of the component. Again fatigue wear can also result in abrasive wear if these micro fractures will get trapped between clearances on mechanical parts. Fatigue fractures are generally considered the most serious type of fracture in machinery parts since they can occur in normal service without excessive loads and under normal operating procedures. There are 4 points to understanding fatigue. First, without stress fluctuations, fatigue cannot happen. Second, fatigue happens at stress levels well below the tensile strength of the material. Third, if corrosion is present, the fatigue strength of metals continuously decreases, and lastly, the cracks will take a measurable amount of time to progress across the fractured surface.

Corrosion: This is a wear process of materials caused by chemical, electrochemical, or other reactions. It is also defined as the destructive and unintentional attack of a metal. It is electrochemical and begins at the surface of the metal. The problem with metal corrosion in industries is of significant proportions. It is estimated that around 5% of the industry's income is spent on preventing corrosion as well as its maintenance and replacements due to corrosive reactions to the environment. There is an actual material loss in metals through dissolution, which is corrosion by forming a non-metallic scale or film. The main difference between corrosion and rust is that corrosion occurs due to the chemical reaction on metal surfaces, while rust only affects iron metals that are exposed to air or moisture. Corrosion is when the metal is converted to a more stable form, such as its oxide, hydroxide, or sulfide state, leading to its gradual deterioration. This usually happens when the metal is exposed to oxygen and moisture, creating red iron oxides, commonly called rust. For parts under lubrication, corrosion is often caused by the depletion of the additive called corrosion inhibitors or rust inhibitors, either due to the extended use of the oil or the oil containing too many contaminants and moisture content. Although corrosion is not catastrophic from a safety standpoint, it can be disastrous if it fractures metal components, especially in vessels that contain hazardous chemicals. Corrosion is visible as surface deterioration caused by the chemical action of active ingredients in the lubricant. It includes acid, moisture, foreign materials, and extreme-pressure additives. During operation, the oil breaks down and allows corrosive oil elements to attack, as in the case of gear contact surfaces. When a material is subject to corrosion, it loses its material strength, hardness, and toughness. A corrosive attack on a metallic structure can cause the formation of small cracks, dents, and pitting, which further leads to deeper cracks over time due to tensile stress acting on the edges of the cracks. These situations can cause catastrophic failure without significant deformation of the structure.

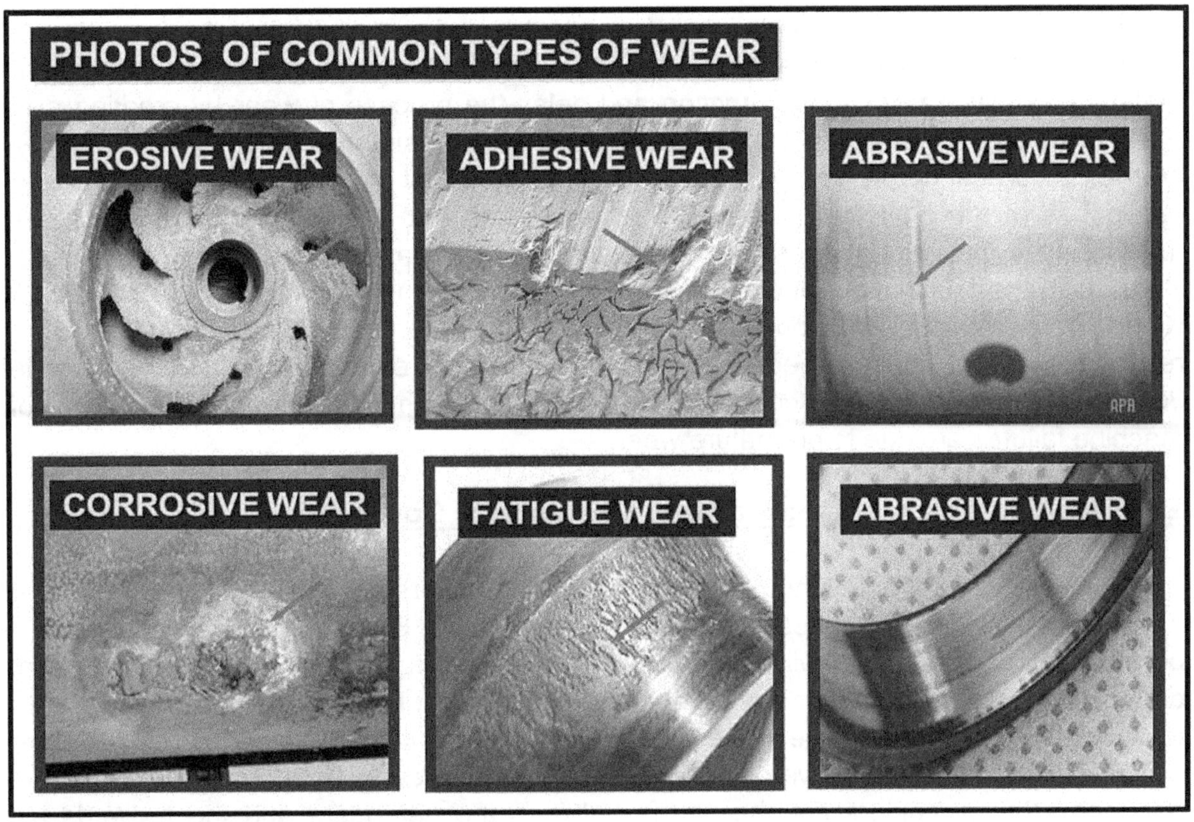

Figure 2.14: Photos of the Different Types of Wear

SUMMARIZING WEAR :

WEAR	PRIMARY CAUSE	HOW TO CONTROL	CAN OCCUR ON
Abrasive	• Particles between two adjacent moving surface Also known as cutting wear, contamination	• Increase hardness of the material • Upgrade filtration by using high efficiency and beta rating	• Rod cylinder wear • Pump Gear, vane, piston • 2 metals with clearances • Bearing raceways
Adhesive	• Loss of surface film • Surface to surface contact due to lack of lubricant which cause metal bonding.	• Application of correct oil viscosity film. • Improve metal surface smoothness or apply coating	• Any two metal bonding together due to the loss of film from the lubrication
Erosive	• Continuous bombard-ment of particles and at high fluid velocity.	• Use of clearance protection protection filter • Minimizing 90 degree pipe to reduce erosion	• High velocity flow on servo, proportional valves, control valves, pump impellers
Fatigue	• Particles damage since surface is subjected to repeated stress and ends with a fracture.	• Control contamination • Correct loading • Reducing machine vibration	• Bearing surfaces • Gear wear • Tooling, dies and punches

Figure 2.15: Summary of Wear

Friction is simply defined as the resistance to its relative motion. It is the force that resists motion. Wear is the loss of materials caused by friction. Lubrication is the use of fluid in solid materials to minimize or reduce friction and delay the process of wear by creating a film to prevent metal-to-metal contact. The subject of Tribology simply tells us that the energy loss due to friction cannot be recovered and is considered lost forever; therefore, the study and application of tribology can prepare us for designing and maintaining better equipment to maximize the longevity and lifespan of equipment's parts and components. In the report by Peter Jost, the founder of Tribology, friction, wear, corrosion, and lubrication cost the United Kingdom over 500 billion pounds per year or roughly around 639,749,500.00 USD. He reasoned out that significant education and research are required to reduce these losses by 30% of the primary energy consumed by friction throughout the world, while 60% of the machine failures are due to premature wear.

2.7: The Maintenance Organization: Everyone is Connected

There are several functions of maintenance in a mature organization, and each of these functions is as good as its weakest link, just like a chain. Each of these functions can never be an isolated group, but it will also depend on the other functions of the entire maintenance organization to be successful. A football team has 11 players. Each of these players will have their own designation and role in the game. The positions include 1 goalkeeper, 4 defenders, 3 midfielders, and three forwards. The goal of the goalkeeper is to block the ball from the opponent, the midfielders are the bridge between the defenders and the forwards usually responsible for carrying the ball, while the forward's role is to score a goal for the team. Just like in a game of football, the different functions of the maintenance organization have a specific role to play and will be needing each other, where the maintenance manager acts as the overall coach. Each of these people involved in the different functions relies on other functions so that together, they can achieve their goals on maintenance. What I am trying to say is that every function of maintenance is connected, and the strength of your entire maintenance organization depends upon its weakest link.

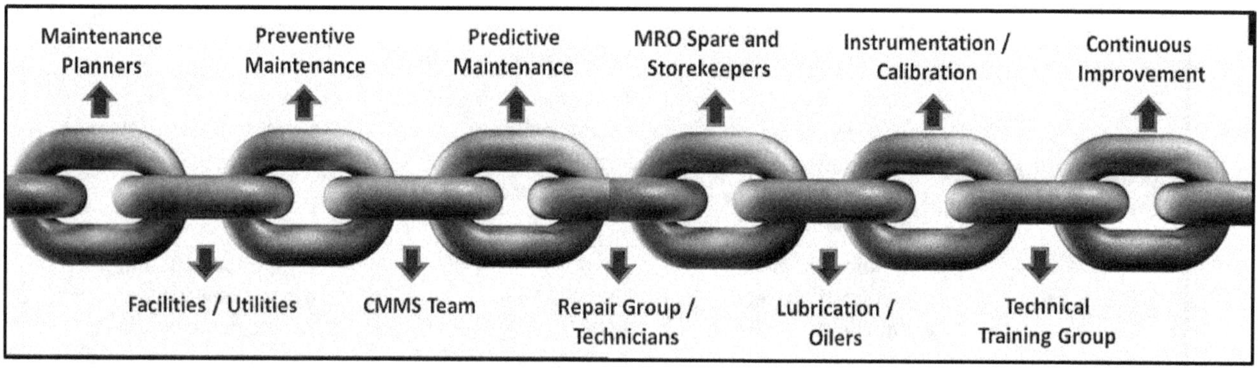

Figure 2.16: Maintenance is a Good as its Weakest Link

- People involved in executing PM depend on the plan of the maintenance planner.
- The planner is also dependent on the feedback of the Preventive Maintenance team.
- The planner is also dependent on the availability of the parts, items, tools, and consumables from the storeroom needed for maintenance.

- The Predictive Maintenance must be in constant communication with the Preventive Maintenance so that only parts with problems will be subject to replacement, or overhaul
- For manufacturing plants, the entire operations and maintenance will depend on the Facilities / Utilities of the plant. These are the people who will provide the power, air, energy, water, and other resources for production equipment and machines to run.
- Instrumentation and calibration people ensure that test instruments, gauges, and protective devices are calibrated and functioning without any deviations from the norm.
- CMMS depends on the output of the entire function on what to automate and streamline. The outcome entirely depends on what we want to populate in the system.
- But all of these functions depend on the very foundation which is training as this function will provide us the knowledge on how to perform and execute our work correctly. Knowledge is the very foundation where we build our skills: The main role of each of these functions includes:

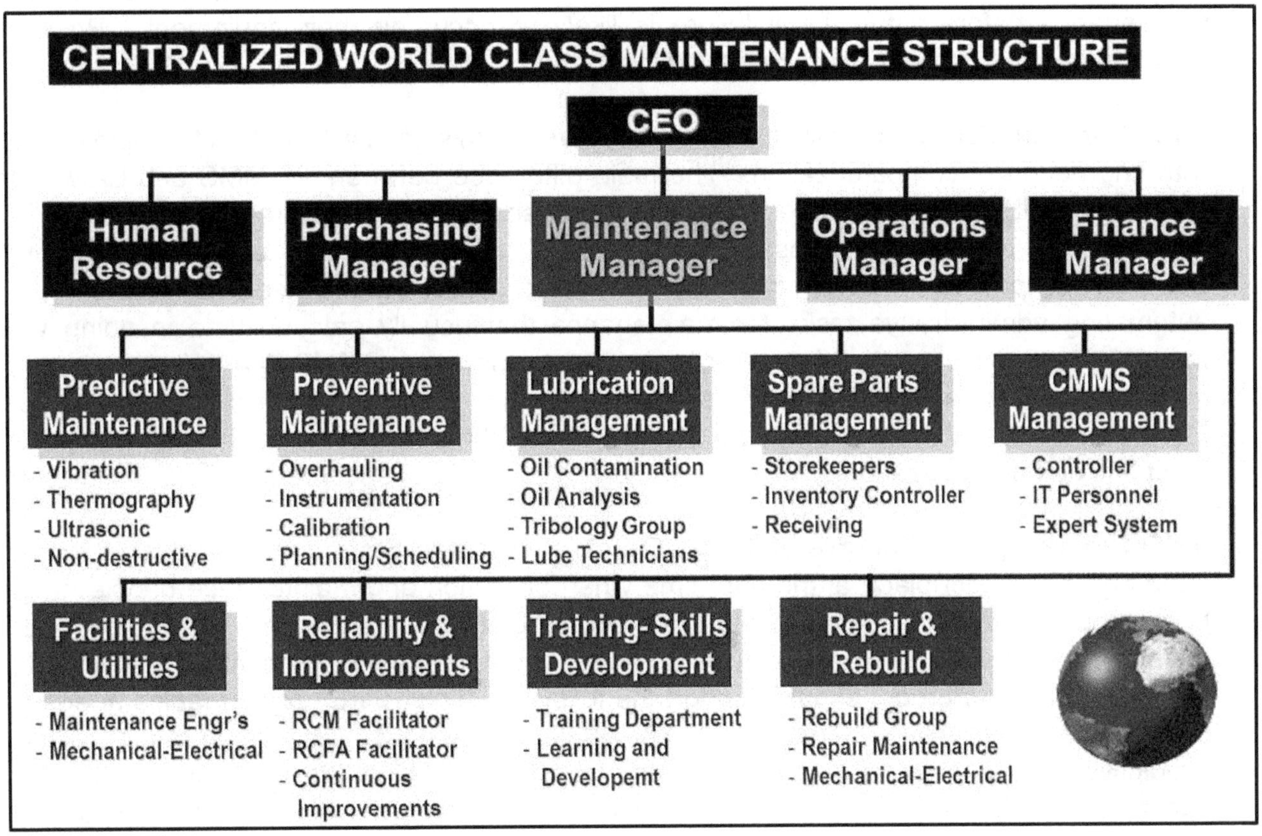

Figure 2.17: World Class Maintenance Organization

Maintenance Planners: Provide the scheduled Planned activities for Preventive, Predictive, and Corrective Maintenance. The planner provides the details of the maintenance tasks to be performed, labor estimates, duration, bill of materials, consumables, and MRO spares needed for a scheduled Preventive Maintenance and Corrective Maintenance from the PdM group.

Preventive Maintenance Crew: People involved in these activities will be responsible for the execution of tasks dictated by the maintenance planner based on the schedule provided. Common activities will include cleaning, routine lubrication, inspection, conducting failure-finding tasks, calibration, scheduled replacements, and overhauls.

Facilities / Utilities Group: The role of the maintenance people from the facilities is responsible for sustaining and maintaining the industry's buildings, facilities equipment, and assets to ensure the production operates without any problems. The role of the people involved in the facilities is to ensure that their equipment and assets are operating daily. People involved in facilities perform routine walk-around inspections, conduct repairs when needed, contract works for third parties, and perform scheduled Preventive and Predictive Maintenance as well. The entire operations machines and assets are entirely dependent on the facility's equipment. If a facility's sub-station fails, then the entire operations will halt.

Predictive or Condition-Based Maintenance Group: The main role of this group is to anticipate or predict failures through the use of their Predictive Maintenance instruments. Their role is to monitor and analyze critical assets in the plant for any signs of Potential Failure. Once a Potential Failure is detected, these people will recommend corrective maintenance before a functional failure is likely to occur on their equipment. Predictive Maintenance can be performed online, offline, or both.

CMMS or EAM Group: Usually the people involved in this case will be the IT people or those with knowledge of computerization. The main difference between a CMMS and EAM is the scope. CMMS or Computerized Maintenance Management Software is designed for the maintenance department whose function is to streamline and automate the entire maintenance process. An effective CMMS should support these needs by capturing and automating administrative tasks for maintenance that usually eats up time in doing when manually, and minimize the chances of human error as well. CMMS is a software design to automate and streamline the process of maintenance, while an EAM or Enterprise Asset Management Software will not only include automating maintenance but also the other functions of the organization.

MRO Storekeepers: The role of the MRO Spare Parts and Storeroom is to ensure the right parts get to the right place at the right time when operation and maintenance people needed them most. People in charge of the storeroom should provide the item or part needed by maintenance, supply them as quickly as possible, yet at the same time control the overall cost of spares. The storeroom should support the following items, parts, consumables needed for a scheduled Preventive Maintenance, corrective maintenance performed as a result of Predictive Maintenance, and parts needed for repairs. MRO Storekeepers should be in constant communication and collaboration with the maintenance function as these are their primary customers.

Lubrication Group: These people are responsible for all lubrication activities in the plant. Although routine lubrication can be performed by operators, especially those involved in TPM's Autonomous Maintenance activities. People involved should be responsible for generating a plant-wide Oil Contamination Control Awareness Campaign, and oil analysis activities. Note that Oil Analysis can also be included in the Predictive Maintenance or CBM group.

Instrumentation and Calibration: Calibration is the process in which instrumentation and equipment used in industries are monitored and maintained to ensure they continue to give accurate and reliable results and ensure that deviations are corrected. The role of these people is to inspect their test instruments on a scheduled basis to ensure that the reading does not deviate and meets expected results. If there is any deviation found, that piece of equipment or instrument will be further analyzed, adjusted, calibrated, or repaired as necessary.

Repair Group: These are people responsible for repairing and restoring the equipment when it failed or breakdown in operations. These people can also be involved in performing corrective maintenance based on the recommendation of the Predictive or CBM group. Their role is to restore the equipment back to an operational stage when it fails.

Continuous Improvement Group: In some industries, a full-time person will be in charge of this group which monitors all the improvement activities of the plant. People involved in the maintenance improvements are usually a cross-selection of people from the maintenance function addressing a particular problem using different analytical problem-solving tools such as FMEA, RCM, RCFA, and basic problem-solving tools. Once the improvement project has been completed, the team disbands and goes back to their usual roles and responsibilities.

Technical Training Group: Although in other industries, these functions may be absorbed by the Training Department, Human Resources, or Learning and Development function. The role of these people is to conduct a Training Needs Analysis for Operators and Maintenance and provide the training needed so that both operators and maintenance can perform their job correctly right the first time around. Training should be based on the needs so that both operators and maintenance can build their skills in performing their jobs correctly. Training is the venue where we absorbed knowledge. Knowledge will provide us the means to develop our skills. Mastery is the point in time where we can transfer this knowledge to others.

2.8: Why Preventive Maintenance is Important

All industries depend on their equipment to operate and fulfill a particular function and that is to produce a product that will generate revenue for the plant. Equipment should operate reliably based on its design life cycle. A piece of equipment that is not well maintained will degrade much faster than we would have anticipated. This is where Preventive Maintenance comes into place. A piece of equipment without any form of Preventive Maintenance will cut short its entire life span by 30% or sometimes even more. The equipment is not a plug-and-play instrument just like a TV or a mobile phone as it contains not only electronic, and electrical but mechanical parts as well. For as long as there are movements inside the equipment, then there are only an (n) amount of cycles before it wears out. The main reason why we perform Preventive Maintenance is to anticipate or prevent the failure from occurring by doing a maintenance task before the failure happens. A failure or breakdown of the equipment will result in downtime and will reduce the income-generating revenue of the industry.

Just like a car, one maintenance performed is to rotate the tires of the car when it reached 8000 to 12000 kilometers. The reason for doing this is that the thread of the tires in front will wear out faster than the rear that is why the front tires will be placed on the rear after reaching this mileage. If the tires will not be rotated this means that we need to purchase two tires for the front which will add to the expense of the driver. Similarly, if we speak about our equipment and assets, performing Preventive Maintenance allows us to inspect the equipment and perform the necessary task needed to keep the equipment running. Although there is a cost in executing Preventive Maintenance, the cost of not doing PM will be greater. The cost of doing reactive maintenance will always be higher if the breakdown happens first instead of preventing it from happening. Unexpected failures can interrupt production and the chances of encountering secondary or tertiary damages are possible. Preventive Maintenance can avoid these situations if done correctly. Problems that are caught early can be addressed which can avoid the chances of catastrophic failures from happening. According to a 1979 Dun's Magazine Review regarding the high cost of bad maintenance, it is estimated that a repair job done on an emergency basis requires three times more manpower, time, and money than a scheduled repair.

Conducting Preventive Maintenance regularly can extend the lifespan of the equipment but the key to making Preventive Maintenance work is to carry out the correct task precisely at the correct interval. As discussed in the previous section that there will be parts that will eventually wear out and these parts should be under the watch of Preventive Maintenance. A boiler may require 25% more fuel to operate if it is not properly tuned. A ¼" leak in a 100 psi system will be a loss of around 100 CFM of compressed air. This will be approximately $12,000 in annual wasted power cost based on 24 hours a day compressor operation with a power rate of $0.07/kW for the plant. Leaks can be a significant source of wasted energy in an industrial compressed air system, sometimes wasting 20-30% of a compressor's output. A typical plant that has not been well maintained will likely have a leak rate equal to 20% or more of the total compressed air production capacity. On the other hand, proactive leak detection and repair can reduce leaks to less than 10% of the compressor's output. In addition to being a source of wasted energy, leaks can also contribute to other operating losses. Leaks cause a drop in system pressure, which can make air tools and pneumatic machines function less efficiently, adversely affecting production.

The bottom line is that the cost of doing maintenance will always have to be lower for a plant with a Preventive Maintenance program in place. But there is a catch to this since whatever task we perform first should address a particular problem or failure mode and second, we need to be precise in how we execute Preventive Maintenance since performing PM too much or too little will likewise have a detrimental effect on the equipment being maintained. Being reactive will cost the industry 3 to 4 times or even more than those with a structure Preventive Maintenance program in place.

2.9: Is Preventive Maintenance Feasible or Not?

We must understand from a maintenance standpoint that Preventive Maintenance activities are most effective only on age-related or wear-out failures; however, we still need to address the most frequent failures, which are random and infant mortality failures. I would also like to emphasize that not all similar parts will fail in the same period. Preventive Maintenance is only feasible when the probability of failure of parts will survive at a specific age, and we are speaking about the useful life of the part or component. Maintenance must understand that infant mortality failures are introduced during overhauls when the equipment is not properly put back together. Overhauls and replacements should only be done by skilled craftspeople equipped with the right tools and knowledge to do the job correctly.

PM is feasible to use when the part or component wears out directly with respect to its operating age, and the part will survive this defined age. Application of Preventive Maintenance is also feasible if the cost of performing PM is lower than the cost of not doing it, which eventually will result in a failure. It is important to conduct a thorough review and assessment of the equipment undergoing Preventive Maintenance and remove intrusive activities being forced on maintenance. Again, I am not against the use of Preventive Maintenance. What I am against are those people who force or add up activities into the never-ending or growing list of activities to be performed on PM because of their ignorance regarding the patterns of failures. Consider the following cases:

Case 1: In a study conducted, 95 out of 100 impellers will reach a life span of 3 years; maintenance decides to replace all 100 impellers before the 3rd year. Is Preventive Maintenance feasible and replace all 100 impellers before reaching the third year of continuous operations? **Answer: PM is feasible** as more than the majority at 95 impellers will survive before reaching its third year of operations.

Case 2: In the mining industry I used to work in, hydraulic pumps wear out prematurely due to abrasion; therefore, the maintenance prepare an MTBF analysis and found out that the average failure is every 3 months. Therefore maintenance decided to overhaul all hydraulic pumps every three months. Is Preventive Maintenance feasible? **Answer: PM is Not Feasible.** In this case, the failure of the hydraulic pumps is random and not age-related. Conducting overhauls and replacements should be based on the useful life and not on the average life. MTBF is about the average life.

Case 3: According to a bearing manufacturer, a ball bearing is calculated to have a life of five years; however, in our experience, similar bearings seemed to fail randomly; others even reach a life span of three months. The maintenance decided to replace all ball bearings every year. Is Preventive Maintenance feasible to replace all bearings yearly? **Answer: PM is Not Feasible** since the failure is random.

Case 4: In case number 3, a Root Cause Failure Analysis investigation was conducted by an RCFA Principal Investigator and his selected evidence gathering group. Strong evidence indicates that both the inner and outer raceway is dried, indicating a lubricant failure is imminent. The RCFA team recommended a different grease with a higher dropping point. After three years had passed, 90% of the bearing have not yet been replaced and are still in working condition. Can we declare that a PM replacement for the bearing will be feasible when more than the majority of the bearing failures are known? **Answer: PM is feasible** since the failure is no longer random; in this case, we shall just wait on when the majority of the bearing will fail to determine its useful life.

Case 5: Many mechanical parts wear out, others gradually, while others will wear out prematurely due to some forcing functions. Suppose more than the majority of the same parts or at least 80% will wear out and reach a specific age for the part or item to fail can we declare that Preventive Maintenance is feasible to use? **Answer: PM is feasible** since, in this case, the majority of the failures are age-related.

Case 6: A vendor performed destructive testing on a newly designed ball bearing where they run 100 bearings till it failed. 15% of the bearing failed prematurely during the first month, while another 25% failed before reaching 6 months. 30 % failed before 1 year, and the remainder of 30% reached a lifespan of 2 years. The vendor recommended replacing all bearings after one year of continuous operation. Is Preventive Maintenance replacement feasible to replace all bearings in this case? **Answer: PM is not feasible** since the bearing's failure is random.

It is important to understand that Preventive Maintenance has its own unique limitations. Understanding when to use and not to use PM will derive the most benefit from its application. When applied correctly, PM will increase equipment uptime; however, if we abuse its use, it will cause us more harm by being reactive instead of being proactive in our assets. Preventive Maintenance is feasible when the part or component will wear out directly concerning age, and its usage will definitely survive this defined age.

I think we need to resound the learning's from Nowlan and Heap, which they discovered in 1960 since it will create a lasting impact if we absorb and apply their discovery that the lessons they provide us are of great value for every person whose responsibility is to take care of their equipment and assets. There is no substitute for training since training is where we acquire this knowledge, and whatever learning we gain should be shared among our people.

Sixty years have passed, yet most industries I have been through have not absorbed the learnings. What Stanley Nowlan and Howard Heap had discovered is not about some rocket science or silver bullet but rather some basic fundamentals and understanding that Preventive Maintenance has its own limitations. And even 20 years before their discovery, Waddington also discovered the same problems especially on infant mortality failures on their RAF Bomber Planes through intrusive maintenance. We must not take this for granted. Preventive Maintenance will always have a place in the reliability world if we understand how and when to use it correctly.

Chapter 3

Building a Solid PM Structure in the Plant

> ***Maintenance must understand that PM overhauls and replacements can introduce Infant Mortality failures. Preventive Maintenance overhauls and replacement should be done by skilled craftspeople with the right knowledge, tools, and skills. If you have the slightest doubt in reassembling the equipment back again in one piece, then think twice or even thrice before dismantling.***

3.1: Phase 1: Preparatory Activities for Preventive Maintenance

There are three main Phases of Preventive Maintenance which will include the Preparatory Phase, the PM Execution Phase, and the Feedback Phase as in figure 3.1. The Preparatory Phase will include activities before executing the PM. It also includes what should be the very basic lists of things to do before executing Preventive Maintenance. The Execution Phase indicates the different available maintenance tasks that are actually performed on Preventive Maintenance. Lastly, the Feedback Phase includes the activities done once the PM tasks are finally completed on the equipment.

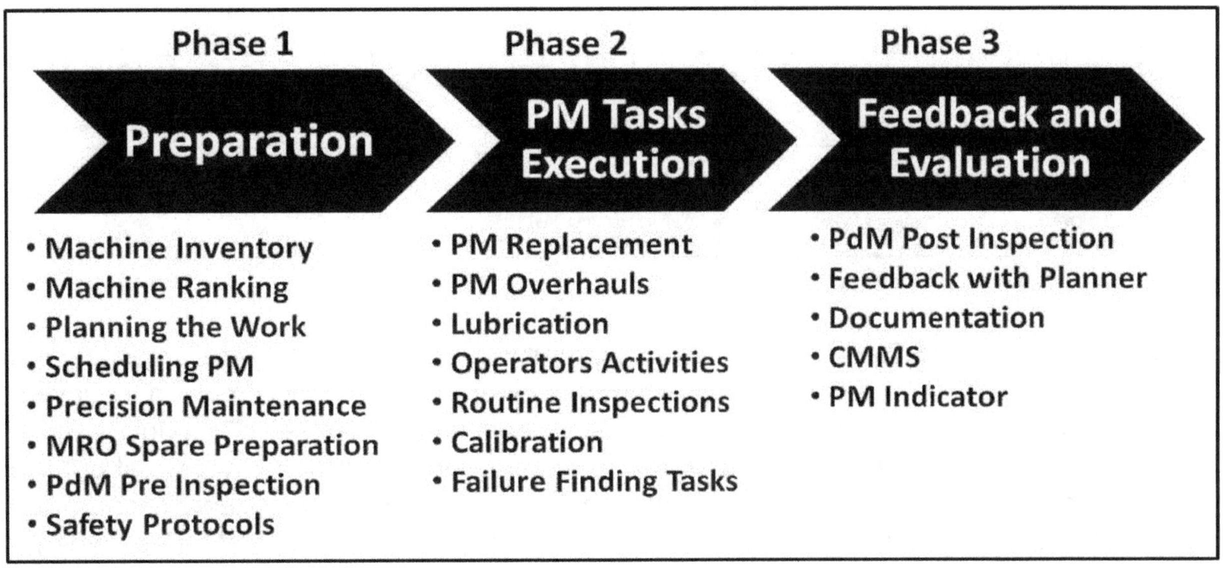

Figure 3.1: Preparation, Execution, and Feedback

Although when we speak about Reliability-Centered Maintenance, Preventive Maintenance will only include those activities that involved scheduled restoration, and scheduled discard. Scheduled restoration includes overhauling activities, while schedule discard includes parts replacement activities. In RCM, inspections are usually included as On-Condition tasks as this refers to any activities which involved inspecting the condition of the equipment using either a Predictive Maintenance instrument or the use of human senses. In this book, we are speaking about Preventive Maintenance in general as it applies to the majority of industries. Hence, the preparatory phase will include the following;

Conduct a Machine Inventory Lists: Although industries vary from one business type to another, the very first thing we need to do is to determine the overall inventory or create an asset hierarchy of all the equipment, machines, and asset operating in the plant. This inventory should include the complete lists of equipment used and operated in the plant. Those machines and equipment that are decommissioned or scheduled to be retired will be excluded from the Preventive Maintenance Inventory lists. Likewise important is to determine the number of maintenance people in the plant who will be involved in Preventive Maintenance. An asset hierarchy is an index of all your maintenance equipment, machines, and components, and how they work and link up together. This explains the parent-child relationship of assets in the plant.

Equipment Inventory Lists and Ranking

Gearing Towards A Pro-Active Mainteannce System . . . . .

Place Company Logo Here

PM Committee : ______________ Area / Division : ______________

No.	Equipment Maker	Equipment Type/Description	Machine Number	Serial No.	Area / Division	Commision Date	Total Years	Machine Ranking			AVE.	Remarks
								A	B	C		

Note : For the remarks, put the following
A - Running
B - Not used for a long time
C - Machine borrowed from other division
D - Decommission / Not running
E - For Phase out
F - Newly purchased (3months)
G - Others not included

Figure 3.2: Inventory Lists Form

Conduct a Machine Ranking for All Equipment: After completing the inventory list of equipment and assets, we need to define the criteria for ranking all equipment in the inventory lists such as in figure 3.4. Once the overall list of inventory is completed, conduct a machine ranking where Rank A will be the worst, Rank B in between, which is not yet worst but is also not considered good or the possibility of transforming into Rank A category, and Rank C should be tagged as good machines. Priority for activities will be given first for equipment and assets categorized as Rank A or worst equipment, then followed by Rank B. Rank C will be considered the last priority. After the inventory has been completed, set up the criteria for ranking equipment and rank all equipment and assets according to the matrix selected. The purpose of conducting a machine ranking is to determine which equipment, assets, and machines are critical and deserves our attention. This will also serve to prioritize the degree of tasks needed on each piece of equipment as well as which equipment will be prioritized in scheduling the PM. Perform this on all assets and equipment lists. Rank A (worse) can be those with the following:

- Considered to be the worst equipment in terms of operations and maintainability
- Possibility of fan-out or horizontal replication is large
- Considered to be the most headache, constraint, or bottleneck equipment
- Considered to be the most problematic equipment in the plant
- Have high costs of operating and maintaining the equipment
- Equipment will not be phased out for several years to come
- Equipment that has experienced Safety or Environmental Consequences in the past?
- Include all Facilities and Utility equipment for manufacturing industries
- Unmanned equipment in the plant
- Equipment that consumes the highest downtime in terms of failures and breakdown

Effects on	Criteria – Severity of Effects	Ranking	
Hazardous without warning	Very high severity ranking – Affects operator, plant, or maintenance personnel, safety and/or affects non-compliance with government regulations	10	
Hazardous with warning	High severity ranking – Affects operator, plant, or maintenance personnel, safety and/or affects non-compliance with government regulations	9	A
Very High Downtime or Defective Parts	Downtime of more than 8 hours or defective parts loss more than 4 hours of production	8	
High Downtime or Defective Parts	Downtime of 4 to 7 hours or defective parts loss of 2 to 4 hours of production	7	
Moderate Downtime or Defective Parts	Downtime of 1 to 3 hours or defective parts loss of 1 to 2 hours of production	6	
Low Downtime or Defective Parts	Downtime of 30 minutes to 1 hour or defective parts loss of up to 1 hour of production	5	B
Very Low Downtime No Defective Parts	Downtime up to 30 minutes – no defective parts	4	
Minor Effect	Process parameter variability exceeds Upper/Lower Control limits. Adjustment or other process controls need to be taken – no defective parts	3	
Very Minor Effect	Process parameter variability between upper/lover control limits. Adjustments or other process controls needs to be taken	2	C
No Effect	Process parameter variability within Upper/Lower Control limits, adjustment or other process controls not needed or can be taken between shifts or at normal maintenance – no defective parts	1	

Figure 3.3: Criteria for Machine Ranking

Set-up of equipment may differ, and consequences of failure may vary accordingly from one type of industry to another. Priority should be given to equipment whose failure modes can affect safety and the environment or have experience failures of this sort in the past. Equipment that will undergo Preventive Maintenance activities must belong to Rank A and B only. New machines and Rank C equipment will just be sustained by performing routine maintenance. All equipment in the plant should be ranked accordingly. You may use any of these matrixes to rank your equipment and assets. Our goal is to convert all Rank A and Rank B equipment into Rank C, which is considered good equipment.

Rank A: Very Severe Consequences: Loss of any one of these components will result in plant outage, total loss of production, severe environmental or safety consequences, and critical, stand-alone equipment. This can also include equipment and assets that consume the most amount of maintenance costs.

Rank B Severe Consequences: Loss or failure will limit production capacity by 30% or more. Also included in this critical classification are machines with chronic failure histories or those that have a high repair and replacement costs history.

Rank C Less Severe Consequences: Failures that do not have a dramatic impact on production but contribute to maintenance costs. An example will be redundant systems. Since the inline spare could maintain production, the loss of any one component would not affect the whole operation.

Department	Station	Rank A	Rank B	Rank C	Total
SOIC	FOL	125	92	206	**423**
	EOL	48	29	5	**82**
MQFP	FOL	117	91	0	**208**
	EOL	5	13	45	**63**
PDIP	FOL	33	30	95	**158**
	EOL	5	15	57	**77**
PSOP	FOL	18	28	54	**100**
	EOL	5	11	13	**29**
TQFP	FOL	11	129	1	**141**
	EOL	2	23	8	**32**
TSSOP	FOL	8	53	157	**218**
	EOL	8	14	30	**52**
SSOP	FOL	5	119	26	**150**
	EOL	0	33	0	**4**
SOT	FOL	0	0	18	**18**
	EOL	4	0	0	**4**
PLCC	FOL	1	66	202	**269**
	EOL	4	46	67	**117**
Hermetics	SSEL/Cerd	25	51	16	**92**
	Analog	24	11	5	**40**
CLF	Plating	4	3	14	**21**
Facilities	---	35	84	80	**199**
Total		486	941	1099	**2,526**

Figure 3.4: Sample of Machine Ranking (FOL – Front of Line, EOL – End of Line)

The ranking will be based on the criticality of the equipment. This may be in terms of the maintenance cost, the number of breakdowns, consequences of failures, or any combination of both. What is important is to establish a Machine Ranking for all equipment on the inventory list so that we can prioritize the maintenance planning based on the most critical equipment in the plant which will be categorized as Rank A equipment, machines, and assets.

Although most industries already have their own Preventive Maintenance, the purpose of this activity is to identify all the critical equipment and asset in the plant as this equipment will be prioritized in developing the tasks needed for Preventive Maintenance. We need to determine the total number of equipment belonging to Rank A and Rank B categories, as these are the equipment that will be prioritized. All Rank C equipment will just be sustained. What is important at this stage is to have an exact figure of how many Rank A, Rank B, and Rank C we currently have in the plant. It is also very important that the machine ranking must be completed first before deploying any Operator Driven activities or Autonomous Maintenance activities if the plant is implementing this.

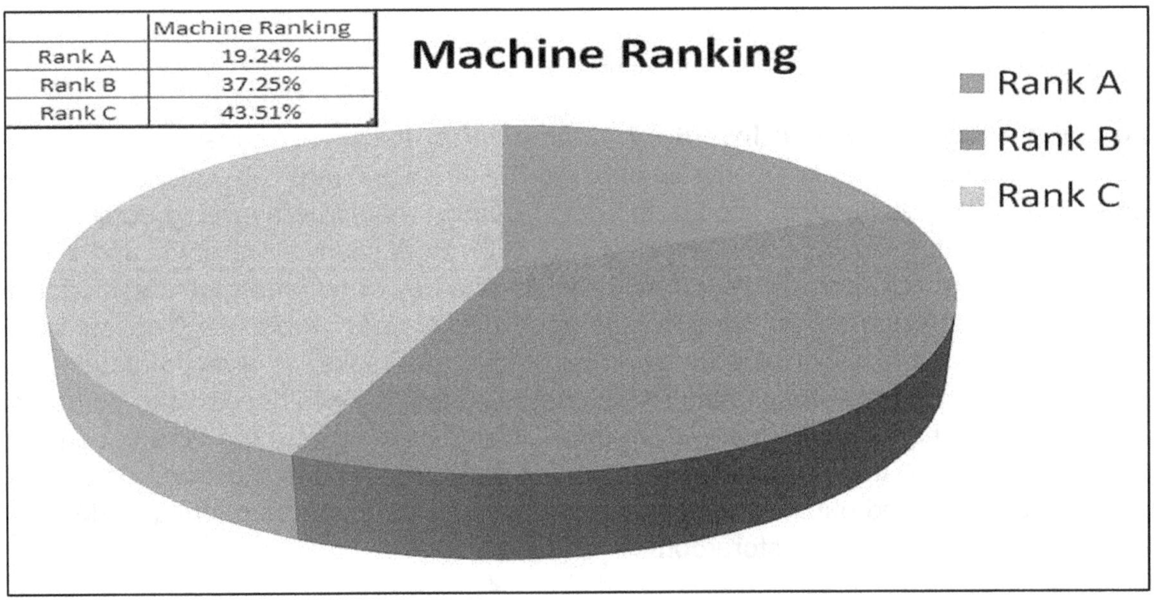

Figure 3.5: Summary of Machine Ranking

Having a summary of machine ranking will provide us a great deal of information on where to focus our efforts on Preventive Maintenance. In figure 3.5, we have a total of 19.4% Rank A machine which is derived from figure 3.4, where we divided all Rank A totaling 486 to the total equipment in the plant which is 2526, and expressed it in percentage. Rank B consists of 941 machines and Rank C consists of 1099 machines. These 486 Rank A machines will be our focus on developing the Preventive Maintenance tasks.

Planning and Scheduling: It is important that all users are well trained in the background theory so that there is a common understanding of not only how but also why certain tasks are necessary to perform in the equipment. The maintenance planning system is defined as the total process set up to ensure that the right resources and materials arrive at the right place, at the right time, and doing the right job in the right way. Planning also has to be based on the scheduled shutdown for the year, the business plan, the marketing plan, the budget, and others. Planned work will include the skills, resources, bill of materials, detailed activities, together with the correct interval that has been identified, documented, estimated in terms of time, and bill of materials, which is due for scheduling. Scheduled work is placed on a formal worklist that will be executed at a prescribed amount period. In summary, planning will include how and who, while scheduling, will determine when the equipment will be due for PM. Note that a separate chapter is written for planning and scheduling which will detail its activities more thoroughly. Usually, the Planning Phase will include the following:

- Detailed scope of the job to be done
- Estimated duration of the maintenance tasks
- Bill of Materials (BOM)
- Estimated manpower needed
- Items to be withdrawn in the MRO Stores for replacement
- Consumables and tools needed for executing the PM
- Safety precautions needed for PM works
- Interval or frequency
- Scope of work for 3rd party contractors if needed

Managing MRO Spares and Inventory: Parts, items, tools, and consumables that will be needed for replacement, and overhauls during a PM shutdown must be scheduled in advance, depending on the lead-time to acquire the part. Spares should be available before the actual PM is due. This must be communicated well with production, storeroom, and purchasing people. Maintenance needs to withdraw these parts a day or two before the said schedule of PM. One of the storeroom functions is to provide parts, tools, and consumables for maintenance that are responsible for executing these PM Tasks. It is recommended that the storeroom people have access to the PM schedule so that the kits needed can be gathered in advance before the actual PM schedule starts. This can save us more time in acquiring the parts to be used for the Preventive Maintenance activities. These must be coordinated well with the storeroom and purchasing people, especially for parts needed during a PM where the item is not stocked inside the storeroom.

Pre - Post-PdM Monitoring: Before executing the actual Preventive Maintenance tasks on the equipment, an actual Predictive Maintenance monitoring should take place to determine which module or sub-assembly possesses problems and potential failure. This will also determine if certain parts need to be replaced or not. This means that if a bearing needs to be replaced and the Pre-PdM monitoring conducted through Vibration Monitoring indicates that the item is still functioning, then the decision is not to replace the bearing or routine lubrication can be performed instead of replacing it. The tasks generated by the planning will be adjusted based on the recommendations of the PdM group. This will also allow the people to execute the actual Preventive Maintenance to focus their tasks on those that are worth doing. After executing the PM tasks, a Post Predictive Maintenance monitoring should once again be conducted so they can compare the before and after data. This means that if the vibration was reduced as a result of conducting PM then the crew has executed the PM correctly, thereby reducing the chances of infant mortality failures from occurring.

Integration of Precision Maintenance: Integrating Precision Maintenance into Preventive Maintenance tasks will reduce problems such as start-up losses and infant mortality failures right after a major Preventive Maintenance is performed especially those that involved direct contact with the equipment such as replacing parts and overhauls. The concept of Precision Maintenance is that whoever is performing or executing the PM tasks should yield the same results whether the person is the most experience or the least experienced in the craft. The reason for integrating Precision Maintenance is to minimize human errors on PM.

Safety Protocols: Complete apparel must be worn during the actual execution of PM tasks such as hard hats, masks, safety glasses, safety shoes, earplugs, and gloves should be worn according to the type and area of work. Provide signs such as slippery floors, adequate lighting, and the use of ladders when performing tasks on higher or elevated portions. When performing Preventive Maintenance the equipment must be completely shut down. Remove the fuse or keys or whatever that can trigger the machine to run accidentally during PM. Never touch moving parts on the machine. Wait for the machine to have a complete stop before performing Preventive Maintenance activities. Use gloves when cleaning metals to avoid cuts and wounds brought about by abrasive particles and materials or sharp corners on the machine. Never use an air gun in cleaning the equipment as the dust will just suspend and return back to the equipment.

Safety on the equipment must be fully observed during Preventive Maintenance to avoid injuries, accidents and near miss situations. Maintenance should wear protective clothing at all times and PM must only be done when the equipment is completely stopped. Remove the fuse if possible to avoid situations on accidentally turning the equipment on.

Figure 3.6: Equipment Safety during Preventive Maintenance

3.2: Phase 2: PM Tasks Execution Includes

This phase explains the different maintenance tasks and activities that are performed in Preventive Maintenance. This is the moment where these tasks will be finally executed on the equipment. Although there is a problem with the nomenclature since according to RCM, Preventive Maintenance will only include 2 activities which are overhauling and replacement, and the rest of the activities will default to On-Condition Tasks. This section will briefly explain the different maintenance tasks performed on Preventive Maintenance.

Routine Inspection and Checking: A routine visual inspection is usually performed on the equipment by using our basic human senses of sight, sound, touch, and the sense of smell. Unless specific parameters have been identified, the inspection quality should be totally dependent on the sensitivity of the inspector in detecting potential equipment problems and abnormalities. Inspection checklists used on equipment are often a list of symptomatic conditions to be looked for where a checkmark is written. Inspection checklists for PM should include a detailed activity, duration, tools or measuring instruments, or human senses to be used. What to do in case there is a deviation from the standard should be clearly specified. The use of visual control management can aid in simplifying inspection techniques. Some inspections will be based on sensors such as temperature, pressure, transducers, thermocouples, ammeters, accelerometers, and others. What is important is whether using human senses or sensors, thresholds, and limits should be clear, and what to do if the limits have been reached be clearly indicated.

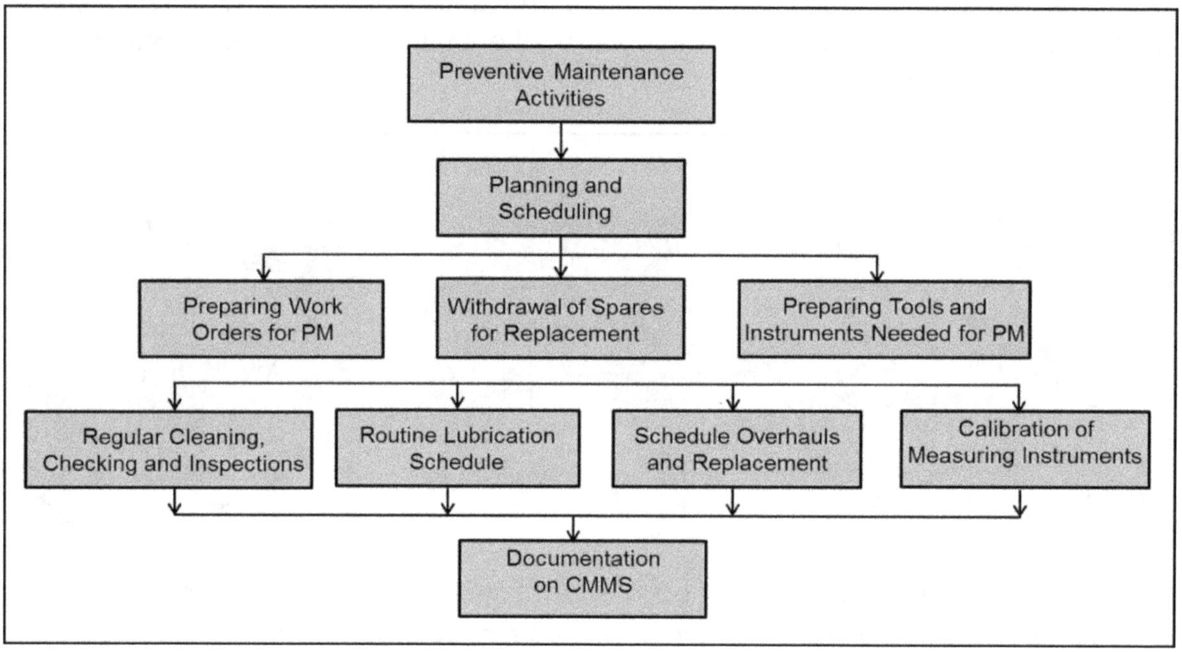

Figure 3.7: Common Preventive Maintenance Activities

Cleaning: Cleaning is the process of removing any form of an unwanted object from the equipment. It consists of removing dirt, contaminants, grime, excess oil, grease, and other foreign objects that can affect equipment parts and components. Cleaning is performed not just to satisfy the cosmetic looks of the equipment but because dirt and contaminants shorten the life of certain parts and components of the equipment. Detailed cleaning of components and equipment is often a no man's island because while everybody agrees that it is important, nobody wants to do it actually. World-Class Companies perform these activities. Components and equipment are kept clean in every detail. Such an organization realizes that inspections cannot be done without this level of cleaning. Cleaning definitely extends the life of parts and components. Through cleaning, we can expose problems to the equipment that has been left ignored for a very long time. Leaks, cracks, and fractures that have been in the equipment for a very long time can be detected. Dirty and untidy equipment will tend to deteriorate more rapidly than equipment that is well maintained and cleaned at all times.

Routine Lubrication: All machines contain lubricants in one form or another. The purpose of lubrication is to reduce friction by creating a film for mechanical parts moving inside our equipment. It is not only important to lubricate the equipment but also to use the correct viscosity and correct amount. Equipment that lacks lubrication or is over lubricated can likewise induce problems. Operators should be taught about the importance of checking the lubrication in their equipment and maintaining the correct amount of lubricant. So many failures can be attributed to lubrication that can be avoided if only these basic equipment conditions can be well established. Maintenance must teach operators not only the necessary points in their equipment to lubricate but also what proper and improper lubrication can do with their equipment. Just like humans, lubrication is the lifeblood of the equipment, and it should be maintained clean and adequate all the time. These things can be avoided and controlled in our equipment if we understand and learn the role that lubrication plays in our assets and equipment.

Overhauls and Replacement: Parts replacement on PM involved the periodic replacement of disposable parts, such as filters, for lubricating oil. Replacement also includes replacing critical components and spare parts that have actually worn out or are considered defective. In RCM parts, replacement means Schedule Discard while Overhauling means Scheduled Restoration. The process of overhauling means dismantling the equipment. Special care should be taken in overhauling the equipment since this can cause problems such as infant mortality failures or start-up failures when endorsing the equipment back to operations. Integrating both Predictive and Precision Maintenance can minimize the chances of Infant Mortality Failures. Note that overhauling and replacement of parts during a Preventive Maintenance shutdown should only be done for age-related failures and not for failures, which are considered random. The aging process simply means the deterioration and wear out of parts and components over time. Any process contributing to age will decline in performance productivity due to the stress accumulated during the aging process. This deserves maintenance attention and intervention. The traditional view is assumed that most parts will operate reliably for a specific period, after which the probability of failure starts to increase rapidly. The analysis of failures will allow us to predict the remaining useful life of an item and take necessary action to overhaul or replace it before it reaches the wear-out mode where the risk of failure becomes evident.

Calibration is the activity of evaluating and adjusting the precision and accuracy of measuring instruments and devices with readings. Calibration is a comparison between a known standard and the current measurement of the equipment. Proper calibration of an instrument allows people to have a safe working environment and produce accurate data for monitoring and inspection. Precision is the degree to which repeated measurements under unchanged conditions show the same result. At the same time, accuracy is the degree of closeness of measurements of a quantity to its actual true value. The purpose of calibration is to identify and eliminate any deviation in the instrument's reading relative to the defined unit of measurement. Thus, the calibration of measuring instruments has two objectives. First, it checks the instrument's accuracy, and second, it determines the traceability of the measurement. In practice, calibration also includes repair of the device if the measurement is out of range.

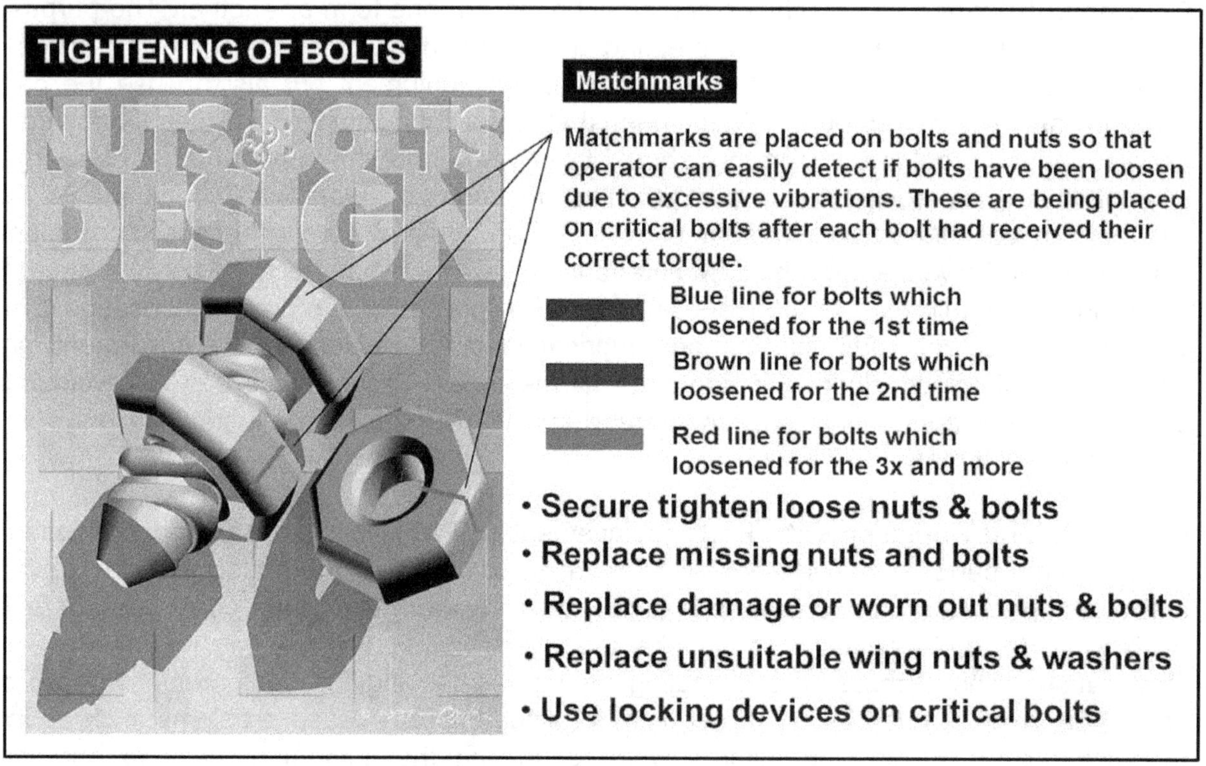

Figure 3.8: The Use of Visual Controls for Detecting Deviations

Operator's Tasks a plant implementing TPM is geared toward developing empowered operators through the 7 steps of Autonomous Maintenance. As the concept of Autonomous Maintenance evolves that operators are important in any reliability and maintenance strategies since these are the people who will witness and experienced the failure first before maintenance. This means that these people are the primary defense against any equipment-related failures. The key is to build up operators who can detect and sense failures before they actually happen in the equipment. Although this will not happen overnight as it takes years to develop an empowered workforce. To be exact, from Step 0 to Step 7, it will take 5 years on average to build a strong and empowered Autonomous Maintenance workforce in the plant. These operators will have their own checklists which will include 3 main standards such as inspection, cleaning, and lubrication standards which will be done on the equipment that they operate. Also, one unique feature of Autonomous Maintenance is the application of Visual Control to make their inspection easier and capture deviations and anomalies on their equipment at an early stage. An example of visual control is called match marking in which once the bolt has finally been tightened with the correct bolt, the operator will place a thin reflectorized sticker on the head of the bolt, and the nut. If the bolt tries to become loose, then the sticker will no longer be aligned on the head of the bolt and on the nut, which can be seen easily by the operator.

Failure Finding Tasks: This is also called functionality inspection or Detective Maintenance. Failure finding tasks applies only to hidden or unrevealed failures or protective devices. The reason for conducting failure-finding tasks is to reduce the risk of multiple failures wherein the protective device and protected function are both in a failed state. Scheduled failure finding tasks entails checking a hidden function at regular intervals to find out whether it has failed. A failure-finding task is technically feasible if first, it is possible to do the job and second, the task does not increase the chance for multiple failures, and third, the task is practical to perform at the required interval. Failure-finding tasks should be considered if a functional failure will not become evident to the operating crew under normal circumstances. Failure is one in which no suitable proactive tasks cannot be found.

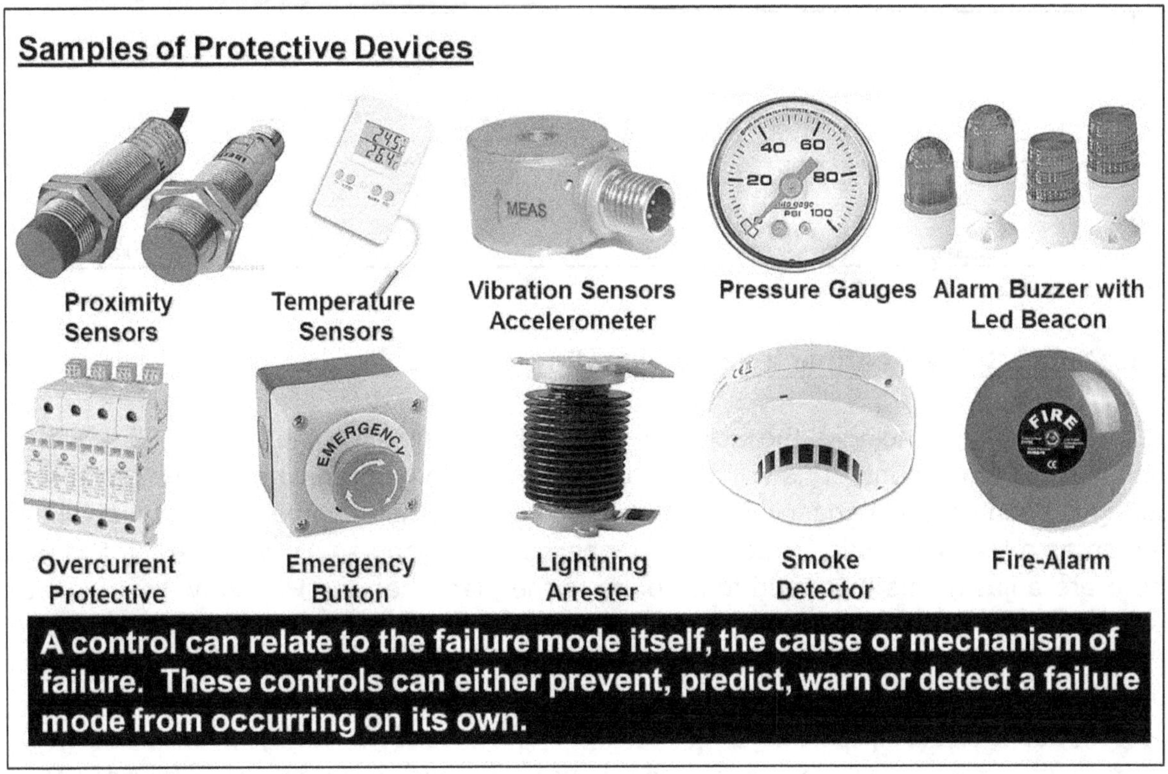

Figure 3.9: Samples of Protective Devices

Note that in highly complex, modern, automated, and industrialized industries expect around 40% of failure modes fall into the hidden category. In other words for equipment full of electronic parts expect more tasks to fall into the category of Failure Finding Tasks. Figure 3.9 are samples of protective devices. These protective devices serve as our defenses not only for failures but also against the possibility of committing a human error. It is important to have a list of all protective devices on every piece of equipment in the plant, and provide the necessary functionality inspection needed as the failure of these devices is mostly hidden in nature. The more critical the protective device, the more frequent it should be inspected if it is still working or not. Note that most failures of these protective devices are hidden. This means that the failure will only be known once the protected function also fails. The succeeding chapter will discuss how to determine the interval for inspecting protective devices.

EQUIPMENT LISTS OF PROTECTIVE DEVICES

Name of System or Asset:

Protective Device	Included in Function		Explain the Multiple Failure in Case it Happens	HIDDEN	EVIDENT	FAILURE FINDING INTERVAL	INTERVAL (ROUND OFF)
	Yes	No					

Prepared by	Checked by	Approved by	Revision Number

Figure 3.10: Form for Listing Protective Devices

3.3: Phase 3: PM Feedback and Evaluation

The last phase includes the activities done after completing the Preventive Maintenance tasks on the equipment. This will also serve as our judgment on whether the tasks executed during Preventive Maintenance are effective or needs to be improved. The feedback portion serves as a learning curve for both the planner and the people who execute the PM tasks. If there are adjustments that need to be done to the plan, then the PM crew who executed the task must provide this feedback information.

- Have the breakdowns and failures been reduced as a result of the Preventive Maintenance just recently done on the equipment?
- Were there cases of infant mortality failures experienced on the equipment after executing PM?
- Have the indices and KPIs on PM improved after performing the PM tasks?
- Was there an improvement in the readings conducted by the PdM group after executing Preventive Maintenance on the equipment?

PdM Post Evaluation: After executing the Preventive Maintenance tasks on the equipment, the Predictive Maintenance group must take a reading to ensure that the equipment is once again in the best shape ever. An example is taking a vibration reading from a multi-stage pump with a velocity reading of 7 mm/second before the actual PM and after the PM had been concluded, the machine was initially run by the PM and the actual vibration monitoring dropped from 7 to 2 mm/second. This will also provide the PM crew the confidence that they actually performed a great job in executing their PM by focusing on the real problems with the equipment. This activity should be done before endorsing the equipment back to operators.

Feedback with the Planner: Although the planner dictates the activities to be done on the equipment, it will not always be a perfect fit. There will be adjustments that need to be done. This should be a continuous improvement effort. The planner must listen to the feedback of the people who actually perform the tasks, or better off if the planner was once involved in Preventive Maintenance themselves. Should the task take much longer as originally been indicated in the plan which is 10 minutes but in reality, it took 45 minutes? Is the interval of performing the tasks too soon, since the task is done monthly which it should be done semi-annually? Are there activities or tasks which should be deleted or added? Was there any part or item that was needed that was not indicated in the plan? It is the responsibility of the people involved in Preventive Maintenance to initiate the feedback with the planner, and it will be the responsibility of the planner to listen and make the necessary revisions in the maintenance planning. Feedback should be constructive.

Monitoring Maintenance KPI and Indices: The primary purpose of conducting Preventive Maintenance on the equipment is to prevent failures and breakdowns. There are several maintenance indices and KPIs that we can monitor to determine if we performed the correct tasks and if the PM conducted was effective. The different maintenance indicators are explained in detail in the succeeding chapter of this book.

Preventive Maintenance Indicators	Formula	Unit	Trend
PM Compliance	PM Compliance = Number of PM Tasks Completed / Total PM Tasks Listed	Percentage	Higher the Better
PM Effectiveness	Compare Number of PM Tasks Completed vs. Corrective Maintenance	Hours	PM High / CM Low
Ratio of PM to Breakdown Maintenance	Compare umber of PM Tasks Completed vs. Number of Breakdowns	Frequency	PM Tasks High / BM Low
Maintenance Backlog	Backlog = Total Sum of All Maintenance Work / Sum of All Man hours	Hours	Lower the Better
Maintenance Cost	Sum of All Incurred Cost and Expenses on Maintenance	USD ($)	Lower the Better
Maintenance Unit Cost	Total Maintenance Cost / Total Products Produced in %	Percentage	Lower the Better
Percentage of Maintenance Cost to RAV	% Maintenance Cost to RAV = (Annual Maintenance Cost ($) x 100) / RAV ($)	Percentage	Lower the Better
Predictive Maintenance Percentage	% PdM = Number of Potential Failures Detected / Corrected in %	Percentage	Higher the Better

Figure 3.11: Preventive Maintenance Indicators and KPIs

Documentation and Closing the Report on the CMMS: Once the Preventive Maintenance tasks have been completed thoroughly, the PM Crew should close the report on the CMMS system. If there are PM tasks that had not been accomplished then it will remain as a backlog. The next machine to be scheduled for PM will undergo the same process. This should be considered as one of the most important parts of PM since this will be the basis for the planner for similar works in the future, especially for equipment with the same operating context or condition. In closing the PM report it should include the following;

- Machine Number, Model, and Vendor.
- Work Order Number.
- Indicate if this is a monthly, quarterly, yearly, or the interval of conducting these tasks.
- Specify and detail the task done on the equipment.
- Actual duration of time each task is performed.
- Specify if there are any other activities done on the equipment besides the scope of work.
- Indicate the spare parts used, part number or codification, and consumables that were withdrawn.
- Indicate if there are special tools, devices, or instruments used to carry out the tasks.
- Indicate if the job was fully or partially completed.

- Indicate if there are any problems or delays in executing the PM and specify it.
- Name the people involved in performing the tasks.
- Indicate if there are third-party contractors involved in the PM.
- Pre and Post Predictive Maintenance Monitoring records

The report should be completed and encoded on the CMMS system or EAM. This must be the standard practice every time a Preventive Maintenance task is completed especially for a major Preventive Maintenance shutdown.

Chapter 4

Planning and Scheduling for PM

> ***The maintenance planner is not always perfect, but the good news is we can always make it better through feedback. What is important is to apply the 4 Cs between the planner and the people who will execute the PM. This means having continuous communication, coordination, cooperation, and collaboration between these two.***

4.1: Needed Non-Productive, Not Needed Non-Productive, and Productive Work

Perhaps you have watched an NBA game in its final seconds, where the coach called his final time-out with 5 seconds remaining in the ball game. The coach gets his tiny whiteboard and marker and discusses his final plan with the players. The opposite team is ahead by 2 points and they have the ball. And the coach said, Scottie Pippen, pass the ball to Michael, you and Dennis provide a screen, I want a 3-point shot. The bell sounds and the crowd is wild on their feet cheering the Bulls. The players are now walking back to the court and Harper passes the ball to Jordan, Scottie and Dennis are on the three-point providing Michael a screen. Michael was expecting a double team and dribbling the ball. Three people were on Jordan, Dennis blocked one of the players with 3 seconds left, and the coach yells, Damn it Jordan, I want my 3. Michael his tongue hearing the voice of his coach without hesitation, focused on the basket, jumps, suspends himself in the air, and shoots the ball, scores and the bell sounds. The Chicago Bulls won by 1 point and the crowd was wild on their feet, and Michael waved his hands to the fans as he chewed his gum. What a game. All went well according to the plan, thanks to the meticulous and precise planning by their coach Phil Jackson.

In preparing for the plan, the maintenance planner must think of all the nitty-gritty details that will be needed for the work to be accomplished. Writing the plan for a specific Work Order is not enough. What we want is that the execution of the work must be precise as possible with little or no unproductive time just like an NBA timeout above. If the maintenance who will execute the job finally finds out that the part or spare needed is not stockable in the storeroom, then there will be delays in the execution of the PM work. Perhaps the job scope will require that a valve needs to be replaced but upon inspection, the job also requires some welding of the base joints which must be done by a qualified welder. A special power tool is needed which was not specified in the job order and once again you need to travel once more to the MRO Storeroom to get the tool but this time you need to queue just like in the airport as there are 6 more technicians ahead of you needing something in the storeroom ASAP. Perhaps the planner thought that the job requires just one person which in reality the job requires 2 or 3 people. All of these are waste or what I called Unproductive Time or those that are considered as Not Needed Non-Productive Work. These are those added activities and resources which are wasted as a result of poor or inadequate planning. This includes wasted time for the wrong part number supplied, lack of manpower, wrong tools, MRO spare part not available, or going back once more to the storeroom since you needed this and that. What is important is that when we execute the task, everything must be planned even up to the smallest detail possible. The maintenance planner must try to list all the details of what it takes for the work to be accomplished.

Figure 4.1: What the Planner Must Do

The main reason for creating the plan is to see to it that our manpower provides more Productive Work than those Not Needed Non-Productive work in the plant. Although we cannot eliminate Non-Productive work 100%, there is also Non-Productive work that we cannot take away from maintenance. Although some books will state that maintenance should only be doing productive work based on what the planners dictate, my stand on this is there are also Needed Non-Productive works are those that are needed by our maintenance and technicians such as conducting improvements, conducting Root Cause Failure Analysis, mentoring operators, conducting TPM, FMEA, RCM, and upgrading their knowledge through training and education. At the moment, they will be considered as Needed Non-Productive works since they are not holding a wrench but this is needed and will be of benefit to the industry in the long run. Hence, we can break down Non-Productive Work into those that are needed and not needed:

Not Needed Non-Productive Work

- Cannibalizing Spare Parts from idle machines (not needed)
- Attending reactive meetings (not needed)
- Looking for Spare Parts in the Storeroom (not needed)
- Requesting for parts on an Emergency Basis (not needed)
- Doing Hero Stuff on Repairs (not needed)

Needed Non-Productive Work

- Lunch and Coffee Breaks (needed)
- Mentoring Operators (needed)

- Conducting Improvements such as RCM, TPM, FMEA, and Others (needed)
- Conducting Root Cause Failure Analysis Investigation (needed)
- Attending training on reliability and maintenance

In this case, I have broken down the Non-Productive Work that is needed and those which are not needed. Let us assume that a plant is operating for 320 days in one year or this will be equivalent to 7680 hours per year. Let us assume that in one day there are 3 shifts. Therefore, in one shift we have 106.67 days or 2560.33 hours in a year.

Case 1: If there are 3 hours per day waste in maintenance (Not Needed Non Productive)

Needed Non-Productive	Hours	Days
Breaktime	133.33	16.67
Mentoring Operators	35.55	4.44
Training	45.00	5.63
Improvements (15%)	128.00	16.00
Total Time	**341.89**	**42.74**
Percentage	**40.06%**	
Not Needed Non-Productive	**Hours**	**Days**
- Waste in maintenance	320.00	40.00
Total Time	**320.00**	**40.00**
Percentage	**37.50%**	
Productive Work (Planning)	**Hours**	**Days**
- Planned Work Orders	191.448	23.93
Total Time	**191.45**	**23.93**
Percentage	**22.44%**	

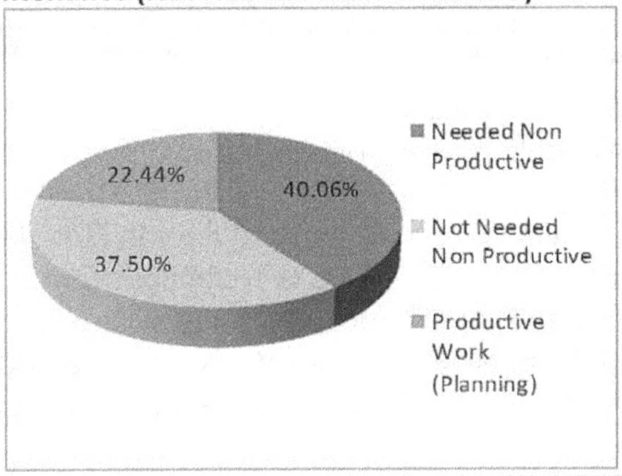

Needed Non Productive	40.06%
Not Needed Non Productive	37.50%
Productive Work (Planning)	22.44%

Case 2: If we can reduce the waste in maintenance from 3 to 1 hour/day

Needed Non-Productive	Hours	Days
Breaktime	133.33	16.67
Mentoring Operators	35.55	4.44
Training	45.00	5.63
Improvements (15%)	128.00	16.00
Total Time	**341.89**	**42.74**
Percentage	**40.06%**	
Not Needed Non-Productive	**Hours**	**Days**
- Waste in maintenance	106.67	13.33
Total Time	**106.67**	**13.33**
Percentage	**12.50%**	
Productive Work (Planning)	**Hours**	**Days**
- Planned Work Orders	404.78	50.60
Total Time	**404.78**	**50.60**
Percentage	**47.44%**	

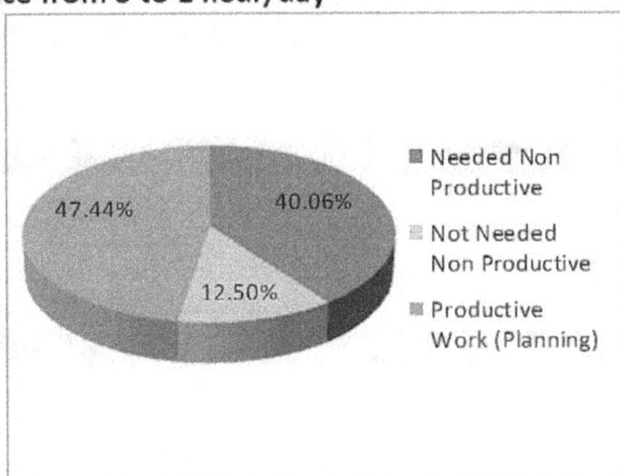

Needed Non Productive	40.06%
Not Needed Non Productive	12.50%
Productive Work (Planning)	47.44%

Figure 4.2: Needed and Not Needed Non-Productive and Productive Work

Let us classify the maintenance time as follows:

- Not-Needed Non-Productive Work: Include all waste on maintenance that must be reduced.
- Needed Non-Productive Work: Includes those that cannot be eliminated and are still needed.

• Productive Work: Those scheduled works that are planned by the maintenance planner.

In figure 4.2, Case 1 indicates that we have 341.89 hours of Needed Non-Productive Work which includes mentoring operators, 45 hours consumed for training as part of their training needs, of course, they need to eat and break which totaled 133.33 hours in one year, and if 15% of the maintenance time is spent on improvements, then this will be 128 hours in a year. The total time spent for Needed Non-Productive Work is 341.89 hours in a year for 1 maintenance as this is per shift. This will be around 40.06%. Assuming that the Not Needed Non-Productive Work is 3 hours per shift or 320 hours per year just for one maintenance person, then this is 37.5%. The total remaining time that can be used for Productive Work which is scheduled work for the planning will be 22.4 %

• Available Time per Head	7680 hours / year	320 days / year
• Available Time per Head per shift	2560 hours / shift	106.67 days / year

In Case 2, if we can reduce the Not Needed Non Productive Work from 3 hours to just 1 hour, then this will be a reduction from 320 to 106.67 hours or equivalent to 12.5%. The Needed Non Productive work is still the same at 341.89 hours or 40.06%. This now increases the Planned Productive Work from 191.45 to 404.78 hours or an equivalent increase from 22.44 to 44.47 % increase in Productive Work.

We cannot eliminate the Needed Non-Productive Work which includes upgrading the knowledge of maintenance through training, doing improvements, coaching operators, and of course taking their lunch and breaks. The details in figure 4.2 are just for a single maintenance person working one shift. If we can reduce the Not Needed Non Productive Work for maintenance from 3 hours to 1 hour, then this will be an increase in the Productive Work from 89.74 to 189.74 hours for 50 maintenance people and still retaining the Needed Non Productive work such as training for maintenance.

Case 1: 3 hours Waste per Maintenance	Percent	1	3	10	20	30	40	50
Needed Non Productive Work	40.06%	3.21	9.62	32.05	64.10	96.16	128.21	160.26
Not Needed Non Productive Work	37.50%	3.00	9.00	30.00	60.00	90.00	120.00	150.00
Productive Work (Scheduled and Plan)	22.44%	1.79	5.38	17.95	35.90	53.84	71.79	89.74
Total Time (hours)		8.00	24.00	80.00	160.00	240.00	320.00	400.00
Case 2: 1 hour Waste per Maintenance	**Percent**	**1**	**3**	**10**	**20**	**30**	**40**	**50**
Needed Non Productive Work	40.06%	3.21	9.62	32.05	64.10	96.16	128.21	160.26
Not Needed Non Productive Work	12.50%	1.00	3.00	10.00	20.00	30.00	40.00	50.00
Productive Work (Scheduled and Plan)	47.44%	3.79	11.38	37.95	75.90	113.84	151.79	189.74
Total Time (hours)		8.00	24.00	80.00	160.00	240.00	320.00	400.00

Figure 4.3: What If We Can Reduce Maintenance Waste from 3 hours to 1 hour

4.2: Concept of Planning and Scheduling for PM

Planning is defined as scheduled work activities on the equipment that will be done and scheduled in the future. This usually defines the what, why, and how their work is supposed to be done. This will involve the following;

• Complete Job Scope of the tasks

- MRO spares and consumables to be used
- Part Number of Spares and Consumables Needed
- Tools, instruments needed for the job
- Consumables needed
- Skills Needed and Number of manpower needed
- Estimated duration of conducting each task
- Bill of Materials (BOM)
- Scope of work for 3rd party contractors
- Safety Protocols

Most industries do not have a plan for their maintenance work; however, what they have is the scheduled work. Scheduling unplanned or emergency work will be very costly since we actually do not know what we need to do and the cost of doing it. Spare parts will be delivered on an emergency basis that will incur additional freight charges by the vendor. In addition, we may be purchasing materials, supplies, spare parts, and providing contracting work that may not be actually required, and those maintenance tasks that are required may not be included in the planning. Whoever will be assigned as the planner should be isolated from other tasks and must work full-time. Maintenance planning and scheduling activities should include the following:

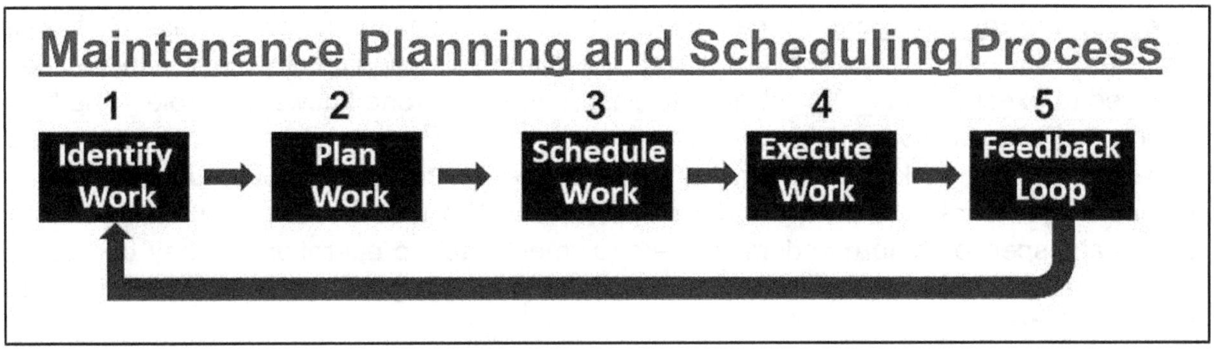

Figure 4.4: Concept of Planning and Scheduling Process for PM

Identify the Work (Input): The planner's job is to gather inputs from the PM crew regarding the maintenance activities and tasks needed to be done on the equipment based on the failure modes experienced in the past. Inputs may also come from the users or operators as a result of records of breakdown, and corrective maintenance done on the equipment in the past. The activities to be done on PM should be based on these inputs. For example, if we are monitoring the difference in pressure on a pressure gauge, we need to know the failure mode that we are trying to address so that we can take action on inspecting it. Perhaps, in this case, we are checking the drop in pressure to determine if the strainer is clogged and needs to be cleaned or not. Inspection should be specific as to what the maximum drop in pressure should be before the necessary actions should take place.

Planning the Work: Planned Work is defined as work that includes preparing the resources needed such as labor skills, spare parts, tools, bill of materials, exact maintenance tasks, and estimated duration to perform these tasks that had been identified, documented, estimated in terms of time, cost and is ready for deployment during a PM shutdown. Planning will answer the question of what and how the PM activities are to be done. PM activities and tasks should be based on the failure modes' experience on the equipment since there might be maintenance tasks listed that seem unimportant or do not address a particular failure mode. Yet, those tasks that must be performed may not be in the actual PM lists. Therefore, both the maintenance planner and those who experience these failure modes must be in constant communication with each other. The planning step identifies all resources required to carry out the work in an orderly, accurate, precise, and cost-effective way. It includes identification, organization, and the procurement of materials needed to execute the PM work, including spare parts, tools, special services or contracted works, work permits, bill of materials, consumables, safety, work procedures, and PM works for 3rd party contractors.

Scheduling the Work: A work schedule will answer the question of who will perform these tasks and when the PM will be done. The work schedule is the process in which all required maintenance-related resources are scheduled to be used within a specified time. A work schedule is approved at a prescribed time or period. The period to execute the activities on PM can be weekly, monthly, quarterly, semi-annually, yearly, or based on other forms and is compiled from a draft schedule and backlog list. There has to be a cut-off for completing these PM task work listed in the initial planning phase. The schedule must be well-coordinated with the operations people.

Execution of the PM Tasks: When the plans have been made, the bill of materials has been projected, and spares are finally ordered, the planner is required to meet with the people involved to execute these PM activities to ensure that everyone knows their role. The highest priority in PM execution must be safety. Once an activity is completed, it will be marked as done in the PM tasks document. Any changes in the plan should be decided upon immediately. Once the PM tasks are complete, the PM team must perform a post-mortem and final inspection before endorsing the equipment back to operations. They can seek the aid of Predictive Maintenance to conduct a post-monitoring to ensure that the PM task is correctly done.

Feedback: This is a way of telling the maintenance planner whether the execution of PM was done accordingly or did we need to make adjustments to improve further the execution. Feedback must be done in a caring and constructive fashion and must not, in any way, insult anyone. Do not mention names in the feedback and address the person in general, such as PM, maintenance, technician, or planner. People need to know how they have actually performed, and any difficulties experienced should be looped back to the planner. The most important aspect of a PM program is giving the people feedback concerning the effectiveness of their activities during PM. If mistakes were made, they should correct them. If the execution of PM was successful, then it should be made known to them. The learnings from feedback are invaluable lessons to people who are actually performing and planning the PM.

4.3: Who Is the Perfect Fit for the Maintenance Planner?

The first thing we need to ask is who should be the right person to sit down as the maintenance planner. Most of the planners that attend my training classes on reliability and maintenance are young people, some of them perhaps I can consider my son or daughter. If this is the case with your plant currently, then I would say that 70% of the time of your planner should be spent not on the computer or desktop writing the plan but involving, collaborating, and interviewing people involved in executing the actual Preventive Maintenance tasks. Let's face it, most of those who wrote the plan are not actually the people who will execute them. This will create a big discrepancy between what is real and not. Some may not even experience the actual execution of these PM tasks when actually performed on maintenance. The duration written by the planner may be too short or too long when actually done on the equipment. A cleaning task may indicate 15 minutes, but several bolts and items need to be removed which might take longer than the duration indicated by the planner. In short, if the planner is uncertain, there is no harm in discussing this with the people who will execute the actual PM or those who have experienced doing the tasks.

Figure 4.5: The Best Planners Will Always Be the Rolling Stones

If I am to select my own planners, I would prefer the old-timers, especially those who were involved previously in the actual Preventive Maintenance execution. These are your plant's Rolling Stones or those with white hair and wrinkles. Let's face it, I have met many of these people in my lifetime and in my training. These people are usually silent as they perhaps practice their right to remain silent. They do not brag about their experience and they just don't talk too much, but seeing their faces, I know that they know things in the equipment that perhaps only a few of them knew. The problem is that when these people reach the age of retirement, 60 years for the Philippines or more in other countries, they will be forced to retire, and the problem is that we have not captured nor documented what they know. In my book on RCM, I indicated that the best people to derive the RCM analysis and be included in the team should be the Rolling Stones. In the book of John Moubray on RCM II, on who should compose the RCM team, he quote, that the objective is to assemble a group that can provide most if not all of the information needed. These are the people who have the most extensive knowledge and experience of the asset and of the process of which it forms a part. In the majority of cases, what happens to industries is that if these good folks retire for good, their experience goes with them to the grave. The industry will be forced to hire new candidates fresh from college, find their way inside the plant, earned their experience by committing mistakes until they grow some wrinkles and white hair, then once again reach the retirement age and retire again for good. What we want is to break this merry-go-round cycle by finally capturing what they know which I truly believe is of great importance and benefit to their industry. We should not let this opportunity slip away and repeat the old cycle. These good old people can leave a legacy and mark in their plant and perhaps write their own book about how they were involved in maintaining their equipment in the plant. The good thing about writing a book is about earning royalties besides their pension of course.

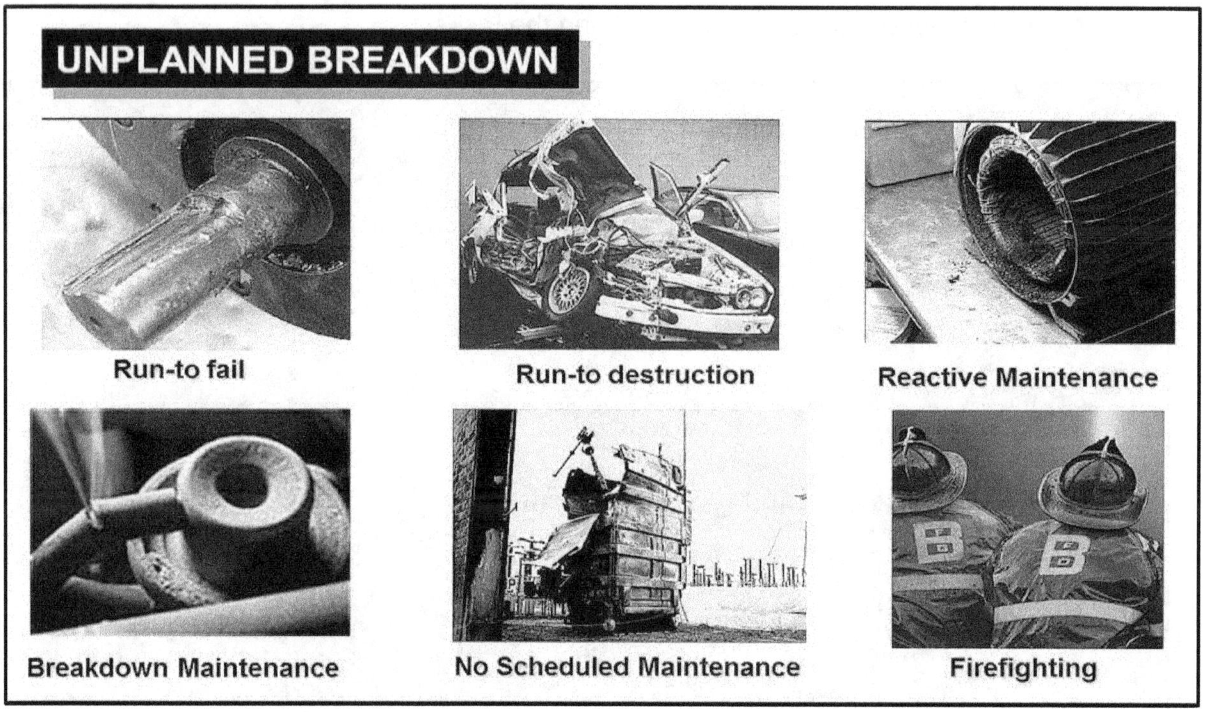

Figure 4.6: Many Terms for Unplanned Breakdown

Usually, if we put things into the correct perspective, the job of the planner is two-fold. This is to generate a Work Order for both Planned and Unplanned breakdowns. Planned breakdowns are those works that will be done in the future to anticipate and manage all possible breakdowns and failures that can occur on the equipment. Unplanned Breakdowns are unexpected failures occurring on the equipment in which the operator will generate a work request where the role of the maintenance or technician is to troubleshoot the equipment until it becomes operational once again. This means that the equipment was in a run to failure mode. Industries have used many terms on this as discussed in the previous chapter, but it all means the same thing which is the equipment failed and the maintenance goes to the equipment to repair the failed equipment.

Maintenance is done at a point when there is repair or actual breakdown. It occurs when repair action is taken on a problem only when the problem results in the machine's failure. Unplanned downtime, in its simplest definition, simply means fixing the machine when it fails. This is usually more expensive than Planned Breakdown since there will always be the possibility of secondary or tertiary damages that can happen to the equipment. In many industries especially if the maintenance culture is reactive, the time of the maintenance planner is being eaten up by requests for unplanned breakdowns which will require emergency repairs. Usually, the ratio will be around 80% of the planner's time will be spent on writing emergency requests, which results in emergency buying, and 20% or less on Planned Work on the equipment. What we need to do is to reverse this situation where 80% of the planner's work is focused on Planned Breakdowns or future work and 20% on Unplanned Breakdowns. If we borrow the case in Chapter 3, figure 3.4, we have the following;

- Rank A: 486 machines = 19.24%
- Rank B: 941 machines = 37.25%
- Rank C: 1099 machines = 43.51%

- Total = 2526 machines

Rank A will consider the worse performers or the critical machines in the plant. The focus of the maintenance planner is to prioritize the scheduled work for all Rank A machines as this will serve as the critical equipment in the plant. If the scheduled planned work will include the following:

• Daily Scheduled Work:	1 point
• Weekly Scheduled Work	1 point
• Monthly Scheduled Work	1 point
• Quarterly Scheduled Work	1 point
• Semi-Annual Scheduled Work	1 point
• Annual Scheduled Work	1 point
• 2 years Scheduled Work	1 point
• 3 years Scheduled Work	1 point
• Total Points	8 points

Hence, the priority of the maintenance planner is to generate (486 Rank A x 8 tasks/machine) which will be equal to 3,888 tasks for all 486 machines. If we estimate that on average, it will take 8 hours for the planner to write the tasks, therefore, we have 3,888 tasks x 8 will be 31,104 hours. If we have 3 overall planners in our plant, then this will take 10,368 man-hours or 1.18 years to complete writing the complete Preventive Maintenance tasks for all the 486 critical machines in the plant. This is of course on the assumption that all the equipment in the plant is not identical which will be unlikely the case. This is just a raw assumption not to mention the revisions in the plan based on the feedback from the PM crew to the planner.

For identical equipment, we can just have a check and balance if there will be fewer or more tasks to be done on each piece of equipment. This means that there will be tasks that will be done on those machines with problems and might not be performed on other machines that do not possess the same problem as in figure 4.7. But this is not yet the end of the work for the planner as they also need to perform Preventive Maintenance for both Rank B and C machines otherwise, if this will not be done, then those Rank B and C can become Rank A machines in the future.

But again, writing the plan is not just a one-time event, as it will be improved continuously either by revising the details, the duration of executing the tasks, the interval which means that if an activity is done monthly, the question is can the tasks be done every quarter and so on. What is important at this stage is to have a feedback session with the people executing the tasks. If the planner is well experienced, he can recommend how the tasks should be done especially where the difficulty will be experienced during the execution. What I am stating is that the maintenance planners should be full-time and should not be pulled out to do other tasks on maintenance.

F	FF	FM	Failure Mode	Pilot Machine	Horizontal Replicated Equipments				
					MC 2	MC 3	MC 4	MC 5	MC 6
1	A	1	Air leak in the system	●	x	●	x	○	○
1	A	2	Excessive Belt Tension	●	x	●	●	○	○
1	A	3	Excessive Vibration on fan-wheel	●	●	●	●	○	○
1	A	4	Soft-foot on base foundation	●	●	x	●	○	x
1	A	5	Worn-out bearing	●	●	x	●	○	x
1	A	6	Worn-out drive coupling	●	●	x	●	○	x
1	A	7	Excessive misalignment	●	●	○	●	○	x
1	A	8	Abnormal end thrust load	●	●	●	●	○	○
1	A	9	Reverse rotation of motor	x	●	●	x	○	○
1	A	10	Excessive oil contamination	x	●	●	x	○	○

Legend : ● Completed X Failure Mode not present ○ Ongoing

Figure 4.7: Performing PM Tasks on Similar Equipment

4.4: Preventive Maintenance Bill of Materials (BOM)

Although, unlike my time when everything was done manually, where we need to check each of the items one by one, today's CMMS or EAM software, can generate an equipment's Bill of Materials (BOM) automatically. Maintenance planners do not have to spend most of their time looking for a particular part, identifying the codification or part number, cost, and other things needed, for as long as the system or CMMS is populated correctly, and closing the PM reports are done completely.

A Preventive Maintenance Bill of Materials will include the part name, codification, part description, items cost, quantity needed, and the total cost. A PM Bill of Materials is a list of materials, items, spares, and consumables needed either for repair or Preventive Maintenance. Having a PM Bill of Materials will allow the people involved to identify easily what items and materials will be needed to conduct a repair or a scheduled routine Preventive Maintenance for a particular asset. This makes the execution fast in which we no longer need to identify each of them individually and manually. Many CMMS software can automatically provide a BOM when parts are issued against work orders. This is on the assumption that maintenance completes their report after executing either a repair or Preventive Maintenance. This means that we need to populate the system first and ensure that the correct information is recorded on the system. System generated Bill of Materials can provide what is needed to perform the maintenance tasks, especially looking for a part number or codification for a particular item which just adds waste and losses in executing the PM task. We do not need to call Jack or Joe who is from the other shift since they are the ones who know the parts. This Bill of Materials is important to every maintenance planner as it will specify the parts of the needed asset to perform a specific work. The planner will usually provide the scope of the work and specify the BOM if something is needed.

4.5: Reducing Breakdowns by Taking Care of the Basics

As we said, we want our maintenance planners to focus 80% of their time on planned future work and not on generating work orders for emergency works and repairs. So how will we reduce the breakdowns in our plant? We must adhere to the principle that *big problems start from small things.* In fact, the majority if not all of the failures and breakdowns we experience in the equipment are simply an accumulation of small problems that have been left neglected in our equipment. It will just take one piece of loose bolt to create a chain of destruction on our equipment as vibration increase incrementally. Cracks and fractures can start to propagate as a result of this. As the machine vibrates, other bolts start to loosen, and that is just the beginning of a bigger problem.

Dirt and foreign matter penetrate rotating parts, sliding parts, pneumatic and hydraulic systems, electrical control systems, and sensors which can cause a loss of precision, malfunction, short stoppages, and breakdown as a result of early wear, blockage, frictional resistance, and other problems ending up in catastrophic failures. Compressed air leaks can contribute to operational problems which included fluctuating pressure causing air tools, and other air-operated equipment not to function properly.

Establishing these very basic equipment conditions simply means eliminating the causes of accelerated deterioration. This is when a certain part or item in our equipment does not reach its useful life. Most books on TPM which were authored by Japanese experts seldom mention the word random failures. They use the term accelerated deterioration. In fact, the goal of Phase 1 of Planned Maintenance is to reduce if not eliminated all accelerated deterioration so that parts will wear out naturally. This means cleaning to remove dirt and sources of contamination, proper lubrication to prevent early wear, and understanding that bolts need to be complete and secure. Parts do not achieve their desired lifespan due to dirt and contamination. The wrong lubricant is poured into the equipment because the guy is new. The machine fails simply because several bolts are missing. The maintenance uses the wrong tools that later damaged the equipment and so on. This is like implementing both Autonomous and Planned Maintenance in a much faster mode since we will not be focusing only on the pilot machine but on all machines in the plant. These small problems that we see daily in our day-to-day operations may mean nothing at the moment, but they will affect the equipment in the not-so-distant future. The problem is that most industries are blind to accepting that these small problems can lead to catastrophic failures in their equipment.

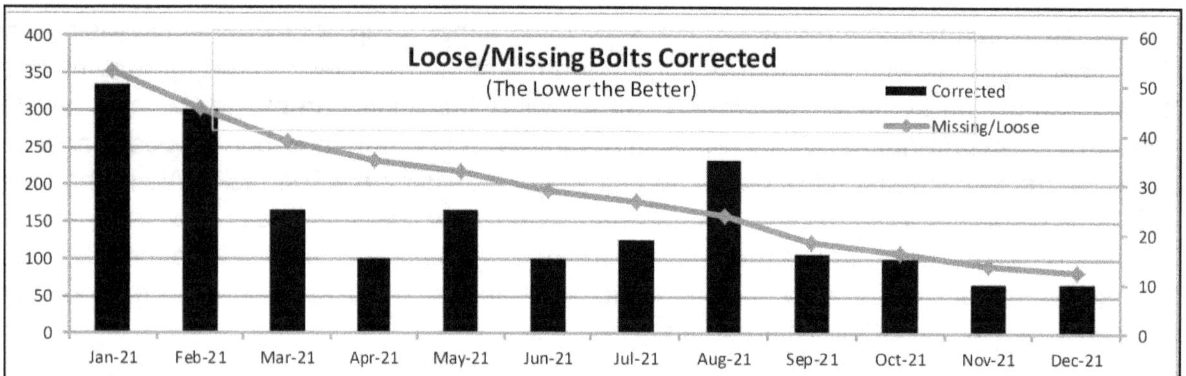

Bolts	Jan-21	Feb-21	Mar-21	Apr-21	May-21	Jun-21	Jul-21	Aug-21	Sep-21	Oct-21	Nov-21	Dec-21
Missing/Loose	352	302	257	232	217	192	177	158	123	107	92	82
Corrected	50	45	25	15	25	15	19	35	16	15	10	10

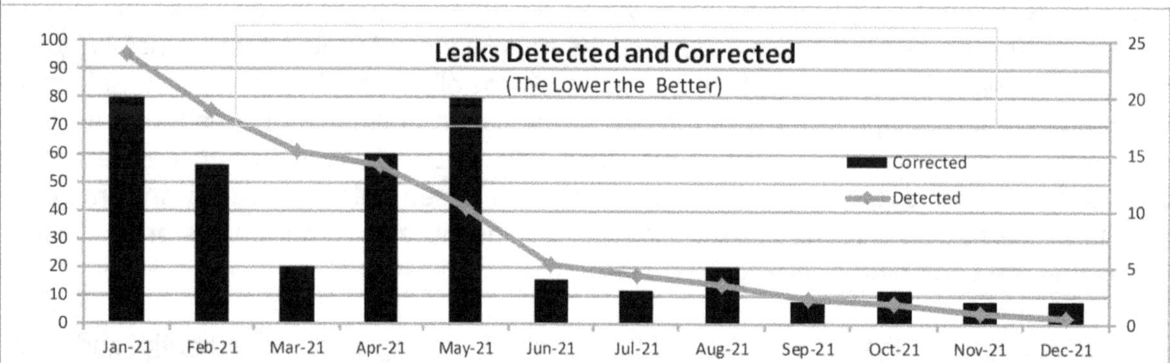

Leaks	Jan-21	Feb-21	Mar-21	Apr-21	May-21	Jun-21	Jul-21	Aug-21	Sep-21	Oct-21	Nov-21	Dec-21
Detected	95	75	61	56	41	21	17	14	9	7	4	2
Corrected	20	14	5	15	20	4	3	5	2	3	2	2

Figure 4.8: Tracking Leaks, and Missing or Loose Bolts

As a starting point in monitoring the basic equipment condition, we can assign a group of people from maintenance to track and tag all equipment with leaks and seek help from operators in locating missing, and loose bolts for all equipment and machines in the plant. As a general rule, the tags will be located exactly where the leak or missing bolt is detected and will only be removed once it is restored and corrected, but not on moving parts. The number of abnormalities and those corrected can be measured weekly or monthly and results are updated not only on the system but can also be placed on their bulletin or activity boards. Tracking the percentage of abnormalities corrected versus detected, the number of leaks, missing or loose bolts perhaps is a unique indicator in the sense that these are the only indicators I know that will measure the failure of secondary functions or function-reduction breakdowns of the equipment. What TPM taught me which I believed with all my heart and soul is that small problems matter most since neglecting them will just result in a bigger problem on our equipment and assets. These small problems are often times neglected thinking that that is the way it is since the beginning of time. Equipment basic conditions will include keeping the equipment clean, correct lubrication, all bolts secure, and equipment with no form of leaks whatsoever. If there are 200 small problems tagged and corrected, these small problems corrected can avoid bigger problems in the equipment.

Cleaning: This includes the process of removing any form of unwanted object from the equipment. It consists of removing dirt, contaminants, grime, excess oil, grease, and other foreign objects that can affect equipment parts and components. Cleaning is performed not just to satisfy the cosmetic looks of the equipment but because this dirt and contaminants shorten the life of certain parts and components of the equipment. World-Class Companies perform these activities. Components and equipment are kept clean in every detail. Such an organization realizes that inspections cannot be done without this level of cleaning. Cleaning definitely extends the life of parts and components. Through cleaning, we can expose problems to the equipment that has been left ignored for a very long time. When operators clean their equipment, they touch parts, and by touching parts, they are actually inspecting them. Leaks, cracks, and fractures that have been in the equipment for a very long time can be detected and exposed when operators touch and clean their equipment. Dirty and untidy equipment will tend to deteriorate more rapidly than equipment that is well maintained and cleaned at all times.

Correct Lubrication for All Equipment: All machines contain lubricants in one form or another. The purpose of lubrication is to reduce friction for mechanical parts moving inside our equipment. It is not only important to lubricate the equipment but also to use the correct lubricant and its correct amount. Equipment that lacks lubrication or is over lubricated can likewise induce problems. Operators should be taught about the importance of checking the lubrication in their equipment and maintaining the correct amount of lubricant. So many failures can be attributed to lubrication that can be avoided if only these basic equipment conditions can be well established. Maintenance must teach operators not only the necessary points in their equipment to lubricate but also what proper and improper lubrication can do. Just like humans, lubrication is the lifeblood of the equipment, and it should be maintained clean and adequate all the time. Many failures are attributed due to lack or inadequate lubrication. These problems can be avoided and controlled in our equipment if we understand and learn the role that lubrication plays in our equipment and assets.

All Bolts Complete and Secure: Machinery contains bolts, nuts, and fasteners as part of their construction, and they serve a particular purpose. The equipment functions properly only if fasteners are securely tightened. All equipment vibrates, but excessive vibration can be destructive as this can cause parts to fracture which can lead to induce secondary damage to other parts and components affected by the vibration. When we touch the equipment, we can feel its vibration. What we feel is simply the sum of all vibrating forces moving inside the equipment. Bolts, screws, and fasteners are placed to minimize the vibration in the equipment. When vibration increases, it can result in cracks and fractures, and these fractures can propagate to the point of rupture as a result of excessive vibration. Remember that it only takes one loose bolt to create a chain reaction of destruction in our equipment. As a result, other bolts become loose, and vibration increases. Match-marks are placed on critical bolts and nuts so that operator can easily detect if bolts have been loosening due to excessive vibrations for some time. Imagine driving your car, and each of your wheels has 3 instead of 5 stud bolts on each of your tire. Would this be all right with you? I guess not. Then why don't we treat our equipment in the same way? We can see many bolts missing or loose as a result of many activities performed previously on the equipment. Why don't we complete them for a start?

No Form of any Leaks: Leaks may be in the form of an oil leak, air leak, vacuum leak, or any other form. One of the most common forms of contamination is process leaks. The simplest method of identifying and correcting leaks is to tag them. Tags will only be removed once the leak had been corrected. Locate the leak and identify its source, type of leak, and severity of the leak. Implementation of a simple, easy-to-use system such as this can save hundreds of thousands of dollars for a plant. Large air leaks can be detected by ear, however, for very small leaks, this will no longer be audible to the human ear. An instrument that can detect these leaks is Ultrasonic Monitoring which can be used to inspect pipes and joints starting from the compressor to the pneumatic equipment. Once the leak is detected, it will be tagged and immediately restored by applying a repair kit to the portion on the joint that contains the leak. An oil leak can also be a source of ingression of dust and moisture that can enter the equipment.

4.6: What Tasks Should PM Include

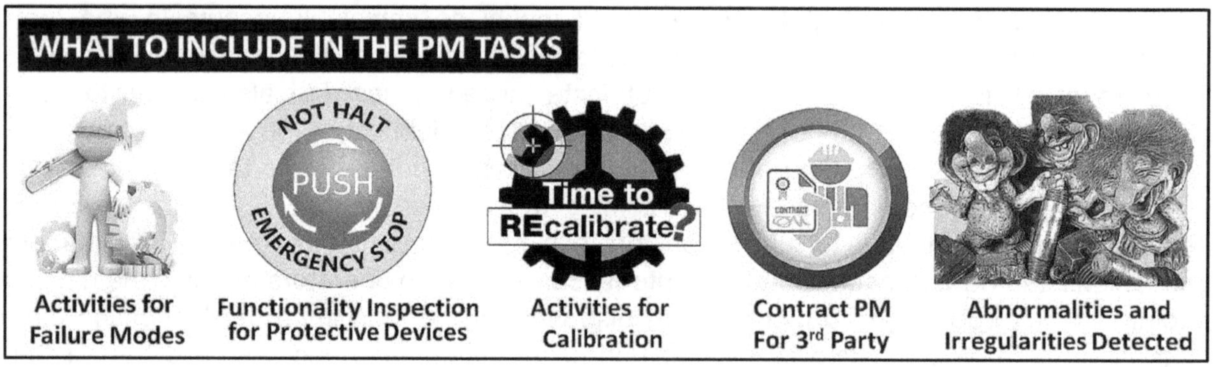

Figure 4.9: What Do We Include in the PM Tasks?

Once the breakdowns are reduced by addressing these small problems, then the maintenance planners can now have the time and focus to develop the plan on what activities should be included in the Preventive Maintenance tasks. The focus should be on all critical equipment categorized as Rank A. What is likewise important is to identify all the MRO Spares and consumable items needed to execute the task. This must be well-coordinated with the Purchasing and the Storeroom people.

Another important point we need to consider is the interval or frequency. The tasks should not be too late or too soon. If the task is performed too soon, then we shall be wasting resources, while if the task is done too late then the failure already occurred. I have dedicated a chapter to determining the interval for the different tasks on maintenance. If we focus our attention on what activities should be included in the PM tasks, we can group this into 5 categories as reflected in figure 4.9. The tasks or activities on Preventive Maintenance should address the following;

1. **Tasks to Address the Failure Modes:** We need to identify and list all the possible failure modes that have occurred in the past and those that are possible to happen in the future. Failure modes should be defined in enough detail for it to be possible for us to select a suitable task to address them. These can be failures that have occurred before in the past or

on similar equipment types and models. If a Preventive Maintenance task is already existing, then we can review the task to indicate if it addresses a particular failure mode or not. Any other failure modes which have not yet occurred and are possible to happen should also be listed. The planner must also understand or have some background on the equipment that is undergoing the maintenance planning process. For mechanical parts, identify those parts that have movements inside the equipment. These parts may either fail concerning their age or they may fail prematurely. Identify if the equipment contains electronic parts and if there were failures of this sort that happened in the past. Do the same thing for electrical, and electro-mechanical parts. Environmental surroundings should also be maintained such as temperature and relative humidity.

Figure 4.10: Equipment Failure Modes

2. **Tasks for Protective Devices:** Protective devices are placed on the equipment for one good reason and that is to protect something. That something is called the protected device. Usually, the failure of these protective devices is hidden unless they can be seen on the computer's display screen. The task performed on these protective devices is called functionality inspection or failure finding tasks. Others termed this Detective Maintenance. The interval of conducting these tasks is explained in the succeeding chapter. The reason for conducting a functionality inspection for these protective devices is to avoid the chances of multiple failures. Multiple failures occur when both the protected and protective device is in a failed state. If this happens, then we experience multiple failures and we are in a dangerous situation in this case.

3. **Tasks for Calibration Activities:** Calibration will be the process of comparing the existing instrument in the equipment with a standard measuring device which will be the basis of known accuracy to detect any deviation and correct for any discrepancy or deviation in the accuracy of the instrument measuring device which is being compared to the standard. The main purpose of calibrating these devices and instruments is to determine their accuracy. If there are deviations or anomalies found during the test, the instrument is adjusted and fine-tuned until the acceptable standard specification is finally met. The main reason for performing calibration is that instruments tend to deviate based on several factors such as the equipment load, vibration, extreme operating conditions, mechanical shocks, or exposure to extreme temperature, humidity, or pressure. The frequency of calibration depends on the allowable tolerance. If the instrument is critical in operations, then the more frequent it should be calibrated. All measuring instruments used on the equipment will require calibration periodically to ensure they are offering accurate results. The following will be subject to their regular calibration

Pressure Calibration Includes

- Analogue Pressure Gauges
- Barometers
- Digital Indicators
- Digital Pressure Gauges
- Test Gauges
- Transmitters

Temperature Calibration Includes

- Laminar Flow Meters
- Rotameters for Gas, and Air
- Thermal Mass Flow Meters
- Turbine Meters
- Temperature Sensors

Electrical Calibration Includes

- Clamp Meters
- Counter timers
- Data Loggers
- Electrical Meters
- Insulation Testers
- Loop Testers
- Multi-Meters
- Oscilloscopes

Mechanical Calibration

- Accelerometers
- Load Cells
- Pressure Gages
- Measuring Instruments such as Micrometers, Vernier, Height Gauges
- Scales and Balances

- Torque Wrenches
- Weight and Mass Scales
- Dial Indicators
- Predictive Maintenance instruments (performed by OEM)

4. **Tasks to Address Abnormalities:** Although this will be a TPM activity, however, if TPM does not exist in the plant, then these abnormalities should likewise be corrected during the Preventive Maintenance outage. For those implementing TPM, Addressing abnormalities will be a joint effort by operators and maintenance. Operators will tag their equipment for any abnormalities, slight or major deviations that they can see during the initial cleaning process. Maintenance will also have its own tags when they performed the Planned Maintenance Phase 1 restoration stage. The goal of both Phase 1 and Step 1 of both Planned and Autonomous Maintenance is to bring the equipment back to its original basic equipment condition. The distinction between Autonomous Maintenance and Planned Maintenance tags is that maintenance will be focused more on the interior part of the equipment, including deteriorated parts that need to be restored, while the Autonomous Maintenance team will focus more on the exterior part of the equipment. As much as possible, it is best for the operators who locate the abnormality to correct it themselves if they can. Abnormalities that the operators cannot correct themselves will be passed on to the Planned Maintenance team. It is highly recommended that when Step 1 Initial Cleaning is performed on the equipment, both operators and maintenance should perform their initial cleaning and restoration activities on the same day so that operators tagging abnormalities can be assisted by the Planned Maintenance teams. One thing important to note is never ever passed all abnormalities to maintenance. There will be abnormalities that can easily be corrected by operators.

The Autonomous Maintenance team should provide a before and after picture of the abnormalities, they have corrected. Hence, when doing step 1 of Autonomous Maintenance, let us say that 180 abnormalities were detected, and 90 of them have been corrected so far, which means that their accomplishment rate will be at 50%. This percentage is what the members of Autonomous Maintenance will be tracking. It is also highly recommended for both Autonomous and Planned Maintenance to have the same pilot equipment at the start so that Planned Maintenance can easily focus on supporting the needs of the operators instead of having different pilot equipment for operators and maintenance. One more important point is that Planned Maintenance activities should be ahead before Autonomous Maintenance is implemented. This means that Phase 0 or the Preparatory Phase of Planned Maintenance should already be completed since Planned Maintenance will be performing the machine ranking on all equipment in the plant. The operator's pilot equipment should belong to the Rank A (worse) category with the possibility of large replication or fan-out. Addressing abnormalities and performing Phase 1 restoration activities for both Planned and Autonomous Maintenance will definitely reduce not only the breakdowns but also minor stoppages on the equipment. In fact, our JIPM consultant once said that restoring the equipment if done religiously must reduce the breakdowns by as much as 80% or even more. We actually experienced what he said and I personally concur with what our Japanese JIPM consultant said. What I love about TPM is its simplicity and one lesson we learned is that small problems are just an accumulation of bigger problems that can happen on the equipment.

7 Types Of Equipment Abnormalities		
1	**Minor Flaws** - Contamination - Damage - Play - Slackness - Abnormal Condition	**Examples** - Dust, dirt, powder, oil, grease, rust, corroded parts - Cracking, crushing, deforming, chipping, bending - Shaking, falling-out, eccentricity, backlash - Belts, chains - Excessive noise, overheating, vibration, strange smell - Over or under pressure,
2	**Unfulfilled Basic Condition** - Lubrication - Lubricant Supply - Oil Level Gauges - Tightening	**Examples** - Insufficient Oil Level, dirty, unidentified, unsuitable or leaking oil - Dirty, damaged or deformed lubricant inlets, faulty lubricant pipes - No minimum and maximum points for lubricants - Loose threads, slackness, missing bolts and nuts, worn out heads, incorrect bolts and missing washers
3	**Inaccessible Places** - Cleaning - Checking - Lubricating - Tightening - Operation - Adjustment	**Examples** - Machine construction, covers, layout, foothold, space - Covers, construction, layout, instrument position & orientation - Position of lubricant inlet, construction, height, lubricant outlet - Covers, construction, layout, size, foothold, space - Machine layout, position of valves, switches & levers, foothold - Position of pressure gauges, thermometers, flowmeter, moisture gauges
4	**Contamination Sources** - Product - Raw Material - Lubricants - Gases - Liquids - Scrap - Others	**Examples** - Leaks, spills, spurt, scatter, overflow - Leaks, spills, spurt, scatter, overflow - Leaking, split, seeping lubricating oil, hydraulic fluid, fuel oil, etc., - Leaking compressed air gases, steam, vapors, exhaust fumes, etc., - Leaking spill cold & hot water, leaks on cooling water, heat exchanger - Flashes, packaging materials, non conforming products - Contamination induces by people, dusty places, leaks on roofs etc
5	**Quality Defect Sources** - Foreign matter - Shock - Moisture - Grain Size - Concentration - Viscosity	**Examples** - Inclusion and entrainment of rust, chips, wire scraps, insects, etc - Dropping, jolting, collision, vibration - Ingression of moisture, defective elimination of moisture - Abnormality on screens, centrifugal separators, compressed air etc - Inadequate warming, heating, compounding, mixing, etc - Inadequate warming, heating, compounding, mixing, etc
6	**Unnecessary / Non urgent item** - Machinery - Piping equipment - Measuring Instruments - Electrical Equipment - Jigs and Tools - Spare parts - Makeshift repairs	**Examples** - Pumps, fans, compressors, auxiliary machines, columns, tanks - Pipes, hoses, ducts, valves, dampers - Temperature and pressure gauges, vacuum gauges, ammeters, etc., - Wiring, harnessing of wires, piping, switches, plugs, octopus wiring - General tools, cutting tools, jigs, molds. Dies, frames, etc., - Standby equipment, spares, permanent stocks, auxiliary machines - Tape, strings, wire, metal plates etc.,
7	**Unsafe Places** - Floors - Steps - Lights - Rotating Machinery - Lifting Gears - Others	**Examples** - Unevenness ramps, cracks, leaks floors, wet floors, peeling, wear - Too steep, missing handrails, wet and leaky steps, dirt - Dim, dirty, busted lights, explosion proofed - Displaced, broken off or broken covers, no emergency stopping device - Unsafe wires, hooks, brakes, and other part of hoisting mechanisms - Special substance, solvents, toxic gases, chemicals etc.,

Figure 4.11: Types of Abnormalities

5. **Third Party Contractor Works:** Other industries will be providing maintenance activities to independent 3rd party contractors, or the OEM of the equipment themselves. The scope of activities may vary from one industry to another. Maintenance or person from EHS can perform a walk-around inspection and identify existing hazards in their plant's vicinity or working conditions that can be considered unsanitary, or hazardous to employees, and recommend a third-party contractor to perform the tasks instead of the regular maintenance employees of the plant crew to correct these hazards. Contract Maintenance can be defined as the contract agreement signed between the industry and a third-party contractor who will perform a maintenance activity based on the terms of the agreement on industries equipment, and assets. Usually, the scope of maintenance work will be dictated by the maintenance head, and performed by a third party and not by the plant's regular maintenance crew. The reason for third-party contractors is that a special skill may be needed that is beyond the skills of the people or it is lesser expensive if this task will be performed by a third party instead of doing it themselves.

4.7: Why Most Planners Cannot Plan Future PM Work

In August 2012, I was in South Africa teaching a course on Lubrication Strategy. In one of my classes, I was informed beforehand that my training will be interrupted for an emergency Safety Meeting for 1 hour, and I said no problem as I can manage my time. They said that I can stay in the room but they told me not to interrupt the meeting which I agreed. Then the meeting came and the safety officers opened their laptops to discuss their recent audits regarding non-conformance. The top of the list includes maintenance working or repairing without a job or work order. After the meeting, the safety people left and we have a break for 15 minutes. I asked the maintenance people while we were taking our coffee break why they do not have a work order. One of the people voiced out and told me that if they wait for the work order and comply, then their downtime will increase. Others told me that there are cases where the work order takes 1 week or longer because the boss is not around to approve the work, or perhaps the maintenance planner is too overwhelmed on what work needs to be prioritized as there is way too much work request and there are just 2 planners for the whole industry. I reflected on that moment and after I left, I had a meeting with the maintenance team and told them why not just eliminate the work order for repairs and emergency work completely and try to think about it.

My point on this is that in most plants I know, the planner can hardly prepare any plans for Preventive Maintenance because their time is consumed by doing daily work or job orders which now becomes a routine job for the planner. When a planner receives a work request, he plans the activities, determines if spares, consumables, or any power tools are needed, and estimates the Bill of Materials, the duration of the work, manpower, and other stuff. If these things are not available on the system, then the planner needs to leave his table and look for the answer himself. He talks to the PM about what materials are needed, goes to the storeroom if the part is available and if not then goes to the purchasing and so on. In fact, if this work is a priority, it will take the planner half his day or one complete day just to complete one job order. The rest of the days will be no different. If there are 10 job orders for the day, then the planer will be delayed in responding to the other 9 requests, as the day ends and as the planner reports to work the following day, then another 10 work request is on his table not to mention the backlogs until it piles up eventually. What happens then is that both planners and maintenance will just be overwhelmed with many unending backlogs.

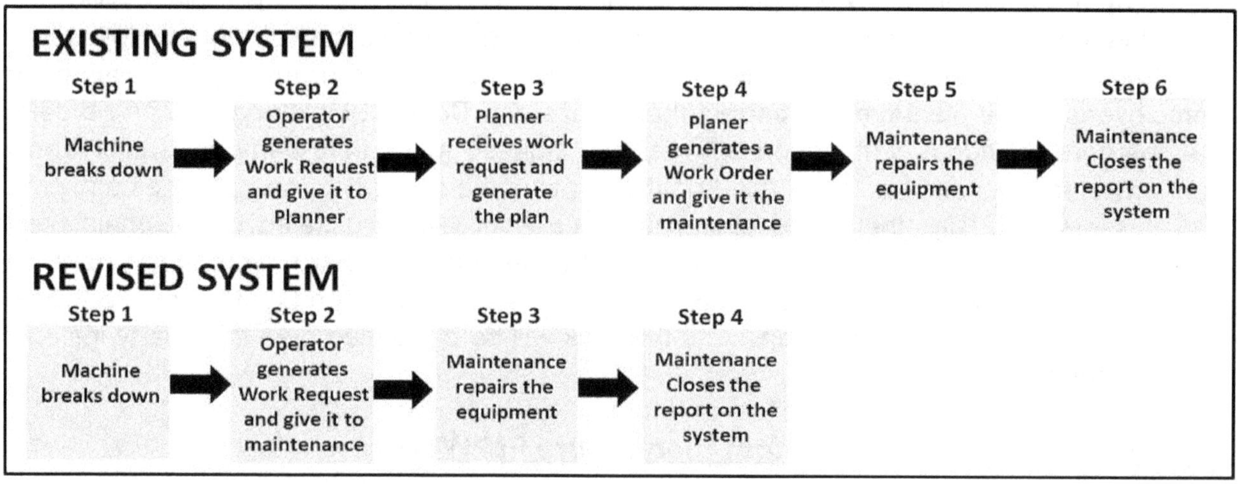

Figure 4.12: Revising the System for Repairs

As we can see in figure 4.12, under the existing system, the planner receives the work request and generate a plan, process the work order, and give it to the maintenance who will execute the tasks. The new system indicates that the planner is removed from doing reactive works and the operator hands out the work request directly to maintenance and maintenance repairs the equipment that failed. What is important in the revised step is for the maintenance and technician to provide the details once he closed the Work Request in the system. Closing the report on the system must include the following:

- Name of Breakdown or Failure Mode
- Machine Number (Machine that was repaired)
- Date, Start Time, and End Time of the Repair (Duration)
- Spare Parts, Consumables, or anything withdrawn from the storeroom. Indicate the parts code
- Provide the details on what was done to repair and restore the equipment.
- Provide Steps if necessary.
- Indicate if there are important points to consider when performing the repair.
- Indicate if there are 3rd party contractors involved.

Example:

- Machine Number: Motor, MTR-0025
- Date: March 1, 2022
- Time Started: 11:45 am, Time Ended, 12:45 am (Duration 2 hours)
- Spare Parts: Bearing Replaced with Parts Code: BRG – 1234J (SKF)
- Steps Involved in the repair:
  - Damage bearing was removed using an Ultra-Torque Bearing Removal tool
  - New bearing was installed with part number BRG – 1234J from SKF
- Were there important points to consider when performing the repair?
  - Use the correct tools in extracting the bearing to avoid further damage

In the revised procedure, everything was recorded in the system completely, hence, if the same failure happens, the maintenance or technician who will be involved in the repair can just check the system if a repair procedure exists and they can just have a print-out on this. The thing is what we want is to free the maintenance planner so that the planner can focus their time more on future works such as planning for routine Preventive Maintenance, and scheduled outages and shutdowns. The planner must focus his plan on Rank A machines or those that are considered worst as discussed in Chapter 3 of this book.

This simple change in the system can ensure that the planner is working on more important work which is the scheduled Preventive Maintenance plan for routine and shutdown works. People involved in Predictive Maintenance can detect potential failures in the equipment they are monitoring several months before the failure happens. This means that at the earliest onset of the problem or once a potential failure is detected by the instrument, the Predictive Maintenance user can coordinate with the planner for a work order and the Predictive Maintenance user can dictate the right moment to decide on when to stop the machine for a scheduled replacement or overhaul once 70 to 80% of the P-F (Potential to Functional) failure have been already consumed thereby maximizing the lifespan of the parts being monitored.

4.8: Do Not Expect the Plan to Always be Perfect

The planner usually determines the scope of the work, estimates the duration for each task, prepares the Bill of Materials needed, and estimates the manpower and skills requirement needed for the PM task. Do not expect the plan to be 100% accurate, unless the planner was once the maintenance who does these things in the past. The planner may estimate a particular maintenance task to be done in 2 hours, but the actual duration almost eat up the entire maintenance shift, or perhaps the planner was referring to Charlie who was the most experience of all the crew who just retired last month. A particular spare was not identified where one of the PM Crew need to once again go to the storeroom only to learn that the part is not being stocked and an emergency request was done by the Purchasing people. A maintenance task actually requires 3 people as they will be removing an entire engine but the manpower only indicates that this will be carried out by a single person unless that person is Superman. Additional tasks need to be done which is not currently included in the job scope or a welding activity needs to be done on a particular base plate, but the welder is currently on leave. There are so many other cases where the plan was either incomplete or incorrect. The bottom line is that planning must be continuously improved for the better.

We also have to consider the human factor and the people who will execute the PM and the working environment. [8]Human fatigue includes the feeling of tiredness, physiological, in the execution of work concerning time. The body temperature fluctuates throughout the 24 hours. This means that our mind and body's energy is not constant. The lowest point or the time humans feel fatigue will be around 0200 to 0300 hours in the morning and the highest is at 1200 to 1400 hours. The difference is 0.5 ° C. This means that the body will fatigue more

[88] Reasons, James and Hobbs Anthony, **Managing Maintenance Errors**, Ashgate Publishing Limited, 2003, page 31 to 32

in the early hours of the morning. This diminishes and the body increases its level of energy until just past noon and decline in the afternoon. These fluctuations are examples of the Circadian Rhythms which means around the day. It comes from the Latin words circa which means around, and the word dies which means a day. These circadian variations are governed by a biological clock located in the brain. Skill-based errors exhibited a significant circadian rhythm, being most prevalent in the early hours of the morning. However, the circadian rhythm can be disturbed by several factors such as a change in time zone, especially when we go to a country with a different time zone or from a shift change. It is much faster to adjust from a day shift to a night shift, rather than from a night shift to a day shift. During the night shift, the body temperature decreases. Alertness begins to reduce. This means that there are much greater chances of human errors to occur during the night shift as it includes working in the early hours of the morning. It is much easier to sleep during the night for workers than for night shift people to sleep during the day. I have experienced this personally. There are a lot of distractions and disturbances sleeping during the day compared to at night. People who will execute the PM must avoid late-night activities and should have a balance of 8 hours of sleep before the actual PM. If possible PM should be performed during the day shift as people on the night shit are more prone to human errors.

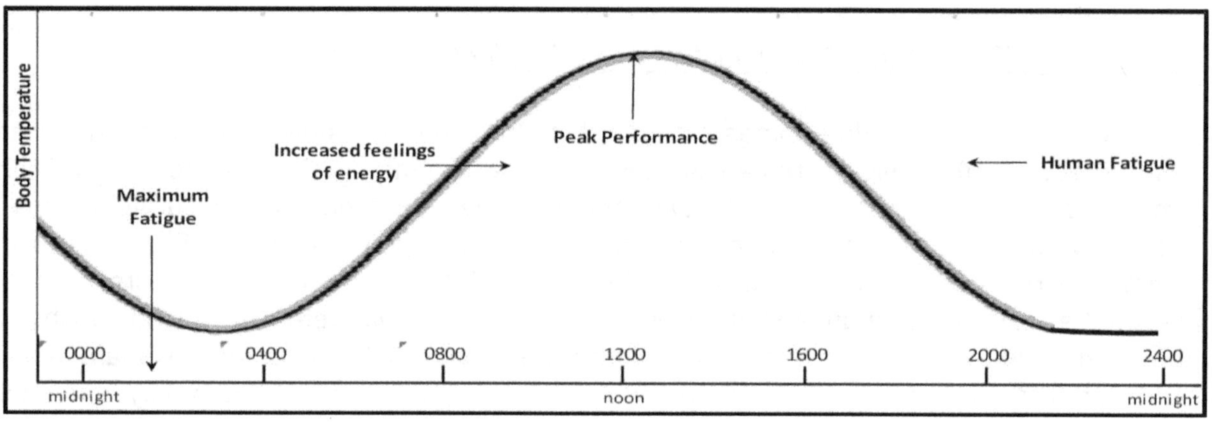

Figure 4.13: The Circadian Clock

Fatigue workers during the night shift can become irritable and may tend to lose concentration during their work especially if the work is routine. Research shows that the lack of sleep or sleep deprivation experienced by these workers is very similar to those produced by drinking alcohol. Both mental and physical performance on several tasks will be affected by a worker who is awake for 18 hours straight. This is like having a Blood Alcohol Concentration (BAC) of 0.05%. It is highly recommended that shift rotation should not go beyond one month for industries since people from night shifts are more prone to human errors. A worker should have at least 11 rest hours between shifts. It is also important for workers to have at least 2 rest days a week. This will allow the worker to start fresh when reporting for work.

The person who planned the work is different from the people who will actually execute the work. This is where the feedback will come in. Whatever discrepancies and inaccuracies should be discussed with the planner continuously. This is not just a one-time event, this should be continuous. People who will be doing their feedback must do it constructively and not sarcastically or should not provide below-the-belt remarks as the planner is also a human being. Treat people professionally with respect and expect the same in return. Planners must be flexible to make changes based on the feedback from the people who executed the PM task. The maintenance planner is not always perfect, but the good news is that we can always make it better through feedback. What is important is to apply the 4 Cs between the planner and the people who will execute the PM. This means having continuous communication, coordination, cooperation, and collaboration between these two. The purpose of the feedback is to correct the fault, recognize it, and become better by learning from the mistakes made so we can further improve future works.

What is important is that the planner must always be asking the people who will execute the task. Although the CMMS will be a good source yet there is no better information than that of the people who will actually execute or have the experience in doing the task. If the PM task will be done annually, the task may be identical next year, but there might be changes or additional tasks that need to be done that were not done last year.

4.9: Scheduling Work for Preventive Maintenance

What is important for the PM crew who will execute the task is to create a PM Master Checklist of everything before the actual PM is due especially if this will be a major Preventive Maintenance shutdown. Figure 4.14 is an example of a Preventive Maintenance Master Checklist when preparing for a scheduled PM. Failure to check these details could prolong the execution of Preventive Maintenance. If a third-party contractor is hired, and a clearance was not provided to the Security of the plant, then there will be delays. This always happens to me when I conduct in-house training only to learn that they forgot to provide me a gate pass where I need to wait outside the plant for several minutes and sometimes hours because they cannot locate the person who made the arrangement for the training. These will cause the contractors a delay in entering the site where you are already being charged for work that has not yet been started.

Many delays can happen during the actual PM, where we need to go back once again to the storeroom to withdraw this and that. A permit to enter was not prepared in advance where the contractors are already at the gate but have not been cleared by the security or we need to go to the other building just to retrieve a document needed for the details on how to perform a particular task. These kinds of waste can be minimized or removed if we can have a list of things that need to be accomplished before the PM such as in figure 4.14. Just like NASA minutes before a shuttle will be launched, every single system will be checked for a go or no go situation.

- Final Check begins - Go or No Go for Launch
- Retro: Good to Go, Flight
- Booster - Go, Flight
- Ecom - Roger Flight Go
- Control - Go, Flight
- Procedure - Go, Flight
- Network - We are Go Flight

- Recovery - Go, Flight
- CAPCOM – We're good to go Flight
- Instrumentation - Roger on that
- Kennedy, you are go for launch.

Checklist for Scheduling PM				
Item	**MRO Parts Checklist**	**Yes**	**Not Yet**	**N.A.**
1	Are parts, spares and items available at the storeroom?			
2	Are those parts not stocked already coordinated with Purchasing			
3	In item 2, if yes, is the part certain to arrive before the PM Schedule?			
4	Are any consumables to be used during the PM and are they available?			
5	Have the storekeepers already informed to stage them?			
6	Are any power tools, instruments or measuring tools needed to be borrowed?			
7	In item 6, if yes, have they been reserved on the day it is needed?			
8	If the time is 24 to 48 hours left before the PM, have all the needed items, tools, consumables, spares and other items been withdrawn, and checked for completeness so that nothing is missed?			
Item	**Manpower Checklist**	**Yes**	**Not Yet**	**N.A.**
9	Have the Manager or Supervisor identified the skills needed for the tasks?			
10	Have the Manager or Supervisor identified and checked the availability of the people needed which means that they are not on leave during PM?			
11	Will the task require 3rd Party Contractors?			
12	Are there any specific skills needed for any of the PM tasks?			
13	Have the PdM Group been notified to conduct a Pre-Post PdM Monitoring?			
Item	**Maintenance Planner Checklist**	**Yes**	**Not Yet**	**N.A.**
14	Have the Work Order been generated?			
15	Is the Bill of Materials checked and complete?			
16	Have Work Permits/Clearances if any been generated and approved?			
17	Is the duration assigned on each tasked reviewed and verified?			
18	If 3rd party contractors are to be assigned, have they been provided security clearance and permit to avoid delays when entering the site?			
	Safety Requirement Checklist	**Yes**	**Not Yet**	**N.A.**
19	Are there any safey kits, or apparels needed for the job?			
20	Have they been reserved during the PM date?			
21	Have safety signs been identified and reserved during the PM date			
22	Are ladders, additional lighting, hoist or chain block needed for the work and have they been reserved during the PM date?			
Item	**PM Documents Checklist**	**Yes**	**Not Yet**	**N.A.**
23	Does the PM Crew have access to the detailed procedure for each tasks on the CMMS system in case they need guidance?			
24	Are any documents needed for guidance such as schematic diagram, PM Manual, and other dossiers?			
25	Are any forms needed to be filled up before executing the actual PM?			
Item	**Other Requirement Checklist**	**Yes**	**Not Yet**	**N.A.**
26	Have operations been notified of the schedule and have inform the operator that will be affected?			
27	Have the equipment been inspected before PM for any additional works that is needed that is not covered in the maintenance plan?			
28	Is forklift needed to move the equipment to the PM Bay? If yes, have they been reserved during the PM scheduled date.			
29	Does the PdM group been monitoring parts that needs to be replaced which can be included in the PM schedule?			

Note: N.A. Means Not-Applicable

Figure 4.14: Master Checklist for Scheduling PM

Make it a habit to have a PM Master Checklist every time there is a PM schedule especially or a major PM shutdown. Another important point to consider is to have a sit-down with the operations people to release the equipment on the given date. Even if everything is prepared if operations won't let go of the equipment, then everything will just be plain useless. Provide operations a PM Schedule for the month so that operations can review their load requirements and make necessary adjustments. Better if we can include operations people in the PM scheduling of the equipment. On the part of the PM crew, they need to ensure and do everything they can to comply with the duration without any further delays so Operations can bring the trust back to maintenance. Operations can also be included in the scheduling meetings as they know what equipment will be loaded and not be. Perform pre and post Predictive Monitoring before and after executing the PM tasks and integrate the concept of Precision Maintenance which will be discussed in the next chapter to avoid or minimize human error that can lead to infant mortality failures.

4.10: Providing a Before and After Photo After Restoration

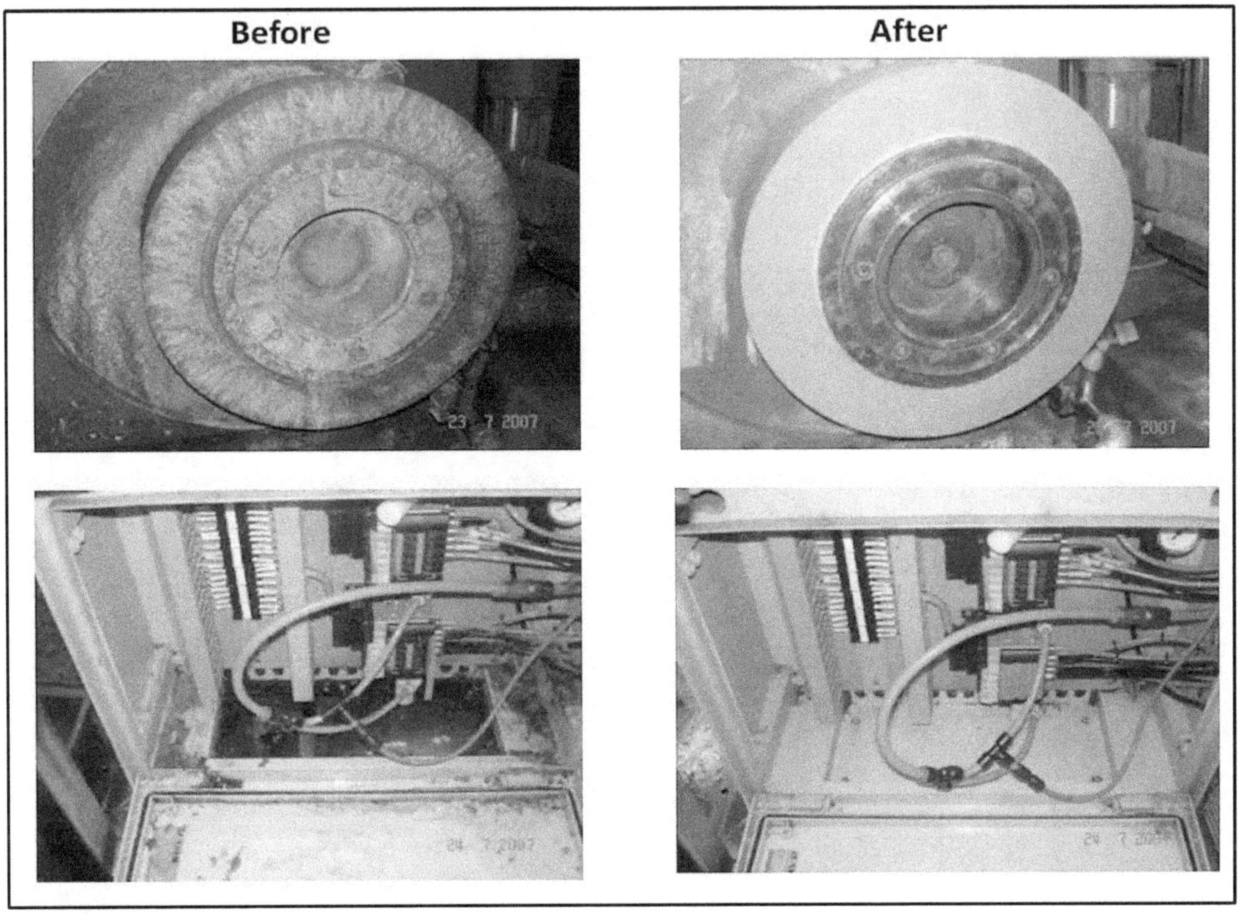

Figure 4.15: Before and After Restoration Photo

Providing a before and after photo on those works that have been completed and placing them on the system or having a printout should also provide future works of the quality of work that was performed during the restoration process. A brief description of the work can be provided. Although this practice has been a tradition for TPM's Autonomous and Planned

Maintenance, we can also apply this practice during Preventive Maintenance activities. This also serves as evidence that the work has been completed based on the Work Order. What is important in taking a before and after picture is that when taking a photo, they should be of the same angle and view. The before photo will indicate the condition before doing the PM works and the after photo shows the same parts after PM has been executed. This can be also applied to contract works as part of the requirement in accomplishing their scope of work in the plant.

Chapter 5

Integrating Precision Maintenance in PM

> ***The most important facet in Precision Maintenance is knowledge. Remember, people are not the company's greatest asset. The right people are the company's greatest asset and we can only have the right people if they are equipped with the right knowledge to do their jobs right the first time around.***

5.1: Human Errors in Maintenance

By definition, human error can be defined as an action planned but not executed according to the plan. Humans make errors since it is part of being human. The principle of error management is that even the best people can make the worst mistakes. Often very good people in established organizations keep on making the same blunders. An error can be defined as an action planned but not executed according to the plan. Research by Dr. James Reasons from the University of Manchester in England found that humans commit an average of 6 errors per week and with all the errors that are occurring and all the ways that we could destroy ourselves, life is still preserved. Why? It is because humans can sense change and break the error chain. Human errors include inappropriate action, or intention to act, given a goal and the context in which one is trying to reach that goal. This is a by-product of poor planned tasking or execution of a task resulting in the failure of a goal. All errors involve some kind of deviation from their intended course. It is the failure of a planned action to achieve their desired goal where this error occurs without some chance intervention. The majority of errors committed in maintenance have been the major cause of major accidents in industries. In summary, most organizations will change the human condition when they should be changing the conditions and environment in which humans work and should treat errors as an expected part of maintenance work.

Human errors are committed daily by physicians, doctors, engineers, designers, carpenters, nurses, lawyers, you name it. Although many people suffered from illegal drugs, many people also died from prescribed drugs. For people involved in the medical field, some errors can be fatal and can compromise the lives of their patients. You have road accidents reported almost daily by the news, errors from management decisions, airline disasters, flaws in systems, procedures, and policies that can have a detrimental effect on its product, and not to mention those cost-cutting schemes that caused a bridge or building structure to collapse, operations deferring Preventive Maintenance to cope with production, outdated procedures, industrial disasters, and you name it. But the most important factor in human error is learning from our failures so we can become better people. If people create problems, then people must also be capable of correcting them, but this can only happen if we look ourselves in the mirror and admit that we, too, are also part of the problem.

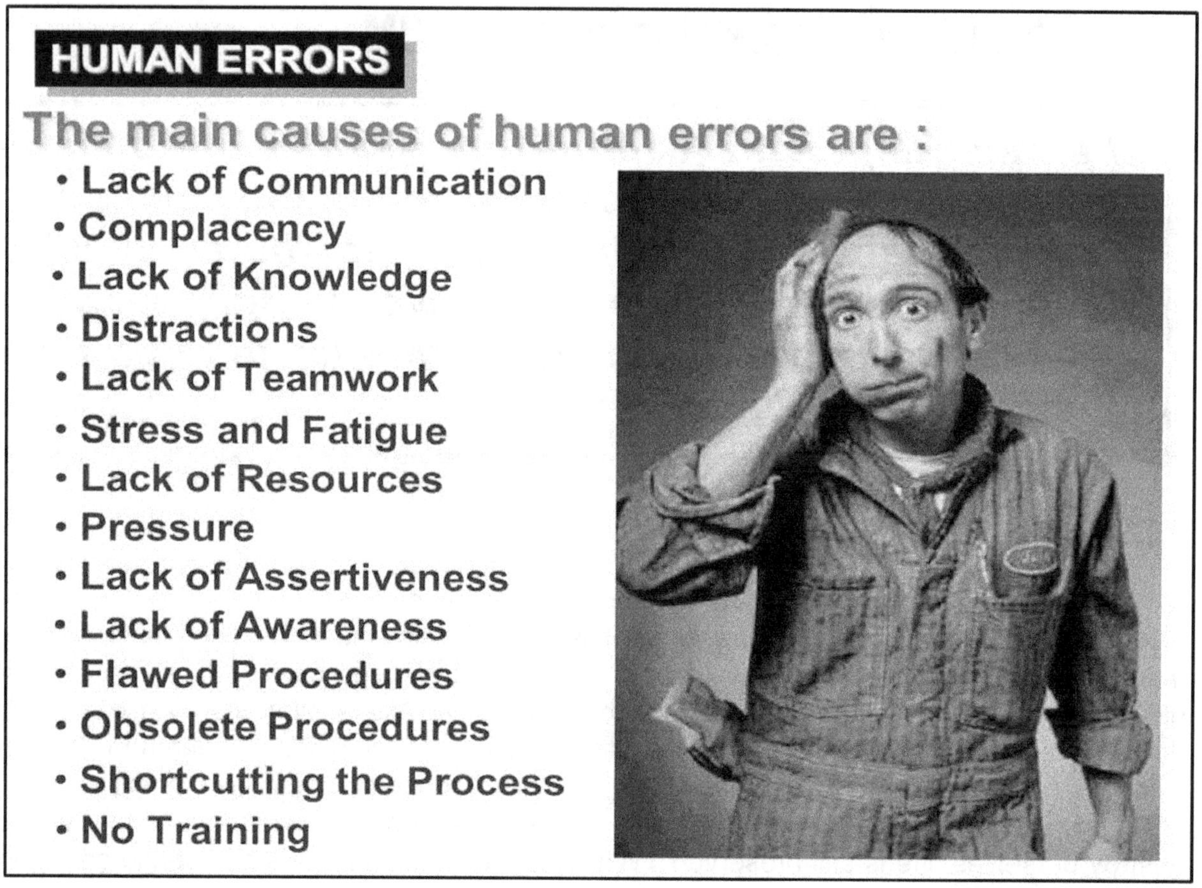

Figure 5.1: Common Causes of Human Errors

Humans play an important role in industries during the design, installation, commissioning, production, and maintenance phases. In maintenance, human error may be defined as the failure to perform a specified task that could disrupt scheduled operations or result in damage to property and equipment. While human error had existed since the beginning of mankind, only in the last 50 years has it been the subject of scientific inquiry. There are various reasons for human errors such as stress, fatigue, working too long without adequate break time, incorrect tools, forgetfulness, complacency, pressure, distractions, insufficient knowledge, lack of training, circadian rhythm, lack of communication, misinterpretation, ignorance, lack of standards, lack of awareness, and the list goes on. However, with every error we make, there is typically an associated change or something out of the ordinary occurring in our environment. The difference between humans and machines is that people can sense change, hear, smell, see, feel, or taste something different and take the necessary actions to correct the anomaly. In industries, human error exists either because we tolerate them or because we are just plain ignorant.

The most common maintenance errors are slips and lapses. As discussed in the previous chapter, a slip occurs when somebody does something incorrectly, for example, an electrician wires a motor that runs backward. Lapse occurs when someone misses out on a key step in a sequence of events or activities. Lapses are also called memory failures. It means forgetting something or simply not remembering. When you are performing a maintenance task and some interruption took place, like a phone call or other distractions, and go back to what you are doing, you simply cannot recall what was the last step you have done on this piece of equipment. Or if perhaps your wife called while you are performing Performing Preventive Maintenance, and your electricity has been cut-off in your home or some personal home problems, then your focus is distracted.

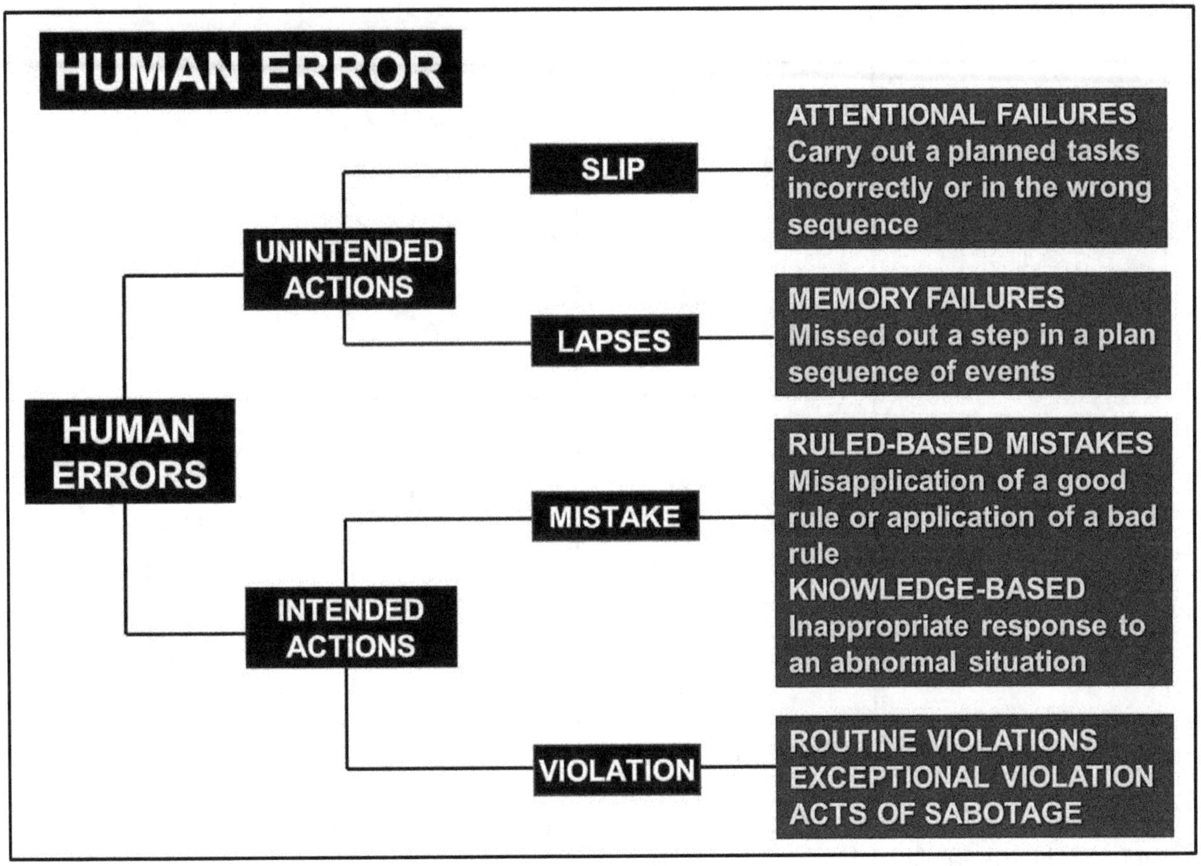

Figure 5.2: Human Errors

Performing Preventive Maintenance includes risks since there is always an assumption that everything will be put back in place well together after a PM initiative. Likewise is the possibility of human errors which leaves to infant mortality of newly installed components which eventually leads to additional failures after endorsing back the equipment to production. Therefore, it is important to conduct a thorough review regarding the activities that are being performed on the equipment during a Preventive Maintenance shutdown and understand that most failures are not related to the operating age of the equipment. The good news is that errors can be managed and human errors in maintenance are not random. According to the book of Alan Hobs and James Reasons on Managing Maintenance Errors, most human errors will occur during reassembly rather than on disassembly. According to their studies on the airline industries, the majority of maintenance errors are associated with reassembly and installation. This is where we can integrate Precision Maintenance on PM to minimize or mitigate the chances of human errors.

5.2: Precision Maintenance Explained

Have you ever bought a piece of knock-down furniture before? If you have, it comes with detailed step-by-step instructions with drawings on how to assemble the furniture. Even without really having the experience, just by looking at step 1, then the next step, and so on, anyone is confident enough to complete the assembly. This is just the same concept as Precision Maintenance.

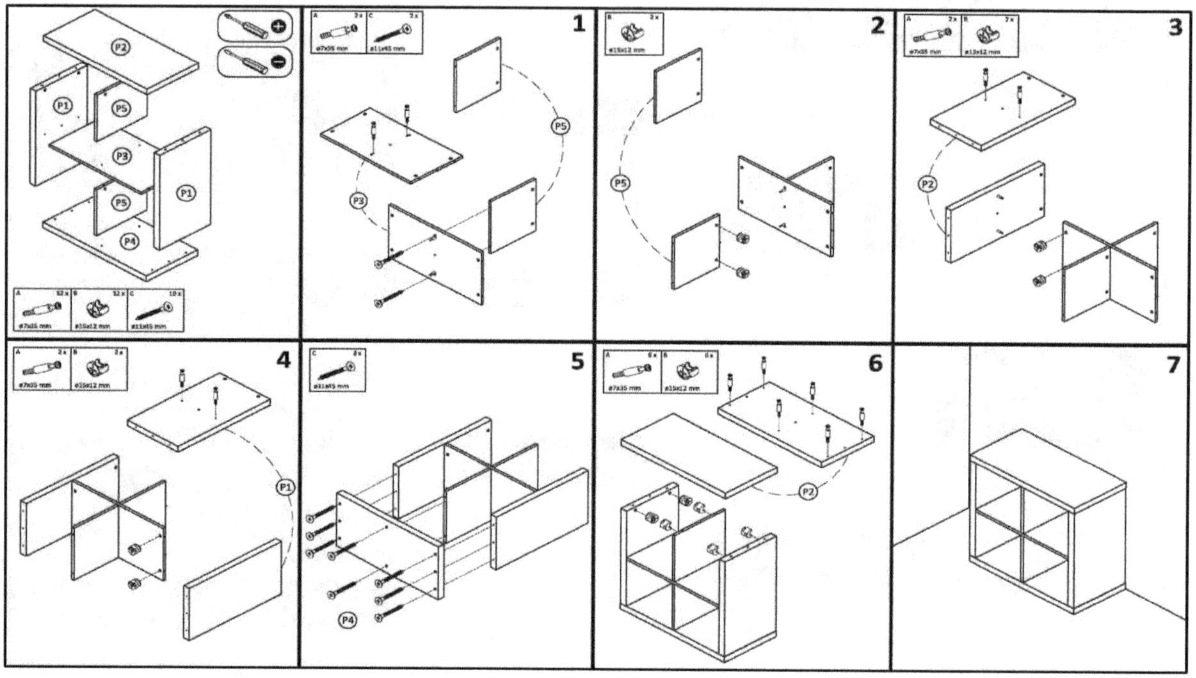

Figure 5.3: Knockdown Furniture Assembly

According to the dictionary, precision means the state or quality of being precise or being exact. It is the ability of a measurement that can be reproduced consistently. Precision Maintenance is the act of maintaining a consistently copied manner. This means that the maintenance that is performed is done the exact same way and delivers the same exact results regardless of the person who performs the maintenance tasks. It is also a method of performing maintenance tasks that must always be done with consistency, precision, and accuracy. Not only do we need to be precise, but also we need to execute the tasks in the shortest possible time precisely.

Precision Maintenance involves performing maintenance work in a consistent, precise, and industry-accepted way. If properly implemented, this means that maintenance should yield the exact same results, no matter who is performing the tasks. Precision Maintenance is much more than merely having procedures on PM. It should be the correct culture of the organization. The good thing about doing Precision Maintenance is that it can even reduce the chances of infant mortality failures seen during the start-up of assets right after a

Preventive Maintenance shutdown or Scheduled Overhauls are performed. It involves performing maintenance work in an accurate and precise manner, which means that the maintenance should provide the exact same results no matter who is performing the work, whether the work is done by the most or the least experienced maintenance craft person in the plant.

When different maintenance people perform the same PM on the same equipment, variations occur. These variations cause problems since they do not meet the requirements for the process. Precision maintenance help to ensure that equipment will be maintained to the highest possible standard so that the variations that can cause defects and failures can be reduced, or eliminated enabling equipment and assets to run at their optimized and peak level of reliability. Precision maintenance rebuilds machines and equipment to the highest standards so that fewer problems can occur during operation. It is a matter of ensuring the important things for equipment and machinery health are done correctly and accurately.

Although during the planning session, the activities on what to perform on Preventive Maintenance were discussed, the spare parts to be replaced were ordered by purchasing, the number of manpower to execute the PM, and the bill of materials also discussed. The PM activities are not really discussed in great detail about how they will be precisely executed from step 1 until the final step. Good experience maintenance people may have no problem with this, but we cannot say the same thing for a less experienced person in the plant.

[9]According to the book of James Reasons and Alan Hobbs on Managing Maintenance Errors, the largest number of maintenance errors in the aviation industry is associated with reassembly and installation. Analysis carries out on their BOEING identified the top seven causes of In-Flight Engine Shutdown (IFDAS) as follows;

- Incomplete installation at 33%
- Damage on installation at 14.5%
- Improper installation at 11%
- Equipment not installed or missing at 11%
- Foreign object damage at 6.5%
- Improper fault isolation, inspection, and test at 6%
- Equipment not activated or deactivated at 4%

There are two ways people can go wrong. First, they can either do something they should not have done, which is called commissioning errors, and second, maintenance fails to do something that they should have done, which is called omission errors. Omission errors during the installation process are the largest single category of maintenance errors. The US Nuclear Power Plant reports that 64.5% of errors associated with maintenance activities involved omission of necessary steps. If the impact and consequences bought about by human error are not acceptable, then we need to do something to not only mitigate but also eliminate the risks involved in this situation completely.

[9] James Reasons and Allan Hobbs, ***Managing Maintenance Errors, A Practical Guide***,(Ashgate Publishing Limited, 2003), Page 5

As you can see in figure 5.4, which is a racing pit crew, each of the pit crew members, which is composed of 20 people, excluding the driver, has a crucial role to perform. Their actions are well synchronized and consistent. The sequence is very crucial, in which they took a considerable amount of time practicing to change the tires in the shortest possible time. The fastest racing pit crew has set a new record beating the old record of 1.91 seconds to a new world record of 1.82 seconds at the Brazilian Grand Prix last November 18, 2019. If you can watch this in a slow-motion process, you can see how synchronized this team of 20 people is on who will be first, who will be next, and so on. Although this is quite an impossible task to perform during a Preventive Maintenance activity as we do not have those many people, I believe that we can still learn some good things from this experience. Being consistent, synchronized, accurate, and precise is what we want to achieve when we conduct a Preventive Maintenance activity on our equipment. This serves well if there are replacements and overhauling activities done on the equipment. What Precision Maintenance wants to achieve is that whoever will be performing the PM activities should yield exactly the same way regardless of the person that is executing the activities. Being precise will reduce the chances of not only infant mortality failures but as well as some cases of random failures.

Figure 5.4: Racing Pit Crew World Record at 1.82 Seconds

5.3: History of Precision Maintenance

The first real evidence of Precision Maintenance was pioneered at NASA by Dr. Wernher Von Braun and his team of rocket scientists in the 1960s. He was a German and was an

American aerospace engineer and space architect. He was the leading figure in the development of rocket technology in Germany and a pioneer of rocket and space technology in the United States. They discovered that for every 20% decrease in vibration, the life of the bearing is doubled. Further reductions produce an exponential increase in the bearing life and all other expensive components that usually are damaged during bearing failure causing a dramatic reduction in costs.

Dr. Wernher Von Braun
March 23, 1912 – June 16, 1977

Ralph Buscarello (The Vibe Father)
Sept. 20,1925 - Sept. 5, 2013

Figure 5.5: Pioneers of Precision Maintenance

In his 80's, Ralph Buscarello also known as the Vibe-Father taught thousands of people in more than 34-plus countries about vibration analysis, including innovations in-phase and resonance analysis. He pioneered the principles of practical prevention through precision and collective teamwork. He is said to be the father of Precision Maintenance. Ralph Buscarello preached that common assembly errors, mistakes, and omissions of essential field installation and rebuild details, along with simple equipment specification issues at purchase are significantly the sources of destructive vibration that fatigues the bearing's metallurgy casing early, causing premature bearing failure. Ralph coined the phrase, Precision Maintenance in the 1960s and spent the rest of his life trying to spread the word to industries. Like Dr. Braun, Buscarello is convinced that the lesser the vibration of the equipment, the lesser will be the cost of doing maintenance. He preached about the relationship between an equipment's vibration and the direct impact it will have on costs and introduced Precision Maintenance's role in increasing machine life and productivity. According to Ralph Buscarello, the higher the machine's vibration, the more prone it will fail and induce more maintenance costs. If the vibration of the equipment can be reduced, so as its operating and maintenance costs.

The following is an extract from his paper on "Proven Benefits of Precision Maintenance," presented by him as his keynote address to the Vibration Association in New Zealand in May 1998. He quotes that in over 46 years of teaching on vibration reduction and conducting seminars in over 34 countries, I could produce hundreds of outstanding financial case histories on the benefits of doing Precision Maintenance, many involving millions of US dollars. However, for those who remain skeptical, I suggest you accept none of my figures. Instead, start with the first idea of measuring the vibration levels taken at the worst point on about 50 ordinary but common machines, such as motor and pump assemblies. Next, plot the trends of their previous year's maintenance costs on the same graph as their vibration amplitudes. Then make up your mind as to whether reducing vibration to precision levels pays off or not.

Machine Type	Highest Velocity mm/s	Dollars Spent in USD	Lowest Velocity mm/s	Doing Precision Maintenancee	Percent Savings
Single Stage Pumps	5.6	$3,200	2	$650	80%
Multi Stage Pumps	4.8	$6,100	1.5	$1,100	82%
Major Fans & Blowers	9	$900	2.8	0	100%
Single Stage Turbines	3.8	$8,200	1	$2,000	76%
Other Machines	7.8	$11,850	3	$3,700	69%

Figure 5.6: Relationship between Vibration and Cost on Breakdowns

5.4: Knowledge and Skill of Maintenance

The key to Precision Maintenance lies in the knowledge and skill of the maintenance craftspeople. Precision maintenance will take time to achieve, but the benefits that can reap are worth doing. Doing Precision Maintenance will not only reduce downtime but will reduce the cost of performing maintenance. Training maintenance to perform, the correct tasks on the equipment will yield outstanding benefits for the plant. Organizations that can build a culture of Precision Maintenance will reap the rewards of a better bottom line, a safer workplace, and more productivity throughout the facility.

To begin with, a skills gap is necessary to conduct and assess the difference between the skills required to perform a specified task and the skills maintenance currently possesses. Precision Maintenance describes the maintenance culture in a facility. This means that the maintenance should yield the exact same results no matter who is performing the tasks. A scoring system will help determine the knowledge and skill maintenance currently has. With these scores, a sensible skills training plan can be developed to transform the maintenance workforce into a highly skilled team that is fully trained and qualified to perform these tasks in a precise state every time maintenance tasks will be performed on the equipment. Achieving a level of Precision Maintenance implementation, success, and sustainability requires a full commitment and advocacy from Top Management, senior leadership, and management roles. In addition, there should be a clear focus on setting written expectations for maintenance, engineers, planners, MRO Stores, and all contributing roles, including operations in achieving Precision Maintenance to reduce human errors, especially during Preventive Maintenance. Maintenance must understand that excessive vibration will reduce the longevity of the equipment in the long run. Unlike Preventive and Predictive Maintenance, Precision maintenance has something to do about the culture of maintenance. If Karl is the most experienced person in the plant, and no infant mortality is experienced if he performs the PM, then we need to know precisely what he is doing differently from others so that others can follow a common procedure or standard. Before we can improve the performance of the equipment, what is needed is to improve the knowledge of the people who will improve them.

Planned Maintenance Skills Evaluation

Gearing Towards A Pro-Active Maintenance System

Division : Central Equipment | Team Nam The Untouchables | Equipment type handled : All Types

Station : PLCC Department | Leade Karl Richards

Legend : LEVEL 1 Knowledge & Skill not Satisfactory ⊕ (0 points) | LEVEL 2 Knowledge Satisfactory ◐ (0.5 points) | LEVEL 3 Skill Satisfactory ◕ (0.75 points) | LEVEL 4 Knowledge and Skill both Satisfactory ● (1 Point)

			PLANNED MAINTENANCE MEMBERS									
Classification	No.	Knowledge / Skill Item	KARL	BOB	CHARLIE	RACQUEL	CAS	JOHN	BUDDY	NENA	FRANZIN	JB
BASIC MACHINE FUNCTION	1	Basic Machine Function	●	●	●	●	●	●	●	●	●	●
	2	Machine Specs, Parts and Function	●	●	●	●	●	●	●	●	●	●
	3	Knowledge in Actual Set-up and Conversion	●	●	●	●	●	●	●	●	●	●
	4	Basic Lubrication Knowledge	●	●	●	●	●	●	●	●	●	●
	5	Bolts, Screws and Fasteners	●	●	●	●	●	●	●	●	●	●
	6	Basic Repair and Troubleshooting	●	●	●	●	●	●	●	●	●	●
ANALYTICAL SKILLS ENHANCEMENT	7	Failure Mode and Effect Analysis	●	●	⊕	⊕	●	●	●	●	●	●
	8	Root Cause Failure Analysis	●	●	⊕	⊕	⊕	⊕	⊕	⊕	⊕	◕
	9	P-M Analysis	●	●	⊕	⊕	⊕	⊕	⊕	⊕	⊕	◕
	10	MTBA Snapshot and Analysis	●	●	⊕	⊕	⊕	⊕	⊕	⊕	⊕	◕
	11	Single Minute Exchange of Dies (SMED)	●	●	⊕	⊕	⊕	⊕	⊕	⊕	⊕	◕
	12	Statistical Process Control	●	●	●	●	●	●	●	●	●	◕
PNEUMATICS & HYDRAULICS	13	Knowledge and use on FRL's	●	●	●	●	●	●	●	●	●	●
	14	Knowledge and use on Pipings and Connectors	●	⊕	⊕	⊕	⊕	0	⊕	⊕	⊕	⊕
	15	Knowledge and use of Cylinders	●	◐	◐	◐	◐	◐	◐	◐	◐	◐
	16	Knowledge and use on Filtration	●	⊕	⊕	⊕	⊕	⊕	⊕	⊕	⊕	⊕
	17	Knowledge and use on Speed Controllers	●	⊕	⊕	⊕	⊕	⊕	⊕	⊕	⊕	⊕
	18	Knowledge on Leaks and Seals	●	●	●	●	●	●	●	●	●	●
MAINTENANCE TECHNOLOGY & OTHERS	19	Bearing Failures and Causes	●	●	●	●	●	●	●	●	●	●
	20	Sensors Technology	●	●	⊕	⊕	⊕	⊕	⊕	⊕	⊕	⊕
	21	Motors and Pumps	●	●	⊕	⊕	⊕	⊕	⊕	⊕	⊕	⊕
	22	Total Productive Maintenance	●	●	⊕	⊕	⊕	⊕	⊕	⊕	⊕	⊕
	23	MRO Spare Parts Management	●	●	⊕	⊕	⊕	⊕	⊕	⊕	⊕	⊕
	24	Reliability-Centered Maintenance	●	●	⊕	⊕	⊕	⊕	⊕	⊕	⊕	⊕
	25	Maintenance KPI and Indices	●	●	●	●	●	●	●	●	●	●
PREDICTIVE MAINTENANCE (Specialization)	26	Basic Knowledge on Vibration Monitoring	●	⊕	⊕	⊕	⊕	⊕	⊕	⊕	⊕	⊕
	27	Basic Knowledge on Heat and Thermography	●	⊕	⊕	⊕	⊕	⊕	⊕	⊕	⊕	⊕
	28	Oil Analysis and Tribology	●	⊕	⊕	⊕	⊕	⊕	●	●	⊕	⊕
	29	Condition-Based Maintenance Tactics	●	⊕	⊕	⊕	⊕	⊕	⊕	⊕	⊕	⊕
	30	CMMS Structure and System	●	●	●	●	●	●	●	●	●	●
S5-03			30.00	22.50	12.50	12.50	13.50	13.50	14.50	14.50	13.50	16.25

Figure 5.7: Knowledge and Skills Assessment Gap for Maintenance

5.5: Requirements for Precision Maintenance

Knowledge is the key requirement for Precision Maintenance. Knowledge is where we develop our skills. Essential to any industry are competent people who understand their assets and equipment intimately. Training is the backbone of any cultural change. It is mainly the missing link ingredient in any change or continuous improvement effort. Skill is the ability to do one's job, and apply knowledge and experience correctly in all kinds of events over an extended time. Skill is the product of personal motivation and thorough training in which the end result is mastery. When mastery is achieved, then maintenance can be capable of transferring their knowledge and skills to others. This is what we want to achieve in our industry, and to enable us to achieve this, companies must develop the most effective training methods. This refers to the skills needed to perform the required work correctly. Define the gap between the current and what is expected from our people. Precision Maintenance involves performing maintenance work in a consistent, precise, and industry-accepted way.

Implementing Preventive Maintenance for Industries the Right Way

Here are some requirements needed to establish a successful Precision Maintenance program:

- Maintenance must be given the training so they can develop the skills they need to perform their maintenance tasks correctly so that they can do their job right the first time around. Likewise, they must also have the correct tools and instruments needed. The tools and instruments used for PM should be calibrated regularly.

- Maintenance should have knowledge of lubrication as well as how oil is contaminated. Oil and lubricants must be properly stored and free from contaminants. Everything will start with how lubricants are stored. Maintenance must understand the effects of storing lubricants in an open and contaminated area. For grease, maintenance must understand that grease should not be mixed with other types to avoid any incompatibility issues with the grease. This problem will shorten both the life of the grease and the part it is greasing. According to Svante Arrhenius a Nobel Price physicist, the life of the oil is cut in half for every increase or rise in its operating temperature. This means that if the lubricants have exceeded their base activation temperature, the oil will degrade or oxidize twice as fast for every 10°C rise in its temperature.

- Maintenance procedures and tasks must not only be written but should be precise and in detail. Use simple words that everyone can understand. Provide drawings, visuals, photos, or, if possible, videos whenever needed. Adopt visual controls to easily spot deviations from normal. The scope of work or maintenance tasks to be done must be made crystal clear to the people who will execute the tasks.

- If bolts are removed during a Preventive Maintenance disassembly, the sequence of which bolt was removed first until the last bolt should be consistent. The correct torque and tension should be known and checked during reassembly and installation. For rotating equipment being dismantled, precision alignment tools such as laser alignment should be performed before reassembling both the driver and driven components. It is important to check the vibration frequency level before and after conducting PM on rotating machines.

- Besides the written procedure regarding the details on doing maintenance tasks, it is more effective to provide videos on how to correctly perform the tasks for easy-to-understand instructions. There are many videos on the Industrial Internet of Things (IIoT) that can be shared with maintenance. Browse youtube.com regarding the correct way to grease bearings, bearing installation, alignment, dismantling, reassembly, and the like. There is free software available on the internet on how to download videos from www.youtube.com. Share these videos during your toolbox meeting, or much better if maintenance can create their own videos and include these as part of their training program for all maintenance as well as operators in the plant.

- Discuss with purchasing regarding MRO spare parts being supplied to them. Only original spare parts should be used on the equipment unless a replicated part has passed an engineering evaluation that complies with the requirements that the part supplied have the same quality, dimensions, tolerances, measurements, and strength of the material.

• Only the correct tools specified are used in performing the tasks. Measuring tools should be calibrated before using them. Any deviations from the standard should be corrected and calibrated immediately.

5.6: Other Applications of Precision Maintenance

Besides using Precision Maintenance on Preventive Maintenance tasks, other tasks where we can adopt the use of Precision Maintenance include the following:

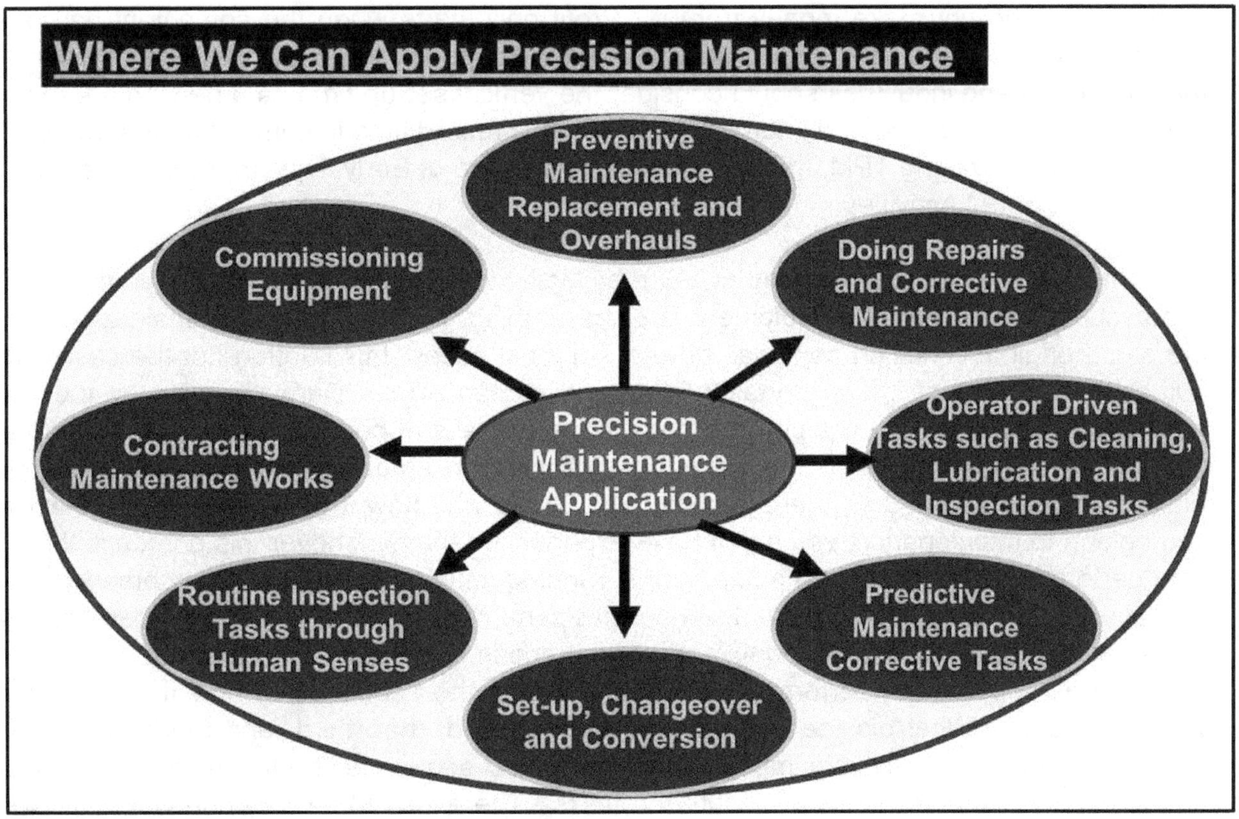

Figure 5.8: Other Applications of Precision Maintenance

Installation and Commissioning of New Equipment: Equipment installation refers to setting the equipment in the plant while commissioning equipment refers to conducting all required tests and procedures required as per set industry standards to show that the equipment can perform for the purpose for which it was installed. For smaller plants, both installation and commissioning can be conducted on the same day, but for bigger plants, the installation will be done first before commissioning. If industries purchase additional equipment in the future in support of their production needs and targets, they should ask for some data regarding the vertical start-up time from the previous equipment installed by the OEM or manufacturers or simply ask how fast they can commission the equipment in the plant. Their previous data can be used as a way of having some benchmark on the process for both installation and commissioning. Determine the activities where time had been spent most. The important part of this is to evaluate their previous activities and provide an improvement in reducing the vertical set-up time or commissioning time needed for the equipment. The vertical set-up time will include the time the equipment was installed, and the commissioning time until the time it is now ready for the start of operation. The goal of equipment manufacturers should be to shorten the vertical set-up time of equipment being commissioned. If the previous commissioning time has been reduced from seven to three days, this must be documented and highlighted on what they have done to reduce the commissioning time. This will be an added value for the organization since the equipment is now capable of operating and adding revenue to the organization. This new record time will now be their benchmark when trying to commission the same type of equipment in the future. Remember that the longer the equipment is commissioned in the plant, the more organizations do not make money from it. Profit only starts when the equipment has been handed over to operations, and production starts to generate products. Both equipment manufacturers and industries should consider the vertical set-up time as a benchmark record. This is like running a race and improving the previous record time in completing that race. For industries implementing TPM, this will be under the scope of Early Equipment Management or Initial Flow Control Activities.

Preventive Maintenance Overhauls: As previously discussed, overhauls contain risks since it is assumed that when the equipment is disassembled, it will be put back again together in the same original condition as it was before. In most cases, this is often not the case. The number one cause of infant mortality failures is human error. During overhauls, the error mostly happens when the equipment is now in the process of being reassembled or put back together once again. When the equipment cannot run smoothly right after a Preventive Maintenance overhaul, then there is something wrong with how it was overhauled. Maybe if you belong to maintenance, you might have experienced that whenever this person is the one who performs the overhaul, the equipment is running smoothly after it was endorsed back to operations. If another person was the one that performed the overhaul, then operations will find a difficult time running the equipment, and it needs to be debugged and corrected. One thing that is important in performing a PM overhaul is the sequence of how the overhauled was conducted. What did the PM did first, then second, then third, and fourth until it was finally completed? This is very important, as the correct sequence should be very crystal clear and in detail to anyone who will execute the tasks. There should be only one way to perform a maintenance task. For example, there is a correct way to remove the cylinder head from a car engine. The unbolting pattern is always fixed to assure no damage is done to other parts of the engine. What is the first bolt to be removed, and then followed by this, and so on? Also important likewise is to provide a thorough list of post-mortem or providing a checklist of all the activities done for a final inspection to ensure that everything has been covered before finally endorsing the equipment back to the operators.

Performing Corrective Maintenance During Repairs: In most cases, when an experienced person retires or leaves the plant for good, his experience goes with him. Most industries have not captured nor documented the experience of these people on how they performed this type of repair for a particular breakdown. What was his first step, then the second step, then the next, and so on? When these kinds of things are in place in the plant, then these can be used to teach other maintenance people who are new to the plant so that maintenance will have a guide on repairing this type of breakdown and not experimenting or doing trial and error all the time. I believe that this is where the true value of MTTR should be reflected.

Most of us in maintenance admit that we still cannot get rid of the hero stuff. A breakdown happens when the product needs to be shipped during the night, and no one in that plant can repair the failure. Finally, the boss called you late at night at your home, telling you to go back to the plant to fix something. For a start, if your plant encounters a major breakdown, call on the most experienced person to repair, then when he starts to repair, try to video the whole process from how he started until he ended. Once the video is in place, set up a meeting with the person who performed the repair, show him the video frame by frame, and let him explain what he did on how he accomplished the whole repair process in detail. Write down what he said systematically. Once the document is completed, have him review it for any more addition or deletion. Once the systematic repair procedure is completed, encode this in your CMMS (Computerized Maintenance Management Software). The good thing about this is that if a similar failure happens in the future, any maintenance on duty can have a printout if the failure is new to them, and the maintenance will be guided accordingly on how to perform this type of repair precisely. For example, if MTTR is monitored in the plant and in a certain group of 10 people, let us say that Karl is said to have the lowest MTTR in performing repair and troubleshooting hydraulic systems, while Joe has the highest MTTR in the same category since he is just new in the plant. What we can do is we can define the proper procedures for repairs and doing maintenance tasks based on Karl's practices. This practice can easily be applied and followed by other people in the plant, thereby avoiding trial and error during the repair process. Another way is that we can collaborate both Karl and Joe to work together during the repair and PM so that Karl can teach Joe from time to time while Joe can assist Karl in performing the repair. After a while, if we can see Joe repairing the hydraulics by himself without Karl it simply means that the knowledge Karl got had already been transferred and passed on to Joe. This is what we actually want with our maintenance people.

Performing Replacement after Detecting a Potential Failure using Predictive or CBM tasks: Once a potential failure is detected on the equipment where a spare part or item is involved, what is important for the Predictive Maintenance user is to estimate how long will the P-F interval will be until the limit has been reached. The user should recommend the precise time the machine will be stopped for Corrective or Preventive Maintenance tasks. The person responsible for executing the Corrective Maintenance activity should be informed by the Predictive Maintenance user in advance so that he can prepare and plan for the replacement activity. This will include having the spare parts available, identifying the tools needed, estimating the time to perform the replacement tasks, and precise detail on how the tasks must be done. Detailed step-by-step activities can be written in this case and for other repairs, which can be made available in the CMMS or EAM software that can easily be printed for reference by other maintenance people.

Chapter 6

PM Interval, Are We Doing PM Too Late or Too Soon?

Determining the correct maintenance interval is equally as important as the maintenance tasks itself. Doing the maintenance tasks too soon will be a waste of money, time and resources while doing it too late or too long will just allow the failure to occur.

6.1: How Did Maintenance Arrive at the PM Interval?

The goal of Preventive Maintenance is to perform these maintenance tasks before the failure happens and one of the questions being raised in this case is when is the correct interval to perform the Preventive Maintenance tasks. There are two ways that we can go wrong with selecting the interval for a specific maintenance task. First, we might be doing the tasks either too soon or too long. If we are doing the maintenance tasks too soon, then we are just wasting precious time, money, and resources. On the other hand, if we are doing the tasks too long, then there is the probability that we might have actually missed the failure. This means that the failure already occurred. Although It is difficult to tell precisely when the best interval for doing the different maintenance tasks for PM, there are basic guidelines and recommendations on selecting the correct interval for the different tasks of maintenance.

Hence, this chapter explains when Preventive Maintenance should be done. There are several tasks involved in doing PM such as human inspections, cleaning activities, routine lubrication, functionality inspection tasks for hidden failures, scheduled replacement of parts, scheduled overhauls, and if we can also include performing Predictive Maintenance inspection. There may be industries that will be using calendar days, running hours, number of strokes, number of volumes produced, number of cycles, or any combination of what was mentioned.

When we ask maintenance people from industries how did they arrive at this interval of conducting a complete overhaul on this equipment? There are two common answers, either it was what the OEM or vendor recommended, or the interval was based on guestimates. In short, we just guess the interval to perform this particular task. The thing is if we rely on the PM tasks on OEM, then the interval might be too conservative. No disrespect, but if I am an OEM, and I know that the life span of this item is 5 years, I will definitely not recommend you to replace this item every 5 years, especially if the item will be purchased with them. Perhaps the OEM might tell you to replace the item every quarter. [10]According to the book of Anthony Smith on RCM – Gateway to World Class Maintenance, quote, yet the basis for many PM programs is the blind acceptance of OEM PM Recommendations as the best course of action to pursue, even though the OEM recommendations are conservative and not necessarily applicable to the plant's operating profile.

6.2: The JIC Concept

One of the common problems in industries experienced is replacing an item or part even when it is still working. The reason for doing this is two folds, first is to comply with the PM task and second, we often use the Just In Case Concept. We replace items that are still working since we assume that if the item will not be replaced then it will fail anyway. But the thing is if the item is still working then it should still be used. In performing part replacements, we only need to replace those parts that are in the process of wearing out, which means that the item to be replaced conforms to either Pattern A, B, or C. These will be the parts that will be in direct contact with the product, and second, these can also include parts that have movements inside the equipment.

A good example of this is the pump. Let us assume that the rate of flow of the pump is at 100 GPM (Gallons per Minute) of water which is used for the cooling system for the heat exchanger. During the pump's operation, it will continuously be bombarded with water which will hit the impeller of the pump continuously. Since the plant is running 24 hours, and the pump will be running continuously for 24 hours, this means that as the impeller rotates, the blades of the impeller will be bombarded continuously with the fluid. After some time, the flow of water hits the impeller continuously, as this happens, there will be tiny bits of the metal fragment that will be chipped off from the impeller. This removal of the metal fragment will now be the start of the wear-out process. As the pump continues to suck the water, this process of continuous bombardment will cause more tiny fragments of metal to chip off from the impeller. As time progress, the operator will now realize that the flow rate seems to drop from 100 to 99, to 98 GPM. This means that more and more tiny metal fragments are removed from the impeller and the process of erosion is now taking place on the impeller. The clearance between the tip of the impeller and the volute of the pump increases, thereby creating more internal leakage, in which the impeller is no longer capable of discharging the rated volume. As the rate of flow drops, the operator will now complain that the discharge is no longer sufficient, thereby maintenance will carry out a PM program and schedule the pump's impeller to be replaced. Hence, the rate of flow can provide maintenance an indication of when to replace the impeller. As we have discussed in Chapter 1 regarding the operating context, the interval of replacement may differ depending on what the pump is suctioning and discharging. For example, the interval for a pump discharging water will be longer compared to an identical pump that is discharging slurry or a more viscous liquid even if both pumps are identical in every way. We also have to take into consideration if the pump has a standby or not. If the pump has a standby, which means that this will be less critical if the pump does have a standby in place.

---

[10] Smith, Anthony, Hinchcliffe, Glenn, ***RCM – Gateway to World Class Maintenance***, Elsevier Butterworth-Heinemann Page 6

Figure 6.1 is a Tool Monitoring Algo we used in the semiconductor End of Line process to dictate precisely when to change the dies and punches. A Tool Algo is an instrument that monitors and records the number of strokes stored on the main computer. This means that if the number of cycles for a die and punch is at 350,000 strokes, (in which one stroke will be equivalent to an up and down movement) when the punch reaches 350,000 strokes, it will automatically stop. The tool algo can also be programmed that when the punch and die reach 320,000 strokes, it will provide a signal that the time to replace is nearing. This will allow the operator to prepare a work request to the maintenance that the end of life for the die and punch is closing in.

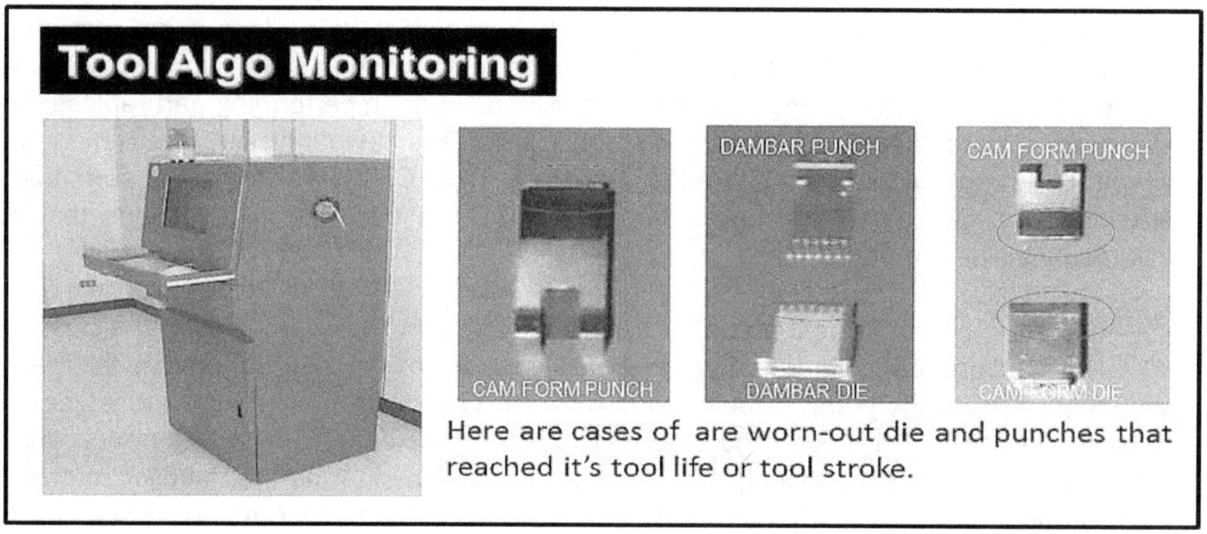

Figure 6.1: Tool Algo Monitoring

Before: Without the Tool Algo

- Hard time to generate cost reduction on tooling.
- Dies and punches are replaced when they failed.
- Difficulties with tooling specs (material strength) which can accelerate the wear and tear process
- Maintenance does not have any idea on when to replace or if the die and punches are now causing quality problems on the product.
- Tool cleaning activity on schedule which is time-consuming.
- Forecast of materials is always run to fail.
- Unexpected breakage of tooling when overused.

After: With the Tool Algo

- High reliability in terms of effective tool replacement
- Better spares management for dies and punches.
- Can provide a data-based tool status report.
- Lesser downtime and higher yield for the products.
- Centralized tool monitoring.
- Can define a Periodic tool for cleaning monitoring.
- Can define a Periodic tool for demagnetization monitoring.
- Real-time equipment utilization tracking.
- Run to fail tool replacements are now converted to stroke-based replacement.

6.3: Determining the Correct Interval for PM Overhauls and Replacement

Determining the correct frequency for replacements and overhauls for Preventive Maintenance activities should be based on the useful life and not on the average life of the part, or item which inhibits an age-related pattern. Parts that do not have a useful life should not be replaced on a time-based dominated frequency because their pattern is different. Only parts that will age because of stress-induced upon the part should be subject to parts replacement or overhaul during Preventive Maintenance tasks activities. Aging is the process where certain parts of the equipment deteriorate because of stress-induced upon the part over a given period. Any part or item that conforms to this process will be under the watch of Preventive Maintenance intervention. Performing PM too often will be a complete waste of time, money, and resources, while performing PM too late will allow the failure to happen on the equipment.

As discussed previously, random failures are failures that occur at any given period. This means that the probability that an item will fail in any one period will be the same as it is in any other given period. This means that the conditional probability of failure is constant or that the part has no useful life. It simply means that failure can happen at any given time. To further illustrate my point on random failures, imagine that you are driving your car on a freeway, and a small piece of rock hit your windshield, which causes a very tiny fracture. My question is, when do you think the second rock will hit your windshield? That will be a very difficult question to answer. To know the occurrence of random failures is like dropping a needle in a haystack and finding that needle when you're blindfolded.

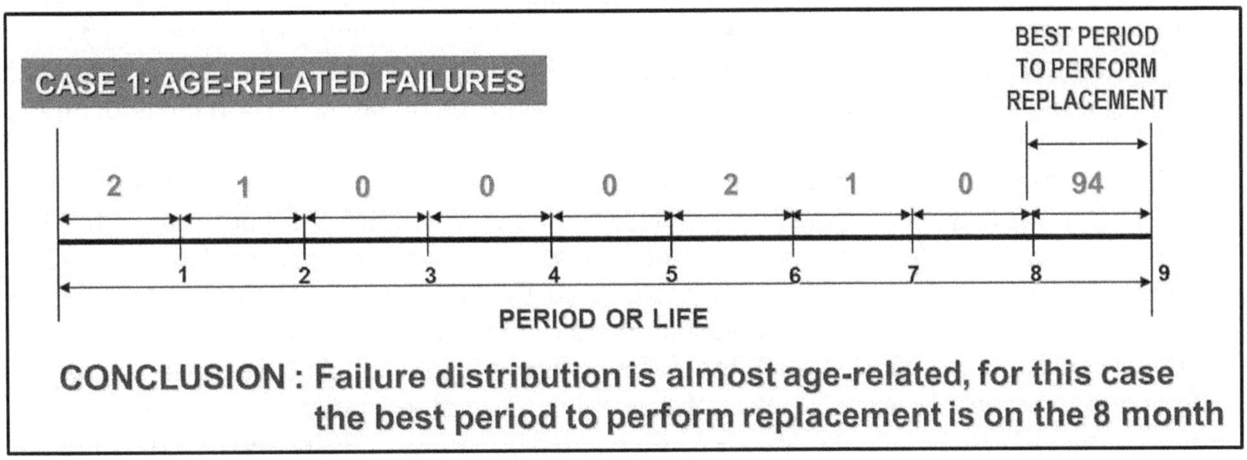

Figure 6.2: The Case of Random Failures

Now let us use this example in figure 6.2; let us say a given plant has several rotating equipment, and every time a bearing fails, the failure is being recorded on a logbook or history book. One reliability engineer decided to study the distribution of the failure and found out that 94 out of 100 bearings reached a lifespan of around 8 to 9 periods while the rest of the bearings failed at different intervals. Let us assume that one period is equal to one year in this case. In Figure 6.2, two failures occurred during the first year, one failure between the first and second year, two failures between the fifth and sixth year, and one failure that occurred between the sixth and seventh year. The remainder of the bearings failed during the eight and nine-year period. In this case, 94 bearings reached that lifespan. If we use MTBF to calculate the frequency of replacement, then we have the following:

- MTBF = [(1x2) + (2x1) + (3x0) + (4x0) + (5x0) + (6x2) + (7x1) + (8x0) + (9x94)] / 100
- MTBF = [(2) + (2) + (12) + (7) + (9x94)] / 869
- MTBF = 8.69 years

It is very evident in figure 6.2 that the replacement frequency should be done before the 8th period or year. If we use the MTBF calculation, it tells us that the replacement frequency should be done at 8.67 years. If we follow this, then we just missed the majority of the failure. It means that if we follow the formula, more than half of the 100 bearings have already failed even before replacing them. In this case, the best time to perform the replacement is in the 8th year or before nearing this period. Therefore, MTBF can be used for inspection purposes but not for PM overhauls and replacement.

In most industries, replacements and overhauls are based on mere guesses, but in most cases, they guess it wrong all along. This means that replacements and overhauls are done because our thinking is that it might fail again sometime if we do not do the tasks now. The problem here is that if a part is still working and it was replaced, then we are throwing something that is still useful. It means that we are throwing money away from our company, which is one of the reasons why doing Preventive Maintenance is costly. This is because most industries are highly too conservative most of the time, in which industries try to select intervals that are way too short or doing maintenance that is intrusive or not needed. This means that we might overhaul equipment after one year when the fact lies that the correct overhauling interval should be done after 10 years. These activities only cause our PM efforts to be more costly than ever. According to a recent survey, around 33 cents out of every dollar is being wasted as a result of unnecessary maintenance being performed on the equipment.

While most industries trust to use the maintenance interval dictated in the equipment manual prepared by the Original Equipment Manufacturer or OEM. The problem, in this case, is that in most cases, the interval may be too conservative, or the conditions in which the equipment is currently operating may not be exactly the same as dictated in the operating manual. In many instances, PM replacements are done too soon, especially if the spare part will be coming from them. I think the reason here is obvious. Another case is that if the equipment operating environment is hot and humid like in the Philippines, then the frequency of performing greasing and the grease to be used might not be exactly and precisely the same grease that is written on the OEM manual. While it will be easy to derive the maintenance tasks for a failure that has a symptom, or what we called the Failure Development Period where we can perform the maintenance tasks less than the Failure Development Period. This means that if the Failure development period is 1 month, then we can perform the task every week or every two weeks for inspection purposes. For PM replacement, we want to maximize its lifespan; hence, the PM replacement will be done when the useful life reaches around 80 to 90%. If the failure development period will be 30 days, then the actual replacement will be done either on the 24th or the 27th day period. In this case, we have more or less maximized the useful life, and the failure has not yet occurred, but the thing is, there will also be failures that will just happen without any form of signs or symptoms.

There are also calculations and formulas to determine the interval or frequency, such as the Weibull distribution analysis. However, the majority of industries have insufficient data, and only a handful of maintenance knows how to use them; in this case, we need lots of data to determine the correct maintenance interval.

In most cases, a car service center will recommend the owner for a car service every six months or 5000 miles, whichever comes first. Depending on the brand of the car, the scheduled semi-annual maintenance will include a change of oil and filter, inspection of brake pads, checking of fluid levels, tire rotation whenever necessary, an inspection of the wiper blades, and others. This will be the standard that most service centers offer. Although car service center checklists may differ from one brand to another or from one service center to another. For example, a Toyota Camry maintenance checklist usually includes a tire rotation and resets the tires to the correct pressure. Scheduled Toyota Camry service may also include an oil change and oil filter replacement. For Toyota Camry owners who prefer to use synthetic oil, they will recommend a longer oil change interval of around 12 months or 10,000 miles, whichever comes first. They will also check and adjust the fluid levels and check the brake pads for both the front and the rear parts of the tire.

A common misconception about MTBF is that it specifies the average time when an item, part, or component will fail. Although this statement is correct, the use of MTBF will not apply to an age-related pattern since MTBF is only speaking about the average interval of failures. Collecting failure data to calculate MTBF to determine the maintenance task interval for replacement and overhauling is entirely wrong and should not be done since MTBF is only about the average. As explained previously, failures fall predominantly into three categories, age-related, which makes up to 20 to 30% of all failures, and the bigger portion will be a combination of both random and infant mortality failures, which can be around 70 to 80% of all failures. For age-related failures, it is not MTBF but rather the remaining useful life that is significant when attempting to determine the best PM replacement or overhaul maintenance tasks on the equipment. Preventive Maintenance replacement and overhauling of tasks interval and frequency should be based on the remaining useful life and not on the average life which is the essence of MTBF. Again, Preventive Maintenance tasks for replacement and overhauls will only apply to age-related failures.

Parts that do not have a useful life should not be replaced on a time-based dominated frequency because their pattern is different. Only parts that will age because of stress-induced upon part should be the subject for parts for PM replacement and overhaul. This simply means that the stress finally exceeds the strength of the material of the part. Any part that conforms to this process will be the subject of Preventive Maintenance intervention. According to the late Anthony Smith in his book RCM – Gateway to World Class Maintenance, [11]in the absence of data and historical records, using our experience to guess the task interval is the only option. One proven method can be adapted to determine the correct task interval for overhauling and replacement, which is called the age exploration method. When we first try to overhaul or inspect the condition of the equipment on parts where aging and wear-out

[11] Anthony Smith, ***Reliability-Centered Maintenance, Gateway to World Class***,(Mc Graw Hill Inc., 1993), Page 126

are thought to be possible and reveal that no wear-out signs exist, we can automatically increase the interval by 10% during the next schedule. Repeat this process until we can finally see the normal wear-out pattern on the part that will be replaced.

This means that if the equipment is due to be overhauled after one year and reveals that upon thorough inspection, there are still no traces of wearing or aging on suspected parts, then we can automatically increase the interval by 10% during the next overhauling schedule. This means the next PM overhaul will happen on the 13th month. If there are still no signs of wearing out, we increase it again by 10% until a wear-out is evident on the parts.

ELEMENTS (PPM)	Hydraulic	Gearbox	Diesel Engine	Gasoline Engine	Transmission	Differential
Iron	**75**	**300**	**80**	**300**	**300**	**1000**
Chromium	**5**	**n.a.**	**25**	**40**	**10**	**n.a.**
Lead	**20**	**n.a.**	**50**	**n.a**	**50**	**n.a.**
Copper	**75**	**250**	**50**	**75**	**400**	**250**
Tin	**10**	**250**	**25**	**40**	**20**	**250**
Aluminum	**25**	**250**	**30**	**40**	**50**	**250**
Nickel	**5**	**n.a.**	**10**	**15**	**20**	**n.a.**
Silver	**5**	**n.a.**	**5**	**5**	**5**	**n.a.**
Silicon	**75**	**250**	**25**	**50**	**50**	**250**

Figure 6.3: Oil Analysis Wear Limit for Metals

Another way for determining when to precisely replace the parts is through the use of the Oil Analysis test for metals. The test for this will include spectroanalysis and analytical ferrography. An FTIR can also be used. This test will indicate the amount of metal debris for each metal element. Figure 6.3 indicates the limits or threshold for each metal. But what is important for maintenance to adopt this method is that maintenance needs to know the different elements that their spare parts are made of. They need to go back to their vendor and ask what specific metal is this spare actually made of. For example figure 6.3 indicates the concentration limits in parts per million (ppm) for different metals for a 100 ml oil analysis sample bottle. It is important to examine past test results to identify potential trends or emerging problems. (Reference: www.nch.com The Oil Analysis Handbook.) This means that whatever mechanical part that this element represents inside our equipment indicates that the process of wear is already starting to happen. Another benefit of this is that there are certain oil analysis laboratories that not only specify the number of metal contaminants per element but also show the shape in silhouette or magnified photo of the metal fragment which provides us an indication of the type of wear occurring on a particular mechanical part. For example, if the metal debris is longitudinal, then we have a case of abrasion or abrasive wear occurring. This important information can allow us to know what is exactly happening inside these parts and provide us an indication on whether to schedule the equipment for a Preventive Maintenance parts replacement. Again what is important to interpret the oil analysis elements is for the maintenance and engineering people to understand what their mechanical parts are composed of or what particular material element is made of. For example, if for the last 3 readings taken for a diesel engine oil, the value of chromium (Cr) is increasing this means that the piston rings are already starting to wear out as most piston rings are made of chromium

6.4: Determining the Correct Interval for PM Human Inspection

Although there will be two kinds of inspection for maintenance which include those that will be done by the Predictive Maintenance group using precision instruments that can be done both offline and online especially with the use of these smart sensors. On the other end, we also have inspection through the use of human senses. An industry implementing Autonomous Maintenance will develop 3 standards which will include the cleaning standard, lubrication standards, and inspection standards. Usually, when Autonomous reaches Step 5, these three standards are now in place and both operators and maintenance will sit down just like the La Cosa Nostra of Mafia and compare their tasks and activities on the equipment and further refine it to avoid any chances of duplication. These two groups will also decide on the best person responsible for performing the inspection task.

Mechanism	Minimum Condition	Inspection should be quantifiable. Precise Condition
	• At least one V-Belt must be installed for operation	• All 3 V-Belts should be installed for operation • All 3 V-Belts should have equal tension • The belts should be free of cracks • Pulley should be free of abrasion • The motor and speed reducer should be aligned properly at all times • Use Tension Checker to check the deflection and should not go beyond 3.5 kg. If tension is beyond make adjustment.
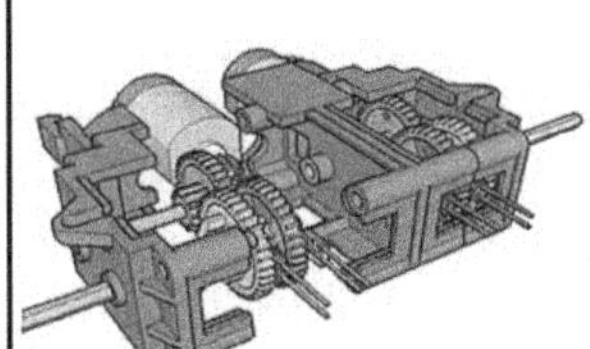	Minimum Condition • Add oil when level of the gear oil is low • Tighten bolts	• No loose or missing fasteners. • All bolts secure with correct torque M27 at 900 Nm (Newton Meters) • Oil level within the min and max level • Oil temperature at 65°C +/- 5 • If gear oil will be added use only ISO VG680 from Shell only, do not use other brand • Ensure EP Additives is present in gear oil

Figure 6.4: Integrating Precision Maintenance on Human Inspection

The interval for Human Inspection will depend on the criticality of what is being inspected. This means that the more critical, then the more frequent inspection will be done on the equipment. But what is the most important part of the inspection process is having a more precise way of conducting the inspection. I would recommend not to use the word check. Check the pressure, check the temperature, check this and that, and so on as this is very subjective. If I get 10 maintenance people to check the gearbox, then they might have 10 different means of checking it. What is important is to apply Precision Maintenance and indicate precisely how to conduct the inspection process and what to do if there is an anomaly or the threshold limit had been reached. In figure 6.4, a weekly inspection will include the inspection of the V-Belts and the Gearbox. What is important in the inspection process is that as much as possible it should be quantifiable. The responsible person whether the maintenance or operator must know what to do if it falls out of the desired or precise condition. Inspection must satisfy its optimal condition, which means that this is the design of the equipment. For example, 1 belt is needed to operate but the precise condition is that there should be 3 belts with the following condition indicated in figure 6.4.

6.5: Selecting the Maintenance Interval for Greasing Bearings

[12]There are several ways to determine the correct quantity of grease, first using a precision grease gun called an ultrasonic grease gun, or by using an auto-lubrication system. If both are not in place, then a meter reading must be installed in the grease gun. We will now discuss the correct interval and frequency for re-greasing the bearing. There are several factors to consider before attempting to determine the correct interval for re-greasing the bearing, which will include the following:

Temperature: Machines that run in a hot and humid environment and plants require more frequent re-greasing because the grease tends to get used up faster and is more likely to run out. There is a greater chance that the thickener to harden.

Contamination: Motors are often re-greased with greater frequency when the risk of dirt and contamination is high to ensure sufficient grease and keep seals from drying out. This means that if the working environment is dusty, greasing will be more frequent compared to an industry operating in a cleanroom environment.

Moisture: This reduces the film strength, which causes hydrolysis of the base additives and thickener, and can likewise increase oxidation rate. Increase the frequency of greasing when moisture is present. Although grease can absorb water, the grease properties such as stiffness and film thickness will be affected and can accelerate corrosion resistance.

Vibration: Excessive vibration can cause the thickener to shear or break off and can shake off the grease out of the bearing. Lubrication will be more frequent to compensate for the vibration on the equipment.

Position: Motors that are non-horizontal or vertical in position require more frequent lubrication compared to horizontal rotating equipment. This is due to gravity, which can cause the grease to flow out of the motor's bearing and the housing, especially if the NLGI range of the grease is from 000 to 1.

Bearing Design: Tapered and spherical rolling element bearings use grease up to10 times fast as radial ball bearings. Spherical roller bearings are most commonly applied in heavy applications where misalignment is prone to happen, and there is high load capacity.

---

[12] Society of Tribologists and Lubrication Engineers, ***Handbook on Lubrication and Tribology***, Vol. 1, Application and Maintenance, (CRC Press, Taylor and Francis Group LLC, 2006), Page 13-16 to 13-17

Condition	Average Operating Range	Correction Factor
Temperature (Ft)	Housing below 150 ° F (65.56 ° C)	1.0
	151 to 175 ° F (66 to 79 °C)	0.5
	176 to 200 ° F (80 to 93 °C)	0.2
	Above 200 °F (Above 93°)	0.1
Contamination (Fc)	Light non-abrasive dust	1.0
	Heavy non-abrasive specks of dust	0.7
	Light abrasive dust	0.4
	Heavy abrasive dust	0.2
Moisture (Fm)	Humidity is mostly below 80%	1.0
	Humidity between 80% to 90%	0.7
	Occasional condensation	0.4
	Occasional water on the housing	0.1
Vibration (Fv)	Less than 0.2 inches per second velocity peak	1.0
	0.2 to 0.4 inches per second	0.5
	Above 0.4 inches per second	0.3
Position (Fp)	Horizontal Bore Centerline	1.0
	45-degree Bore Centerline	0.5
	Vertical Centerline	0.3
Bearing Design (Fd)	Ball Bearings	1
	Cylindrical and needle roller bearings	5
	Tapered and spherical roller bearings	10

Figure 6.5: Table for Re-greasing Interval from Handbook of Lubrication

Where:

T = Time until the next re-greasing in hours

K = Product of All Correction Factors from Figure 6.5

K = Ft x Fc x Fm x Fv x Fp x Fd

n = Speed of the motor in RPM

d = Bore diameter

Note: 1 inches per second = 0.5 millimeter per second

$T = K\,[(14{,}000{,}000) / \{(n) \times (d^{0.5})\} - 4d]$

Case 1: Very Harsh Condition

- Temperature = 105 ° C
- Contaminant = Heavy Abrasive dust
- Moisture = Heavy condensation
- Vibration = Above 1.0 mm/second
- Position = Vertical
- Bearing Design = Tapered

- n = 2500 RPM
- d = Bore Diameter = 100 mm

$T = K\ [(14{,}000{,}000\ /\ (n^{0.5})\ x\ (d) – 4d]$

K = Ft x Fc x Fm x Fv x Fp x Fd

K = 0.1 x 0.2 x 0.1 x 0.3 x 0.3 x 10

K = 0.0018

$T = 0.0018\ [(14{,}000{,}000)\ /\ \{(2500)\ (100^{0.5})\} – 4\ (100)]$

T = 0.0018 [(14,000,000) / {(2500) (10)} – 400]

T = (0.288 hours) x (60 minutes / hour)

T = 17.3 minutes (The bearing on the motor needs to be re-greased at this interval)

It is recommended to have auto-greasing lubrication in this case.

Case 2: Less Harsh Condition

- Temperature = below 65 ° C
- Contaminant = Light Non Abrasive dusts
- Moisture = Humidity below 80%
- Vibration = Below 0.5 mm/second
- Position = Horizontal
- Bearing Design = Ball Bearing
- n = 1000 RPM
- d = Bore Diameter = 100 mm

$T = K\ [(14{,}000{,}000/\ (n^{0.5})\ x\ (d) – 4d]$

K = Ft x Fc x Fm x Fv x Fp x Fd

K = 1 x 1 x 1 x 1 x 1 x 1 = 1

K = 1

$T = 1\ [(14{,}000{,}000)\ /\ \{(1000)\ (100^{0.5})\} – 4\ (100)]$

T = [(14,000,000) / {1000) (10)} – 400]

T = (1000 hours) x (1 day / 24 hours)

T = 41.6 or 42 days; we can round this to every 40 days

6.6: Selecting the Maintenance Interval for Changing Oil

Although most OEMs will have their own recommendation on when to change the oil in the equipment which is in the form of running hours. A more accurate way to determine the change in oil interval is through Oil Analysis. In fact, several instruments can dictate when to change the oil precisely. There are several oil analysis instruments such as RULER or Remaining Useful Life Evaluation Routine which is an oil analysis instrument that measures the level of remaining antioxidant additive levels in lubricating oils, including turbine oils and hydraulic fluids. Another way of determining when to change the oil is the TAN (Total Acid Number) and TBN (Total Base Number) Crossover. When the oil is new, the base was high (TBN) and the acidity content (TAN) level is low. Let us say TBN is 12 and TAN is 0.2. This is the case when we purchase the new oil. If we do not understand how to store lubricants correctly the TAN will increase from 0.2 to 1 for example. As the TAN increase, the TBN will decrease automatically just like during my childhood times, computer games were not around so we have a see-saw, which means that if you are down, then the other player is up. As we use the new oil in our equipment, the TAN increase, and the base decreases. When the temperature inside our equipment increase, the TAN increase from 2 to 3 and the base decrease from 10 to 8 to 6, and so on. When the value of TAN and TBN becomes the same let us say at 4, then we have reached the crossover which signals to us that the oil needs to be changed.

STANDARD BASE CLEANLINESS TARGETS

MACHINE / COMPONENT	ISO RATING
• Roller Bearings	- 16/14/12
• Journal Bearings	- 17/15/12
• Industrial Gearboxes	- 17/15/12
• Mobile Gearboxes	- 17/16/13
• Diesel Engines	- 17/16/13
• Steam Turbine	- 18/15/12
• Paper Machine	- 19/16/13
• Servo-Valve	- 13/12/10
• Proportional Valve	- 14/13/11
• Variable Volume Pump	- 15/14/12
• Fixed Piston Pump	- 16/15/12
• Vane Pump	- 16/15/12
• Gear Pump	- 16/15/12
• Ball Bearing	- 14/13/11
• Turbine	- 17/15/12

Offline Filtration Equipment

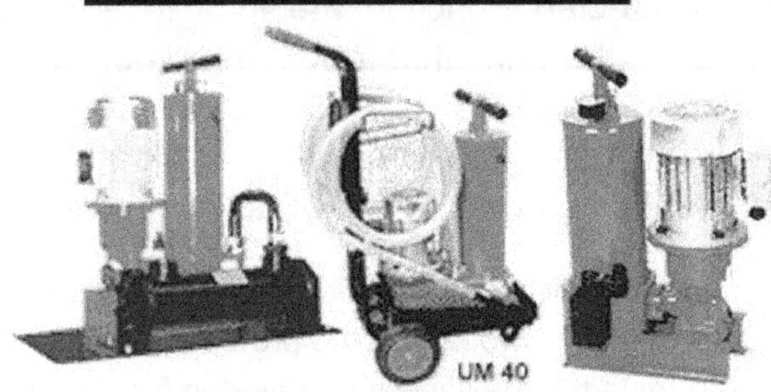

CCS2 Particle Counter

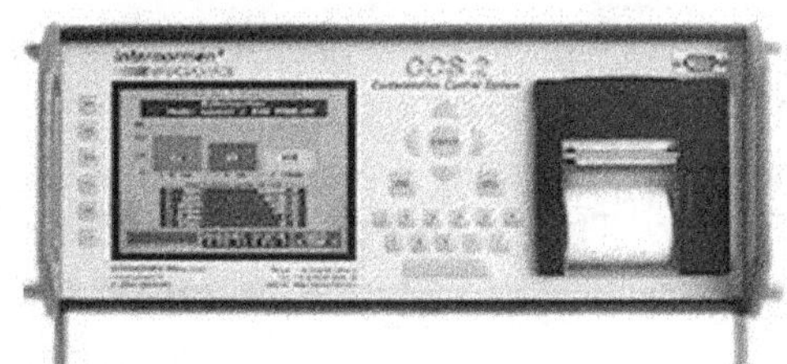

- **Determine fluid cleanliness of oil in ISO and NAS Standards**
- **Best used for hydraulic oil applications**
- **Determine particles in different size distribution**

Figure 6.6: ISO 4406 Cleanliness Standard for Different Lubrication Systems

While there are a few industries that understand lubrication and have matured in this process. Their change oil is no longer based on OEM recommendations but on the ISO Cleanliness Codes. These industries either use offline or by-pass filtration or both to clean and remove the contaminants in the oil. This means that if the oil is maintained clean then the additives will last much longer. When a certain hydraulic system reached its ISO cleanliness standard, then a portable offline filtration using high Beta Rating filters with high efficiency will be used to clean the oil from solid contaminants. For as long as they hit the codes, then the oil will not be changed. Or they may use a technique called sweetening in which 30 to 40% of the used oil will be removed and will be replaced with new oil so that the additives of the new oil will be mixed with the existing oil. An oil analysis instrument called a particle counter will be used to check the oil for its cleanliness and provide either an ISO 4406 Cleanliness or NAS Codes, but I prefer to use the ISO Codes. The ISO 4406 Cleanliness Codes will compose of three digits, where the first digit is those contaminants equal to or greater than 4 microns but less than 5. The middle contaminants are mid-range in size equal to 6 but less than 14 microns. The last include those big contaminants that are equal to or greater than 14 microns. Another method is the use of an Oil Analysis test called RULER or Remaining Useful Life

Evaluation Routine ASTM D-6971-04 3 and ASTM D-6810-02. This oil analysis test measures the lubricant's remaining useful life. This test can be used to determine whether the oil needs to change or can be extended. The RULER oil analysis test will measure the remaining level of antioxidant additive in the lubricating oils.

6.7: Selecting the Maintenance Interval for Failure Finding Tasks

Functionality Inspection or Failure Finding Interval also called Detective Maintenance should only be done for hidden failures such as protective devices and redundant functions to avoid the chances of multiple failures. Multiple failures occur when both the Protective Device and Protected Function are in a failed state condition. Hidden failures will only become evident if the ones that they are protecting also fail in operations. The interval for determining Failure Finding Tasks or Functionality Inspection will depend on two factors, the desired availability and the average life or Mean Time Between Failure (MTBF). The question of how frequent should we perform Failure Finding Tasks should depend on the following:

- What probability can we accept for multiple failures?
- Determine the probability the protected function will fail.
- Determine the availability of a hidden function that must be achieved to reduce the chances of multiple failures from occurring and will the risks be acceptable?

If the failure of a protective device is less critical then, we can aim for a lower availability but for failures of protective devices that are considered critical, then we should aim for much higher availability. Note that the higher the availability, the more frequent will be the Failure Finding Tasks or Functionality Inspection. Use this table in figure 6.7 in determining the availability and FFI Interval for protective devices.

Aim for higher Availability when critical and lower Availability when less critical							
Availability for hidden function	99.99%	99.95%	99.90%	99.50%	99.00%	98.00%	95.00%
FFI Interval (% of MTBF)	0.02%	0.01%	0.20%	1.0%	2.0%	5.0%	10%

Figure 6.7: Table for Failure Finding Tasks Interval

Let us assume that an emergency stop button has an average life (MTBF) of 10 years. Then, what will be the interval of performing a Failure Finding Tasks or Functionality Inspection to inspect if the emergency stop button is still working or not?

If Availability = 95%, Failure Finding Interval (FFI) = 10%, MTBF = 10 years
FFI = (10 years) x (365 days / year) x 10 / 100
FFI = 365 days or Once a year it should be inspected

If Availability = 99.90%, Failure Finding Interval (FFI) = 0.20%, MTBF = 10 years
FFI = (10 years) x (365 days / year) x 0.20 / 100
FFI = 7.3 days or Once a week it should be inspected

If Availability = 99%, Failure Finding Interval (FFI) = 2%, MTBF = 10 years

FFI = (10 ~~years~~) x (365 days / ~~year~~) x 2 / 100
FFI = 73 days or can be rounded to every two months

Note here that if we go for the lowest availability at 95%, the emergency stop must be inspected once a year. This means that inspecting the emergency stop once a year will guarantee us 95 % availability with a 5 % chance of failure. However, if we increase the availability to 99.9%, the chances of having the emergency stop available will be very high, with a 0.1% chance of failure, but the inspection will be more frequent. In this case, if we go for 99% availability, the FFI will be every 73 days with a 1% probability of failure. I think I will be happy with this. Use this table to calculate the frequency of inspection for protective devices and redundant functions. Again, Failure Finding Tasks and this table will only be used for hidden failures.

The reason for inspecting these protective devices is to ensure that they are still functioning. These protective devices are considered our defense on equipment against failure. Since in most cases their failures are unknown or hidden. The only time we know that these protective devices fail is when the protected function fails. The frequency or interval of conducting a Failure Finding or Functionality Inspection depends on the criticality of the ones they are protecting. For critical protective devices, aim for higher availability in table 6.7.

Protective devices for electrical circuits primarily serve as protection by discontinuing the power supply caused by overcurrent, which can cause fire hazards and electrocution. These protective devices for electrical usually protect the circuits from extreme voltages, currents, or short circuits. Just like in our home, a fuse or circuit breaker is usually installed to protect the circuit from overcurrent. The fuse may serve as a self-sacrificing device that can fail once it absorbed an overcurrent as further damage can be expected once this overcurrent is absorbed by our household appliances. While other homes are equipped with a circuit breaker which is an electrical switch that is used to protect the electrical circuit against a short circuit or an overload which is usually caused by excess current supply. The function of a circuit breaker is to stop the flow of current by disconnecting it once a fault has occurred.

6.8: Selecting the Maintenance Interval for Redundant Components

While in most industries, an alternative of switching the standby and duty is done weekly. The problem with this case is that the number of start-ups rather than the duration of operation influence the number of seal failures. This is because seals that are running are well lubricated and generally show low wear rates. The worst time for a seal to fail is during the start-up condition. During start-up, the seals are dry, and it will take a bit of time to build up the fluid film between the clearances. Therefore, frequent start-ups such as weekly is a major cause of wear in seals. By reducing the number of start-ups, we can reduce the number of failures for the seal. Another issue will be the impeller. If the switching is done evenly; this means that the rate of erosion on the impeller for both the duty and standby will happen at the same time, which means that both the duty or running pump and the standby pump can fail at the same period.

For a standby pump, the most important part is to start on-demand, or when it is needed in the case, the duty pump fails. Depending on historical failure rates relating to this failure mode, a test start can be organized at a suitable frequency to ensure its desired availability. The solution is to run the standby for a reasonable length of time, say 8-24 hours, with the duty equipment shut down. In this case, the duty pump can be run either between 3 to 5 months, and the standby equipment runs for 1 month. What is important is that the interval of their operating time should not be the same. The advantage of this policy is that it produces the lowest number of start-ups for both pumps in a year while allowing a long duration of the test run of 1 month for the standby and a frequency of 3 to 5 months for the duty equipment. Besides, take note that even if both duty and standby pumps are of the same type and make, their maintenance tasks will be entirely different. One of the maintenance tasks to perform on the standby pump is to rotate the shaft manually once or every two weeks at 360 degrees + 60 degrees or 720 degrees plus 60 degrees. Just make sure that the balls of the bearing will not end up in the same exact position as the outer raceway. The reason behind this is that we are avoiding the chances of false brinelling, which mostly occurs on the standby pump and motor bearing. If a bearing is subject to vibration when stationary, fretting damage occurs at the points of contact, giving rise to what is known as **false brinelling**. Standby components inside the storeroom should likewise be rotated once a week if the vibration is heavily felt inside the storeroom.

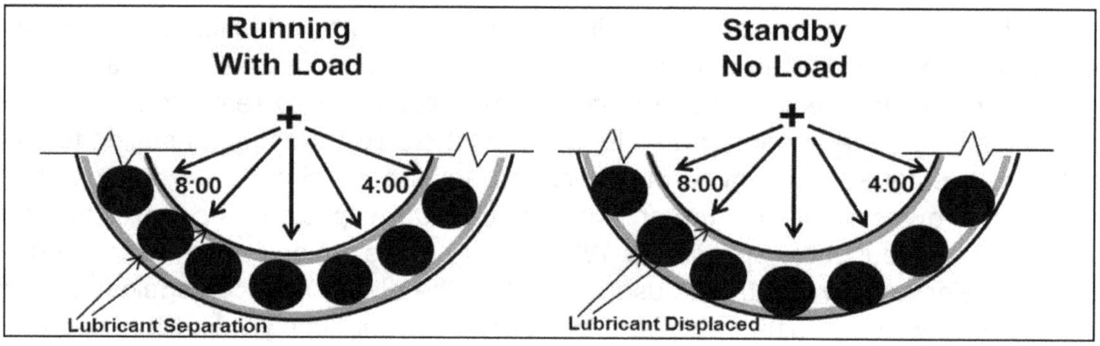

Figure 6.8: False Brinelling

When the duty pump is running, the forces of vibration are not only confined to the duty pump but also to the standby pump. This means that the balls inside the raceway of the standby pump and motor are vibrating concerning the inner and outer raceway of the bearing. When a component is on standby, the balls or rollers are usually in contact with the outer raceway of the bearing. This means that the lubricant is displaced which creates a metal-to-metal contact between the balls and the outer raceway of the bearing. The vibration will not be confined only to the duty pump, but in this case, the standby pump is likewise vibrating too. If the ball's position concerning the bearing's raceway remains unchanged, this will create some fretting and can cause an early fatigue fracture once the standby pump is used. That is why the shaft should be rotated weekly or every two weeks.

6.9: Selecting the Maintenance Interval for Predictive Maintenance Tasks

Unlike Preventive Maintenance in which the schedule can be made fixed, an initial monthly interval can be set for both vibration and thermography monitoring. This means that the user can take measurements on their equipment. The initial interval of monthly will only be valid for as long as the measurements are within their limits and range. Once these measurements start to increase and near their limit or threshold, the frequency or interval of conducting the tasks should be shortened. In this case, the frequency will be shortened from monthly to every two weeks. If the vibration continues to increase, then two weeks will be shortened to weekly once again until the user finally provides his recommendation to stop the equipment and replace the part that is on the verge of failing. The goal of the Predictive Maintenance user is to estimate the duration of the P-F interval so that a suitable proactive task can be done before the failure happens. The P-F curve is simply the failure development period.

Another important thing to consider for Predictive Maintenance is for the user to coordinate with both the planner and purchasing regarding the item being monitored especially if this part is not held in stock in the storeroom. For Predictive Maintenance activities, the decision to stock the part or not in the storeroom will depend on the lead time of the part or item. This means that the item will only be stocked if the lead time for the item to arrive is longer than the failure development period.

Oil Sampling Frequency	Hours	Days
Diesel Engine	150	6.25
Transmission, differential, and final drives	300	12.5
Hydraulics for mobile equipment	200	8.33
Hydraulics for industrial equipment	700	29.17
Hydraulics for aviation	150	6.26
Industrial gas turbine	500	20.83
Steam turbines	500	20.83
Aviation turbines	100	4.17
Air or gas compressors	500	20.83
Chillers	500	20.83
Gearboxes for high speed / duty	500	20.83
Gear boxes for low speed / duty	300	12.5
Bearings, journal, and rolling elements	500 - 1000	20.83 – 41.67
Aviation reciprocating engines	25 to 50	1.04 – 2.08
Aviation gas turbines	100	4.17
Aviation gearboxes	100 – 200	4.17 – 8.33
Aviation hydraulics	100 – 200	4.17 – 8.33

Figure 6.9: Frequency for Conducting Oil Sampling for Oil Analysis

[13]This table is extracted from the book Oil Analysis Basics with permission from Drew Troyer, which can serve as a baseline for industries for determining the interval of conducting oil analysis. This will only serve as an initial interval, except for aviation reciprocating engines; if the reading of the elements is nearing the limit point, then the sampling frequency must be shortened.

[13] Drew Troyer and Jim Fitch, ***Oil Analysis Basics***, (Noria Corporation, Oklahoma USA 2001), Page 56

Motor Criteria: Temperature rises are placed into four categories for Thermography:

- Under 10 ° C will be repaired at the next scheduled maintenance.
- 10 to 20 ° C will be corrective measures will be a Priority 2 basis.
- 20 to 30 ° C will be corrective measures will be a Priority 1 basis.
- Above 30 ° C will be corrective measures required now.

Bearing Failure Based on research by NASA, for Ultrasonic Monitoring

- An 8-dB gain over baseline indicates pre-failure or lack of lubrication.
- A 12-dB increase establishes the very beginning of the failure mode.
- A 16-dB gain indicates advanced failure conditions.
- A 35-50 dB gain warns of a catastrophic failure.

6.10: Decision to Perform Corrective Maintenance on PdM

Although the Predictive Maintenance group will also be conducting monitoring on the equipment, what is important is to coordinate this with the PM Group and Planner as well. If a particular part is monitored and shows an indication of a potential failure, the question is when will the PM schedule be for this particular equipment? If this particular failure can stop the whole process then the machine should be scheduled for corrective maintenance. Questions to be asked will include the following:

- Is the part or item with Potential Failure currently stored in the storeroom?
- If not, can the part or item be delivered before the functional failure occurred?
- Does the equipment have a standby in case it fails?
- If the equipment does not have a standby, will the failure affect other equipment or the entire process?
- Is the P-F curve long enough to reach the next PM Schedule?
- Is it economically feasible to advance the PM Schedule so that the item that needs to be corrected can be carried out before the functional failure?

Case 1: Let us assume that a manufacturing firm has a unit cost of $ 150.00. The machine that is being monitored by the PdM Group can produce 1000 units/hour. The item monitored is a bearing that is non-stockable. If the bearing fails it will take the vendor 24 hours to deliver the bearing on an emergency basis according to the purchasing. The PM schedule is due in 6 months, but the Predictive Maintenance is only giving 1 month for the bearing to seize. It will take 2 hours to replace the bearing if the part is available. **Question:** Is it economically feasible to advance the PM Schedule if the Bill of Materials is $10,000.00?
Given:

- Unit Cost = $ 150.00
- Machine Capacity (UPH) = 1000 units per hour
- Repair Time (Downtime) = 2 hours if part is available
- Revenue Loss = $ [150.00 per unit x 1000 units per hour x 2 hours] + $ 10,000.00
- Revenue Loss = $ 310,000.00

If the equipment will be allowed to fail which it will take purchasing 24 hours to acquire the part or item and 2 hours will be the repair time, then;

• Revenue Loss = $ [150.00 per unit x 1000 units per hour x (24 + 2) hours]
• Revenue Loss = $ 3,900,000.00

The decision, in this case from the economic point of view, is to advance the Preventive Maintenance Schedule and purchase the bearing so that the bearing that Predictive Maintenance is monitoring can be included in the PM task.

6.11: Refining the PM Interval

Once the PM tasks have been used for a year, we need to have a review of whether both the PM tasks and the interval we indicate are effective. Our main goal is to provide the necessary PM tasks. There are two ways to refine the PM Tasks. The first is to review all the existing tasks if it really needs to be done or not on the dictated interval. The second is to review the frequency or interval if we are doing the PM tasks too soon. Ask the following questions.

ORIGINAL PM INTERVAL

Interval	PM Tasks	Conversion	Total PM
Shiftly	24	1,095	26,280
Daily	20	365	7,300
Weekly	15	52	780
Monthly	25	12	300
Bi-Monthly	25	6	150
Quarterly	30	4	120
Semi-Annually	35	2	70
Annually	40	1	40
TOTAL ORIGINAL PM TASKS			35,040

REVISION 1 PM INTERVAL

Interval	PM Tasks	Conversion	Total PM
Shiftly	12	1,095	13,140
Daily	32	365	11,680
Weekly	10	52	520
Monthly	30	12	360
Bi-Monthly	15	6	90
Quarterly	45	4	180
Semi-Annually	25	2	50
Annually	50	1	50
TOTAL ORIGINAL PM TASKS			26,070

REVISION 2 PM INTERVAL

Interval	PM Tasks	Conversion	Total PM
Shiftly	9	1,095	9,855
Daily	35	365	12,775
Weekly	10	52	520
Monthly	20	12	240
Bi-Monthly	25	6	150
Quarterly	35	4	140
Semi-Annually	35	2	70
Annually	50	1	50
TOTAL ORIGINAL PM TASKS			23,800

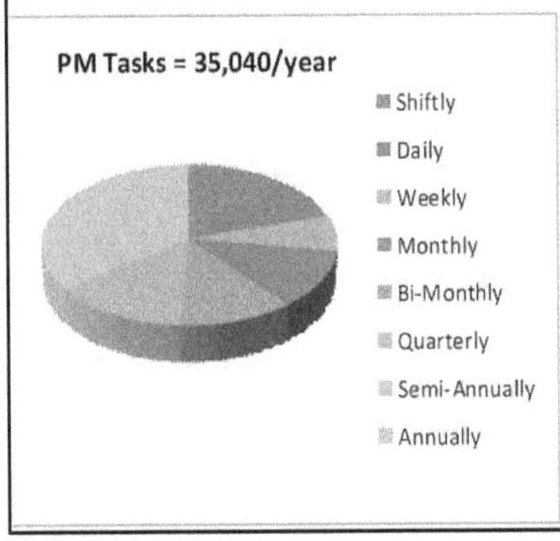

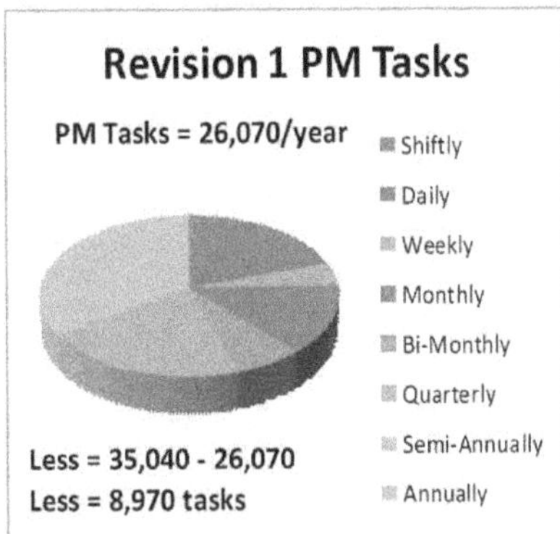

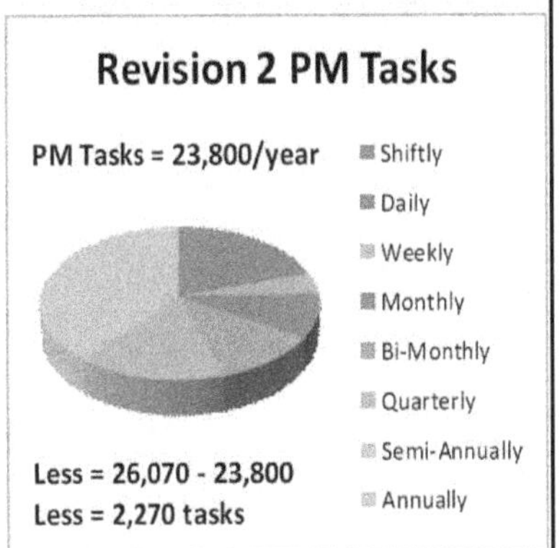

Figure 6.10: Reviewing the PM Interval

• Are there any PM tasks done every shift that can be done daily?
• Are there any PM tasks done daily that can be done weekly?
• Are there any PM tasks done weekly that can be done monthly?
• Are there any PM tasks done monthly that can be done every bi-monthly?
• Are there any PM tasks done bi-monthly that can be done quarterly?
• Are there any PM tasks done quarterly that can be done semi-annually
• Are there any PM tasks done semi-annually that can be done annually and so on?

In figure 6.10, adjustments have been made by the maintenance planner together with the PM crew who are responsible for executing the PM tasks. The original total PM tasks done on certain equipment totaled 35,040 activities yearly. The following changes have been made in the interval which is designated as Revision 1 of the PM Tasks Interval.

For Revision 1 on the PM Interval

- 24 tasks performed every shift were reduced to 12, increasing the daily tasks from 20 to 32 tasks.
- 15 weekly tasks were reduced to 10, increasing the monthly tasks from 25 to 30 tasks.
- 25 bi-monthly tasks were reduced by 15, increasing the quarterly activities from 30 to 45 activities with additional 5 new tasks.
- 35 semi-annual tasks were reduced to 25, increasing the annual activities from 40 to 50 tasks.
- Performing this revision decreased the number of PM tasks from 35,040 to 26,070 PM tasks which eliminated around 8970 tasks.

A second revision was made once again by the PM Planner together with the PM crew and the final revision includes the following changes from Revision 1 to Revision 2:

- 12 shifting tasks were reduced to 9 tasks, increasing the daily tasks from 32 to 35.
- 30 monthly tasks were reduced to 20 tasks, increasing the monthly tasks from 15 to 25.
- 45 quarterly tasks were reduced to 35 tasks, increasing the bi-monthly tasks from 25 to 35.
- Again, performing the revision reduces their annual tasks from 26,070 to 23,800 tasks which reduced around 2,270 tasks in a year.

Chapter 7

The Role of MRO Spare Parts on PM

> ***Maintenance, reliability, engineering, and technical people must have a voice in the process of selecting the right parts for their equipment. It should not be the sole discretion of Purchasing to whom to award the part. Maintenance are the users of these parts and they know what is best for their equipment.***

7.1: MRO Spare Parts for Preventive Maintenance

When a machine fails, it is mostly caused by a part that fails to fulfill its function. There are mechanical parts that wear out and need to be replaced and what is important is to keep downtime to a minimum. A storeroom is a place to store parts that we need to keep our equipment running. But managing spare parts simply means how fast we can respond in acquiring the right part during the time when maintenance and operations needed them most. The role of maintenance is to make the equipment available when needed. If the equipment fails and the part is not around, downtime occurs and the operation is halted. This means that no revenue is generated by the plant. Therefore, the goal of any MRO spare parts or Materials Management is to create a balance by minimizing the inventory cost of spares as well as providing all needed items required to keep the plant operating. Although the MRO goals are quite contradictory, still a well manage MRO Spare parts can provide and carry on this goal.

MRO Spare Parts are items held in inventory that are used to replace failed parts or components. This could be anything from a drive belt or bearing through an entire component such as a pump or motor set. In MRO Spare Parts, we simply cannot stock everything in the storeroom. The requirement to balance keeping the parts in inventory with the need to control spending is the reason it is important to have a clear, rational, and well-understood policy on Spare Parts. This means that we need to manage and control our spare parts inventory.

What MRO spare parts to stock in the storeroom are mostly unknown, especially if the plant is reactive. Emergency breakdowns and corrective maintenance usually result in the absence of parts when needed. Some say that spare parts are those parts or items that are available when not needed and those that are not available when needed. It does not have to be this way. Although it is not required for a storeroom to hire someone with psychic or clairvoyance capabilities like Nostradamus and place his predictions in quatrains on which equipment and what part of the equipment will fail and when. A good maintenance planning and MRO decision diagram or algorithm can help maintenance people decide on what parts need to be kept in stock in the storeroom

The decision on what spares to stock or not to stock should not be based solely on the OEM recommendations alone. If an industry is implementing RCM or FMEA, maintenance can have some background on what particular failure modes are mostly experienced in operations as well as their consequences. We also need to consider the carrying or holding costs of the item, lead-time to procure, and its impact on operations if spares are not held in stock in the storeroom. Having an MRO Spare Parts Decision Diagram or algorithm can guide us in making sound decisions on whether to hold the part or not in the storeroom. Following OEM recommendations every time will just increase the chances that the part will become a non-moving part in the future. Yes, of course, there will be exemptions to this, especially if the lead-time is long and perhaps may take years to reach the storeroom. There are 4 groups that the MRO Storeroom should support which include the Preventive Maintenance, Predictive Maintenance, Repair Group as well as those Improvement groups that Modify or Redesign certain parts or spares with inherent design weaknesses. Among these 4 groups, the Repair Group is the most difficult to support because in reality no one actually knows what part or item will fail and when it will fail.

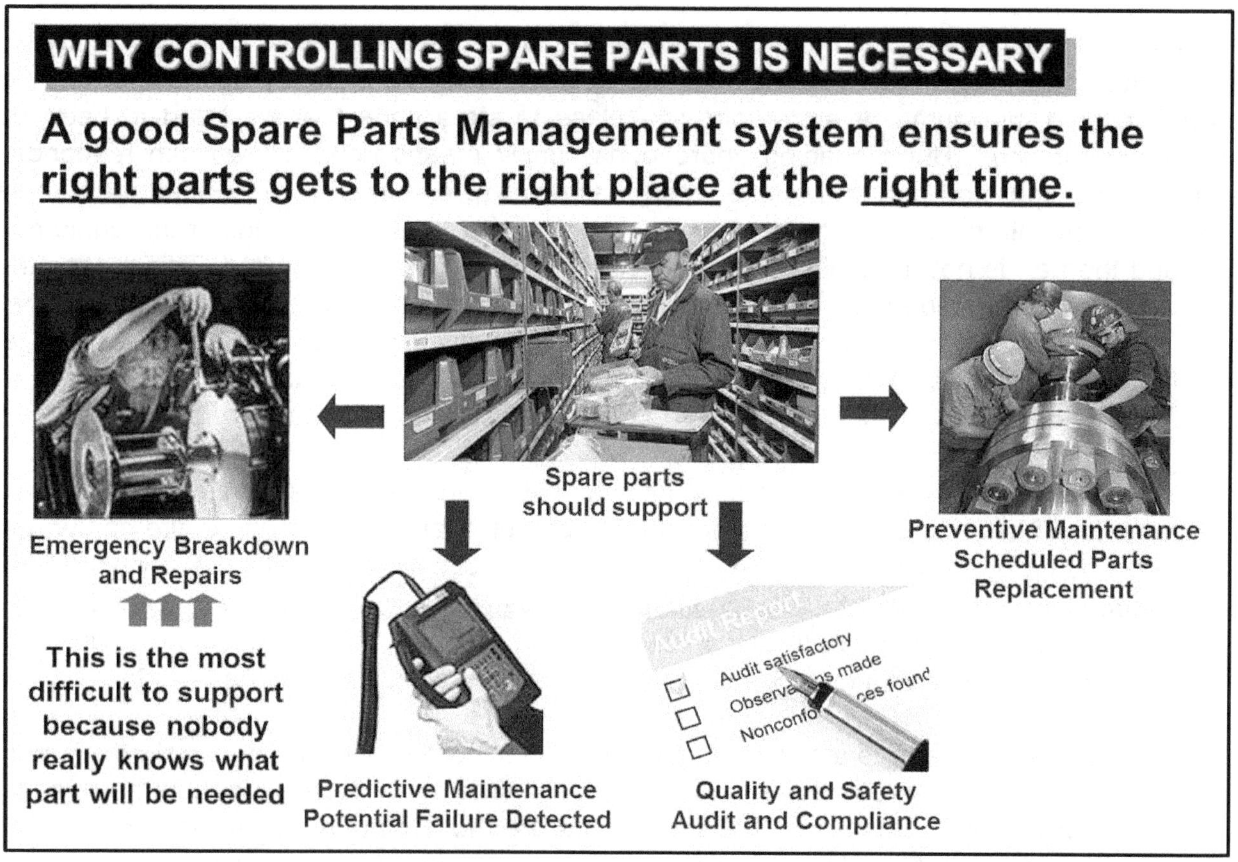

Figure 7.1: What MRO Storeroom Should Support

Preventive Maintenance: For spare parts under the watch of Preventive Maintenance, we can stock the parts with a long lead-time while we do not need to stock those parts that will wear out at a given time and whose wear-out rate is directly proportional to age. The parts can be ordered when the time comes. If a Preventive Maintenance shutdown will happen next year, it would be feasible not to store the parts in the storeroom and order them 2 to 3 months before the actual Preventive Maintenance Schedule shutdown. Exemption to this will be parts whose lead-time will take longer than a year to arrive at the storeroom. If the Preventive Maintenance is done monthly, then we need to stock the parts. In this case, what is important is to determine if the part can arrive before the PM shutdown will be executed. If this is the case, then we don't need to stock the spare or items in the storeroom.

Predictive Maintenance: For spare parts that are continuously monitored by Predictive Maintenance, we do not need to stock the part if the failure development period (P-F curve) or the lead-time to failure is longer than the lead time to deliver the spare part in the storeroom. For example, if the user of a Vibration Monitoring can predict the failure of a bearing with part number BRG-2406J three months or 90 days before it will fail. The longest lead-time to deliver based on Purchasing record is 40 days. Definitely, I will recommend not to stock the bearing in the storeroom.

Standby / Redundant Components: for those redundant components where the default task is Run to Fail, the decision to stock or not to stock will depend on the criticality and lead-time of the spare part to arrive. If the spare part can arrive in less than a fortnight (1 fortnight means 2 weeks) or less, then I recommend not to stock the part in the storeroom. That is why as previously discussed, the switching interval of the duty and the standby should not be the same since some parts of the component can fail at the same time.

Repair Group: As said previously that this is the most difficult to support because no one knows actually what part will exactly fail and when it will fail. The good news is that we can reduce the number of breakdowns occurring by implementing Preventive and Predictive Maintenance correctly. Addressing the basic equipment condition and performing a thorough Root Cause Failure Analysis Investigation will minimize the chances of these failures from occurring. My take on this is that if the lead time to acquire the part would take long, then we need to stock the part, especially for critical and bottleneck equipment that will also affect other processes in the plant.

Modification or Redesign: For those parts subject to modification or redesign, the storeroom people should be notified in advance regarding the plan for modification so that the ordering of the part can be controlled and adjusted. This is frequently the problem with people involved in improvements, redesign, and modification. They plan everything perfectly, but they miss out on one important part, which is the storekeeper. Remember that most modifications result in having the lifespan of the part increased, which means that the interval for replacement will definitely have to be longer. The withdrawal will be less in the future. Another important thing to consider is that, if a part or spare is modified, the original part will no longer be used, which makes it obsolete. A feasibility study must be done to determine if the existing part will still be consumed or if the modified part will be used immediately. If the original part will not be consumed, then it becomes obsolete and should be removed from the storeroom.

7.2: The Lead Time to Order

The most important factor in determining the availability of MRO Spare Parts for Preventive Maintenance is the lead time. For Preventive Maintenance tasks that will involve the replacement of parts, and the use of consumables such as changing oil, it is important that what is needed from the storeroom should be available before the actual execution of Preventive Maintenance. This must be well-coordinated and communicated with both the Maintenance Planner, PM crew, the Storeroom, and the Purchasing. If a needed spare part or item is unavailable during the actual execution of PM, then the PM will be delayed and prolonged. Both the planner and PM crew must check the availability of everything they need before the actual PM date. Parts and items that are not stocked in the storeroom should be well-coordinated and communicated with the purchasing to ensure that these items will be available before the actual PM execution. The decision on whether to hold the part needed for Preventive Maintenance will be based on the lead time to acquire the part. If the PM will happen a year from now and the longest delivery time among all the items needed for PM will be 8 months, then it is not advisable to store the item immediately inside the storeroom. Perhaps if one or two items will have a lead time of 8 months, then 9 months before the PM, these items will be ordered by the Purchasing.

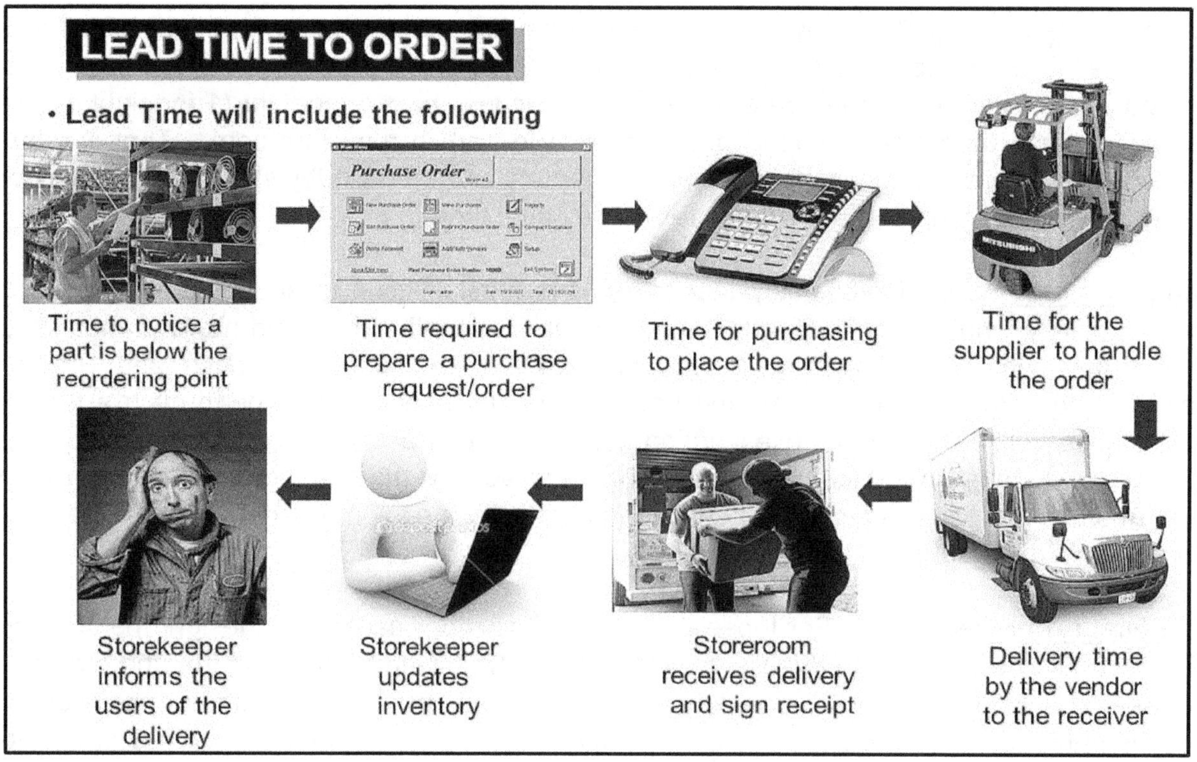

Figure 7.2: Lead Time to Order Includes

By definition, lead-time is the time the storekeeper noticed that the part is below the reordering point and makes a request for the part until the part is finally delivered and received in the storeroom. Lead-time information is a very important piece of information. It is one of the key factors in considering whether to stock the parts or not in the storeroom. The inventory stock shouldn't reach zero. The lead-time should give us some forecast that the item must reach the storeroom before the inventory reaches zero. This is important, especially for fast-moving items. The reason for this is due to the lead-time, which is the time required to purchase and receive the re-order quantity of the item. This is also called replenishment time. Our goal is to ensure that the stock item's inventory does not reach zero levels before the replenishment of the stock arrived. An order is placed at an inventory level called the reordering point. The reordering point is the sum of the minimum stock, and in certain cases, we add the safety stock. Others called this the buffer stock. Safety stock is an addition of items, which is applied mostly to critical parts for the storekeeper to increase their confidence in avoiding stock out. The Reordering Point = Minimum Stock + Safety Stock.

There are 2 ways to determine the lead time to order, the first is by getting the average dates and the second by doing it exponentially. Getting the lead time exponentially is more accurate as it focused more on the most recent lead time. Note that those parts or items that are purchased on an emergency basis should not be included in the lead time. Storekeepers must seek help from their IT or CMMS vendor so that the lead time of each part stocked in the storeroom can be reflected on the system instead of calculating each part individually. This will provide benefits to the users on whether to stock the parts or just purchase them before the PM date.

LEAD TIME TO ORDER

Action	Date	Lead Time
Ordered Received	1/30/2014 3/05/2014	34 days
Ordered Received	3/05/2014 4/11/2014	37 days
Ordered Received	4/11/2014 5/06/2014	25 days
Ordered Received	5/17/2014 6/26/2014	40 days
Ordered Received	7/14/2014 8/12/2014	29 days
Ordered Received	8/13/2014 9/12/2014	30 days
Ordered Received	9/13/2014 10/04/2014	21 days
Ordered Received	10/11/2014 11/17/2014	37 days
Ordered Received	11/18/2014 12/10/2014	22 days

Parts Codification

BRBL502SSR

Getting the Average

$$\text{Average Lead Time} = \frac{34+37+25+40+29+30+21+37+22}{9}$$

$$= 30.55 \text{ or } 30 \text{ days}$$

Computing Exponentially

(34.0 + 37) / 2 = 35.5 days
(35.5 + 25) / 2 = 30.3 days
(30.3 + 40) / 2 = 35.2 days
(35.2 + 29) / 2 = 32.1 days
(32.1 + 30) / 2 = 31.1 days
(31.1 + 21) / 2 = 26.1 days
(26.1 + 37) / 2 = 31.6 days
(31.6 + 22) / 2 = 26.8 days

Average Lead Time = 26.8 or 27 days

Figure 7.3: Calculating the Lead Time Using Average and Exponentially

SDE Analysis, classification of spares is based on the lead time to acquire the part. This classification is carried out based on the lead time required to procure the spare part. The classification is as follows; **Scarce (S)** are items that are imported and those items which require more than 6 months of lead time. **Difficult (D)** are items that require more than a fortnight but less than 6 months of lead time. Note: One fortnight means 2 weeks or 14 days. **Easily Available (E)** are items that are easily available, usually less than a fortnight's lead time. SDE Analysis and classification help in reducing the lead time to acquire the parts. This can reduce downtime in case of stockout. The storekeeper will know which parts need to be stocked or not based on the lead time to acquire the part.

Note that reordering or replenishment of parts will only be applicable for fast-moving and slow-moving items. In figure 7.2, the lead time to order the parts will start when the storekeeper finally noticed that a part is below the re-ordering point. In most MRO software, these items can easily be flagged down in the system for automatic replenishment. The storekeeper, or purchasing, will be responsible for generating the Purchase Requisition. In other cases, such as Preventive Maintenance, the maintenance planner or PM group will be responsible for requesting the parts needed for the scheduled maintenance. For a scheduled Preventive Maintenance, the maintenance planner will prepare the Bill of Materials (BOM), which includes the labor, materials, spare parts, consumables, and other items required for initiating an upcoming Preventive Maintenance schedule. The storekeeper needs to have access to the CMMS or EAM system regarding the Preventive Maintenance schedule so that the parts needed can be prepared in advance days ahead before the actual Preventive Maintenance shutdown will be executed. Once the Purchase Requisition is finally approved, a Purchase Order will be generated by the Purchasing or Procurement staff. Then it is time for the Purchasing to call the vendor. If automation is in place such as Electronic Data Interchange (EDI), the system will automatically be hooked up with the vendor or supplier of the part through a direct communication link in which software is required making everything paperless, including calling and payment to the vendor. It adopts, and standardized purchasing data from any CMMS or EAM system into a format that is sent and recognized by the supplier, and then the supplier directly bills the customer into the system.

After the vendor had confirmed the Purchase Order, the vendor handles the order and conforms to the agreed date of delivery. Usually, when the plant wants to expedite the delivery, the plant will incur an additional cost, which is usually the case for emergency buying. Finally, the item is loaded into the delivery truck, and now the truck is on its way to the plant to complete the order. Finally, the delivery truck reaches the plant and drives through the receiving area of the storeroom, where the storekeeper receives the deliveries.

7.3: Storekeeper Access to PM Schedule

Both the maintenance planner and PM crew need to finalize what parts, consumables, power tools, solvents, cleaning materials, and safety supplies are needed for a particular Preventive Maintenance job. Once these parts are finalized and identified, this must be well communicated with the storeroom people. These parts must be arranged by the storekeeper in advance a day or two before the actual Preventive Maintenance schedule. The storekeeper can just call the PM crew that their kit is already available at the gate and the PM crew must withdraw these parts 24 to 48 hours before the actual PM is due. Getting the parts during the actual PM day would add up to the time in executing the PM especially if several spares, consumables, and tools will be needed. If CMMS or EAM is in place, the storekeeper can just click on the PM schedule for the week and have a print-out of the things needed. The storekeeper can retrieve the needed parts, and stage them in a kitting area. For bulky items, the storekeeper will use a forklift or lifter. Once everything is staged at the kitting area, the storekeeper calls the PM crew that the items needed for their Preventive Maintenance are now available in the kitting area of the storeroom.

What is important is for the Storeroom people to have access to what is on schedule for PM as well as those needed from the storeroom. The planner usually will include those items needed and the PM crew will prepare a withdrawal form for all those items needed to execute their Preventive Maintenance work efficiently. Although not all items will be stored in the MRO Storeroom, both planner and PM should coordinate with the Purchasing people regarding the items needed to be delivered before the PM date is due. Failure to do this will just cause a delay or prolong the execution of their Preventive Maintenance. MRO Spare Parts, consumables, power tools, and safety signs, should be listed, counterchecked, and well-coordinated with the storeroom and Purchasing people if the part is not stocked in the storeroom. A PM Master checklist must be generated so nothing will be missed out during the actual PM execution.

Another important thing to consider is that when the list of items and spares are identified, the PM crew can seek the help of Predictive Maintenance to determine if the part really needs to be replaced. The PdM instrument will detect a potential failure when it is nearing its threshold or limit. There might also be items or parts that the Predictive Maintenance side of the house is monitoring and can provide accurate recommendations on whether to replace them or not. Remember that when we replace parts that are still working then this just adds up to the maintenance cost.

7.4: Shortening the Travel Time to the Storeroom

Walking back and forth to the storeroom for an emergency repair or when we need something may take around 30 minutes if the storeroom is a few hundred meters. If several parts will be withdrawn from the storeroom, then the time to get the parts may take 10 minutes or longer for the storekeeper. Let us assume that it takes 10 minutes to walk to the storeroom, 10 minutes for the storekeeper to retrieve the parts, and another 10 minutes to go back to the equipment so you can start the repair, then the total time consumed will be 30 minutes which definitely adds up to the repair time. If the plant operates for 365 days a year and if there are 20 maintenance in the plant and on average, each of them goes to the storeroom once a day to withdraw a tool, spare, consumable, or whatever, then in one year, this will be equivalent to as follows:

Time Consumed by Walking

= (20 transactions / day) (0.5 hours / transaction) (365 days / year)

= 3650 hours

When Using a Bicycle

= (20 transactions / day) (0.27 hours / transaction) (365 days / year)

= 1971 hours

Difference = (3650 hours – 1971) = 1679 hours

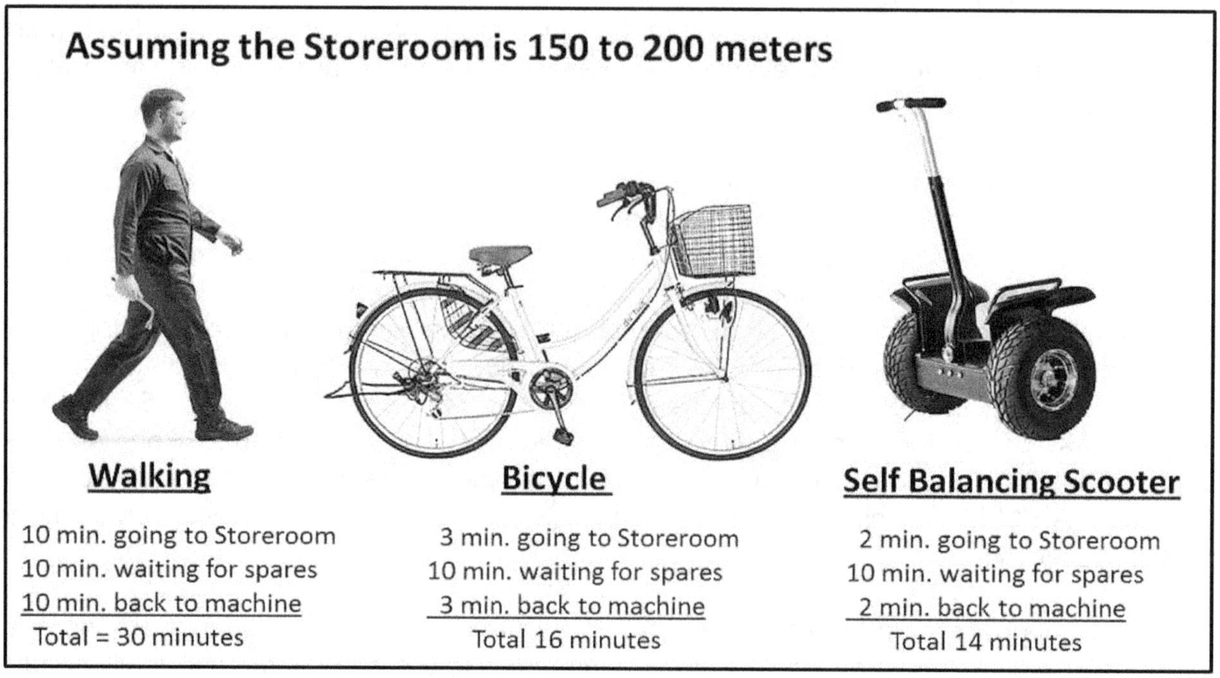

Figure 7.4: Shortening the Travel Time to the Storeroom

Using this simple example, in figure 7.4, taking a bike back and forth to the storeroom, will take 16 minutes compare when walking which takes 30 minutes. For 20 maintenance, this can be a savings of 1679 hours wasted by just walking to the storeroom which can be used for more useful activities for maintenance such as performing more scheduled work. A bicycle must also be equipped with a compartment for the parts, a functional braking system, safety rules, a bike helmet, and parking when not in use. This will only serve for parts that can be hand-carried by the maintenance. For bulky items, either a forklift or lifter will be needed to transport this to the machine. These are just some minor losses but when we add the total maintenance manpower and see the figure in a year, then everything adds up. What if we can reduce these losses and use them where our maintenance people can be more productive in the plant.

Although the calculation above is just a very rough estimate, what is certain is that there will always be waste and losses in maintenance, and the need to reduce them is essential. Purchasing 10 bicycles perhaps for a crew of 20 people is a very small investment compared to the time to acquire the parts in the storeroom. When we look at the number of hours that can be saved in a year converting it to productive work then would make a huge difference in the maintenance organization.

7.5: Do Not Believe Everything the OEM Tells You to Stock

The traditional thinking for maintenance is to keep stock of the part just in case we might need it someday. If the machine fails and the stock was kept, maintenance sentiments will be, “See, I told you we might need it someday.” However, if not, and the part is still around for years, again, the maintenance sentiment is to just hang on to the part, just in case we might need it someday. This is mostly the case when the industry is reactive because maintenance does not know which equipment will fail and what parts are needed. It is not practical to store every single spare in the storeroom. If the plant does this, then expect the plant to close in just months or just even a few days. In the absence of information and data, most industries rely too heavily on their OEM and vendors on what parts of the equipment need to be stock or not, which most frequently results in the part becoming a non-moving part.

In most cases, the OEM will tell us that the part is critical, and the lead time is long. This is usually where the JIC (just in case) concept usually scares maintenance most of the time. Here is the thing, OEM wants us to stock parts that we are not sure of whether they will be used or not. In most cases, they will end up as non-moving parts inside the storeroom. On the other end, if the equipment fails and the part is not around, then we are in big trouble. The point I would like to make is that, before we make our final decision on whether to stock or not to stock parts inside the storeroom, we need to take into consideration several factors before making our final decision. While many industries are blinded to accepting the fact that parts need to be stocked if the OEM tells them to stock, we also need to consider the cost of carrying the part in the storeroom, the lead-time to procure, and its impact on operations if spares are not held in stock. Here are some excerpts from other books regarding OEM recommendations to stock parts in the storeroom.

[14]OEM manufacturer usually exploits the user's ignorance by dumping slow-moving parts, resulting in obsolete and project surplus items at the user stores.

[15]Yet the basis for many Preventive Maintenance programs is the blind acceptance of OEM PM Recommendation as the best course of action to pursue, even though the OEM recommendation is conservative and not necessarily applicable to the plant's operating profile.

Many years ago, when the equipment was commissioned in the plant, the OEM told you to stock several items because they told you that they were critical parts and you agreed with them. Today, the parts remain untouched inside the storeroom. Mostly, those parts are big in size and in cost. Do not believe everything your OEM tells you to do. Remember that OEM is in the business of making money, and the author of this book is in the business of saving you money. When I posed a question to my class regarding if critical parts of the equipment need to be stocked in the storeroom? Almost all will agree and say yes. That is also what the OEM will recommend us to do. Before we answer this question, let us discuss figure 7.5 for a while and analyze each of the three items. Let us find out which part or items we need to stock or not.

Not all critical parts need to be stocked in the storeroom. When the machine was newly commissioned in the plant, there is a tendency for the OEM to convince us to stock parts in the storeroom since they were considered critical parts. Most of them right now have not been moving in the storeroom for a couple of years or even more. The decision or logic tree diagram can provide us some basic guidelines in making our decision on whether to stock or not to stock

---

[14] P. Gopalakrishnan and A.K. Banerji, ***Maintenance and Spare Parts Management***, (New Delhi, Published by Asoke K. Ghosh, PHI Learning Private Limited), Page 214

[15] Anthony Smith, ***Reliability-Centered Maintenance, Gateway to World Class***,(Mc Graw Hill Inc, 1993), Page 11

parts in the storeroom. The more we can control the timing of failure, the better understanding we have of the decision-making process. The traditional thinking for most maintenance people is to keep stock of the part just in case we might need it someday. Therefore, the next time your OEM tells you to stock this part, consider the following factors first and ask your team to make sound decisions on whether to stock or not to stock parts in the storeroom. An MRO Decision Diagram can guide both the storeroom and maintenance to make better decisions on whether to stock the parts or not. With a sound understanding of MRO Spare Parts Management, we can make better decisions instead of always following OEM recommendations all the time. Let us take a look and analyze figure 7.5.

Figure 7.5: Which of These Items Do We Stock?

For the cooking gas, when this runs out we have several options to choose from. So the decision is not to stock this item. For the spare tire, we always keep a spare tire in our car just in case of a random flat tire which seldom happens. The decision is to stock a spare tire in our car. Although the consequences of the failure of the brake pads are more dangerous than the spare tire and may lead to harm or death we do not keep stock of this because not all drivers know how to replace them. The decision is not to stock the brake pads. This is where reliability and maintenance come into place. Although it is not necessary to stock the brakes, what we need is a constant inspection of the brake pads and fluid and inspect the brake pads from time to time for any traces of abrasion on the brake disk and the depth of brake pads as well. These pads gradually wear in time depending on the usage so knowing how many km the car had run would allow us to decide on when to replace these brake pads. Not all critical parts need to be stocked inside the storeroom especially if the lead time to acquire the part is short.

7.6: Factors to Consider in Stocking Parts or Not In the Storeroom

In industries, the approval for the addition of a new item to stock is sometimes a complex decision to make, time-consuming, and subjective. At extreme times, bureaucracy, politics, and red tapes are involved. Worse if corruption is involved. Sometimes it is a disputable issue if the people involved come from different functions of the organization. There is a rational and smarter way of deciding on whether to stock or not to stock parts in our storeroom. To perform this logically, good background and understanding of reliability principles are required. The best people to make decisions on this will be our humble and down-to-earth maintenance people, but of course, with the inclusion of other functions in the organization like the storekeepers and purchasing people as well. My point is simple, not all critical parts need to be stocked in the storeroom. Here are some factors to consider when deciding on whether to stock or not to stock parts in our storeroom.

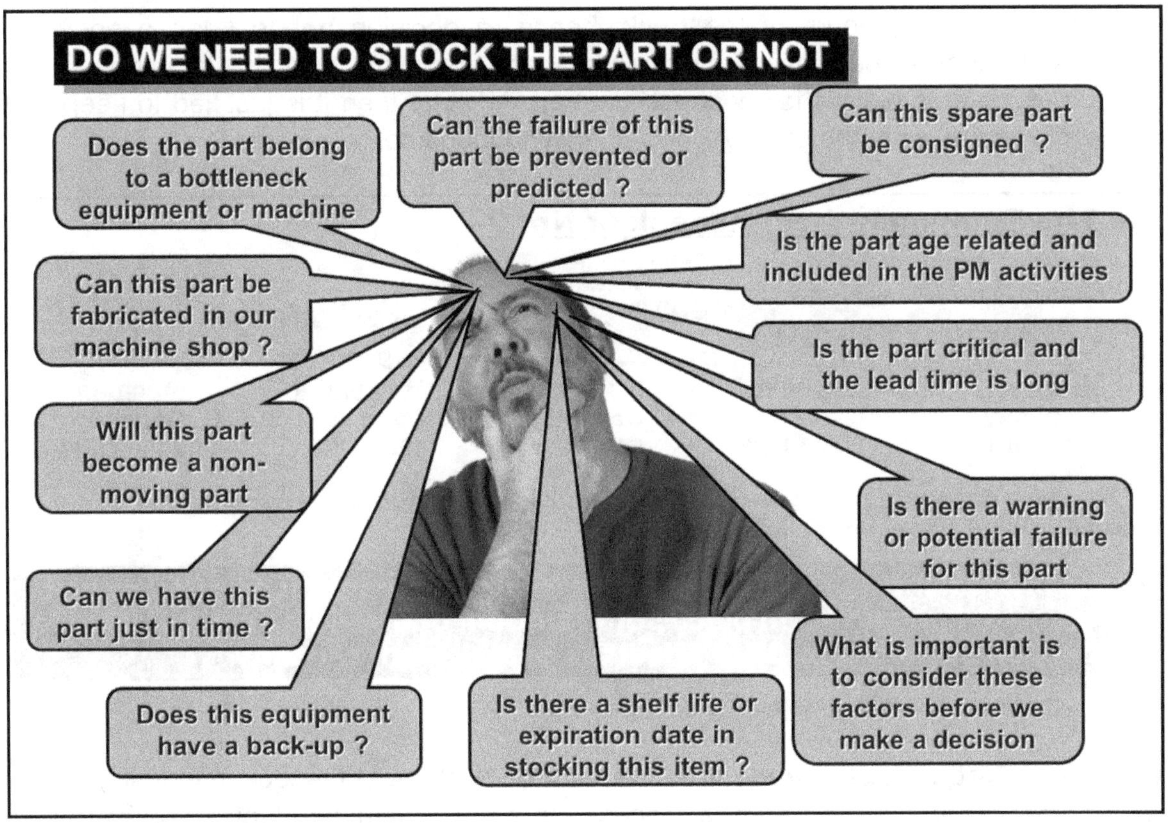

Figure 7.6: Do We Need to Stock the Part or Not

1) Type of Failure Modes: Some failure modes will provide us some signs or warning that they are on the process or verge of failing. One of the key factors in the decision to stock is the warning period of the parts they provide before the breakdown. The failure pattern can be classified as infant mortality, random, or wear-out failure. Many failure modes will give some sort of warning that they are in the process of occurring. The P-F (Potential and Functional Failure) curve shows how a failure will start and then deteriorate slowly to a point where it can be detected, and if the failure is not detected will deteriorate at an accelerating rate until the functional failure is finally reached. Hence, the P-F interval is the interval between the occurrence of a potential failure and its decay into a functional failure. The failure development period can take as little as hours, days, weeks, months, and sometimes a year. Determining the potential failure can only be possible if maintenance people are currently using Predictive Maintenance technology and instruments. Although human senses can also be used, these PdM instruments will stretch it further. Hence, if the failure development period is longer than the lead-time to purchase the part, then it is a smart and good decision not to stock the part. Let us say that the vibration analysis record for a bearing indicates that the failure development period is 3 months, checking the Purchasing records indicates that the longest lead-time it took for this bearing to be delivered is one and a half months. In this case, it is a good decision not to stock the part in the storeroom since the failure development period is half the lead-time to purchase the part. For parts with a very short warning or potential failures such as fuse, shear pins, and electronic parts the decision will be to keep stock of the part. A normal failure mode for the tire of a truck will be a wear-out pattern. For as long as the tires are aligned, properly camber, and the road the truck traveled is asphalted, then the tire will wear out gradually concerning the running hours of the truck; hence, a decision not to keep a spare in the storeroom will sound to be feasible although a spare tire in the truck is still needed. On the other hand, if the tires always fail because of sharp objects, then it is justified to keep a spare tire in the storeroom, as in the case of open-pit mining industries.

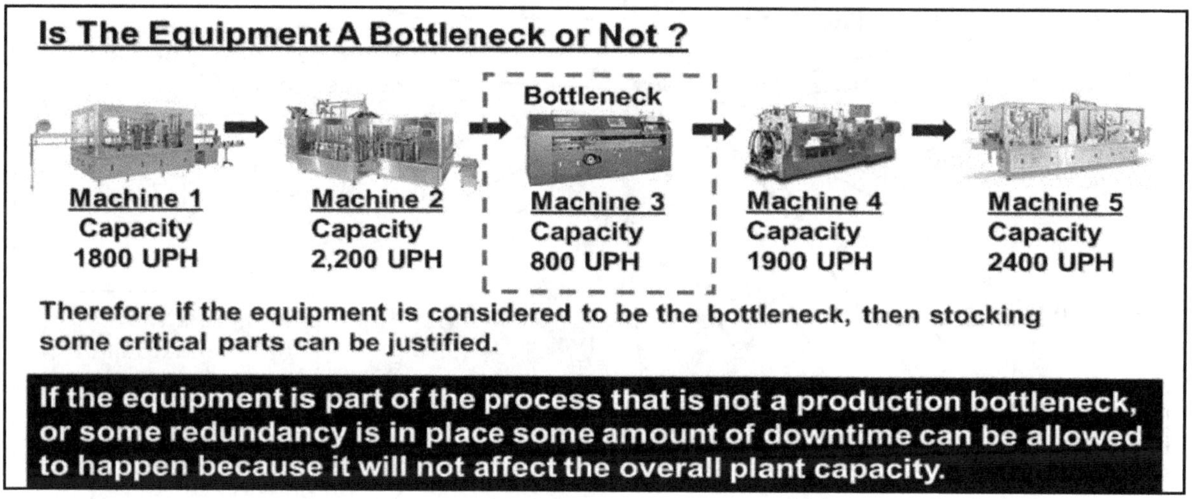

Figure 7.7: Is the Equipment Part of the Bottleneck?

2) Is the Item Part of the Equipment Bottleneck or Not? A bottleneck in industries refers to equipment that constrains the flow of a process. In a manufacturing plant, if you have a series of equipment that produces a product, the machine with the lowest output per hour produced or units per hour (UPH) would be considered as the bottleneck equipment since the inventory of the product is expected to build upon that piece of equipment. Therefore, if the equipment is considered a bottleneck, then stocking some critical parts can be justified. Still, on the other hand, if the equipment is not considered a bottleneck since some redundancy or standby is in place, then some amount of downtime can be allowed to happen because it will not affect the overall output for the day.

3) Shelf Life of the Spare Part or Item: Some items have a shelf life or are similar to an expiration date (best if used before) and are better obtained fresh from the vendor, rather than stocking them in large quantities in the storeroom. Examples include paint, chemicals, batteries, fire extinguishers, lubricating oil, grease, and so on. For lubrication, oxidation occurs in oil that is in direct contact with air, including stored lubricants. The base oil and additives combination affects the rate of oxidation. For grease, the presence of thickener increases its rate of degradation. Environment and storage conditions have the greatest effect on the rate at which the oil will degrade. If we increase the temperature at which the lubricant is stored by 10°C (18°F), the oil will double its oxidation rate, which cuts the useful life of the oil in half. The presence of moisture, usually introduced as a result of temperature variations, and condensation will definitely increase the rate of oxidation in the oil. The point here is that, if the item has an expiration date or is best used before and this date will be reached before it can be used, then it is better not to stock the part and just purchase the item when we need it.

4) Does the Equipment have a Back-Up or Redundancy? A run to fail situation is sometimes valid if some of the equipment has some form of redundancy or backup, reducing the consequences of failure to non-operational consequences. When equipment has a standby or backup in place, it makes the consequences of failure less critical. It can allow the plant or system to continue operating. The duty equipment that fails will then be switched on to the standby or redundant component. The duty equipment that failed can be fixed, and parts that have been affected can be ordered during the time the equipment actually failed if the lead-time for the part is short. In this case, we don't need to stock them in the storeroom. What is critical in this setup is for maintenance to perform a functionality test on the standby component to ensure that it will run when we need it. If the lead-time for the parts affected is long, then it is justified that we need to keep a stock of the part in the storeroom. For backup components held inside the storeroom, they should properly be maintained by the storekeeper especially if the vibration is felt inside the storeroom. For components that have not been moved for a long period, the storekeeper can seek the help of maintenance in checking the lubricant and re-grease them whenever necessary. This will only apply if both the standby and duty components are not run at the same interval.

5) The Use of Alternatives: Substitute or alternative parts for less critical parts can be used, and retrofitted in the equipment. There will be parts where the lead-time to acquire may take quite some time. If a part or item can be substituted or if there is an alternative part that will not compromise the quality of the equipment's performance, then it is possible not to stock the part. Purchasing people are good at finding alternative suppliers or substitutes for parts, or in most cases, they already exist in the storeroom. What is important in this case is to communicate with the maintenance people so that these people can evaluate the parts for a go or no go situation. What is important is that these alternative or substitute parts should be reflected in the system. Some plants have fabrication shops or machine shops, where some mechanical parts can be fabricated locally by their own shops. Likewise, there are independent machine shops that perform this type of work. The decision on whether to stock or not to stock a spare part will always depend on the lead-time to fabricate the part.

6) The part Is Age-Related Based on Preventive Maintenance Replacement: If the part is age-related and the wear-out process would be longer than the lead-time to acquire the part, and the failure is consistent and will survive to that specific age, then it is justifiable and feasible not to stock the part in the storeroom. Usually, wear-out is a gradual process and does not happen overnight. Maintenance can experience parts that tend to deteriorate or wear out concerning a specific age, and the wear-out process is consistent for the part, which means that most of the parts will survive this period. In this case, maintenance can re-order the part on that fixed period when the part will start to deteriorate. What is important is to identify the parts that will wear out concerning their operating age. Examples are brake pads, clutch, liners, tires, batteries, impellers, crusher jaws, punches, tooling, dies, or parts that are in direct contact with the product such as blades, tooling, dies punches, and so on. For example, if the wear-out pattern for an impeller will take 2 years consistently or around 85% of the time, the time to fabricate and lead-time for this impeller to arrive is 1 year, then we don't need to stock this impeller in the storeroom. The question to ask is if this part can be delivered before the actual PM execution, then it is feasible not to stock the part inside the storeroom.

7) Parts Monitored by Condition-Based Maintenance: Although this is related to item 1, parts that have a potential failure or failure development period will be under the watch list of a Predictive or Condition-Based Maintenance program. Since the failure development period is consistent and can be known by the user of these instruments, these people can predict not only when the part will fail, but also the best time when to requisition the part or item that will be affected. With this, operations people can be notified in advance on the time the equipment will be stopped for a corrective maintenance replacement activity. Both the Predictive and Preventive Maintenance groups can advise purchasing to buy the parts they need where the parts will arrive before the actual failure takes place. Once the part has arrived, then the replacement can be done on the equipment. In this case, there will be no surprises on the equipment since the failure modes can be predicted, and repair can be planned in advance. What is important is that if the failure development period is longer than the lead time for the part or item to arrive, then it is feasible not to stock the item in the storeroom.

8) Market Demand is Low: In some manufacturing plants, there are times when the capacity or volume is high that you cannot afford a single operator or maintenance to go on vacation leave because of the demand for the product. For example, during summertime, the demand will be high for soft drinks, mineral water, canned juice, and ice cream. During this period, machine uptime will play a vital role; hence, some fast and slow-moving spares and items need to be stocked in the storeroom. Likewise, in the semiconductor industry, there are periods where the demand is high, as well as tough times where the demand is low where skeletal force is usually used since the plant is not operating at its peak. Instead of their regular 24-hours operation, the plant may only run for two shifts where most operators and maintenance will be forced to use their leaves. In this case, not all equipment in the plant is operating, and the storekeeper, purchasers, and maintenance must adjust their re-ordering and stocking levels just enough to meet the demands they need.

9) Consignment: Some vendors and suppliers are willing to accept some form of consignment for spare parts. They can keep the parts in their storeroom, and the industry only pays them when the part will be used. Still, the vendor will be adding some carrying costs, which is quite normal for storing the item with the vendor. Some vendors can offer to hold some of your critical spares inventory of more expensive items on their site where you pay an annual fee, which is the holding or carrying cost. Another method of consignment is that the supplier will loan you expensive items, which you can place in your storeroom, and you will only be charged if the item has been actually used. The storeroom dollar value remains at zero. The consignment will be a good strategy, especially if the cost of the item is high. This is practical for expensive items. Remember that these vendors want to remain in business with big industries for their survival as well. This means that if a spare part can be consigned, then we actually do not need to stock the part in the storeroom.

7.7: MRO Decision Diagram on Whether to Stock or Not to Stock

In section 7.6, we have covered the different factors that we need to consider before making our final decision on whether to stock or not to stock parts in our storeroom. Shown in figures 7.9 to 7.10 is an MRO Decision Diagram or algorithm I developed that can help us decide more rationally on whether to stock or not to stock the parts in the storeroom. This is an enhancement of the MRO Decision Diagram included in my second book on Maintenance Roadmap to Reliability. Everything will start from the first question by asking if the failure of the spare part is age-related or random. Let us explain the algorithm in detail so that the reader can understand how to use it. In figures 7.9 to 7.10, the MRO Decision Diagram contains four columns, namely:

Age-Related or Random: First, let us define age-related and random failures. Age-Related Failures mean that the part or item will eventually survive to its guaranteed age and eventually wear out. This means that the part or component has reached its remaining useful life. If we speak about people, then these will be our grandparents and those living in the home for the aged. Once the part reached its useful life, then it will be replaced. Age specified may be in the form of running hours, calendar days, number of strokes, number of revolutions, number of stress applied, amount of volume, or any other form depending on the plant. The best maintenance strategy to use for this type of failure will be to identify when the majority of these parts will start to wear out and apply Preventive Maintenance by replacing them. This is the easiest failure pattern to address. Age-related failures refer to failures that are consistent with reaching their remaining useful life in the process. When the majority of the failures reached a certain period consistently, then we can declare that the failure of the part is age-related. For example, a study on 100 impellers indicates that around 85 % of the pump's flow rate starts to drop after two years of continuous operation, in this case, the impeller is starting to erode, and the rated capacity drops until it is no longer acceptable to production. In this case, the impeller will be scheduled for replacement after two years of continuous operation. This will be done at a point where the rated flow is no longer acceptable to the user. Hence, we can declare that the failure of the impellers is age-related. Random failures are failures that occur at any given period. This means that the probability that an item will fail in any one period is likely the same as it is in any other given period. This is also called chance failure. The conditional probability of failure remains constant. If the failure is random, then the part will not wear out.

[16]**Potential Failure** is a failure mode that provides signs or symptoms that they are on the verge of failing. This type of failure mode is often addressed by doing Predictive Maintenance tasks.

---

[16] John Moubray, ***Reliability-Centered Maintenance II***,(Great Britain: Butterworth Heinemann, 1997), Pages 145-149

An ailing bearing will provide signs or symptoms such as an increase in temperature, noise, or an increase in vibration. Predictive Maintenance instruments such as Vibration, Ultrasonic Monitoring, Infrared Thermography, or Oil Analysis can be used to capture the symptoms. Predictive Maintenance tasks include checking the equipment for any signs of potential failures so that decisions and actions can be taken to prevent the functional failure or to avoid the consequences of a functional failure from occurring. In this case, we are not giving any chance that the functional failure will happen, since maintenance will be one-step ahead by replacing the part before nearing its actual failure. A potential failure is defined as an identifiable physical condition that indicates that a functional failure is about to occur or is in the process of occurring. It means that the failure is just beginning. Functional failure is when the equipment is already in a failed state. When a potential failure is detected, that is just the start of the failure process. The functional failure will not immediately happen, as it will take days to several months before the functional failure finally occurs. The P-F curve or P-F interval is the interval between the occurrence or start of a potential failure and its decay or rupture into a functional failure. Many failure modes will provide a symptom that they are in the process of failing. These are good candidates for Predictive Maintenance monitoring.

Operational Consequences: The primary function of most equipment in the industry is connected in some way with the need to earn revenue to support revenue-earning activities. A failure mode has operational consequences if it has a direct adverse effect on its operational capability. Operational consequences may affect the total output or increase production costs through reworks or overtime. It may also affect the quality of the product manufactured or may affect the company's relationship with its clients and customers. In other cases, failure with operational consequences may increase the cost of failure for both operating and maintenance costs. Most failures in manufacturing plants will default to operational consequences and may stop the production process. If a part or item will affect the operations and other processes then it is feasible to stock the item in the storeroom.

The initial code R stands for random, PF will be the potential failure and LT is for the lead-time. A series of questions on the MRO decision diagram will be asked to the team, which must be answered by a simple yes or no. Indicate Y if the answer is yes and N if the answer is no on the MRO to Stock or Not to Stock Form in figure 7.8.

- **R1, PF1, LT1** questions are the same, which ask us about the lead-time to acquire the parts. If the lead time to acquire the part will just take a few hours to days, then we do not need to stock the part. However, if the lead time for the part or item takes several months, then we need to stock the part.

- **R2, PF2, LT2** questions are the same, which ask if the item to be stocked has a shelf life and if this item has a chance to expire before it can be used. The start of an expiration date will begin once the item had been manufactured, and they will be stored at the vendor for some time. The storekeeper needs to understand what items inside the storeroom have a shelf life. Usually, this can be traced to the package, just like when we buy medication, food, and so on. I would advise that the storekeeper inspect new deliveries and check those items with shelf lives as they might be accepting items that are either already expired or will expire very soon.

Do not accept items with shelf lives of less than one year to be on the safe side. For oil, make it 2 years as the expiration. As for grease, this is about 5 years.

- **R3, PF3, LT3** questions are the same, which ask us if the part can be consigned. Some vendors accept this type of arrangement that allows consigning the parts with the plant. The only time that the industry will be charged is, if the part will finally be withdrawn, and used in the equipment. Consignment can be done where either the vendor can keep the item in their warehouse, or the plant will keep it in their storeroom. In most cases, the vendor will be charging a holding or carrying cost when it will not be used per year. This would be appropriate for items with high costs and value.

- **R4, PF4, LT4** questions are all the same, which ask if the items or spare parts are currently being used for a Preventive Maintenance shutdown. If the parts can be available in the storeroom before the PM shutdown, then we definitely do not need to stock the parts. Exemption for this question will be imported items or parts that will take years before they can finally arrive in the storeroom. This means that for items whose lead-time will take years to arrive, the parts need to be stocked in the storeroom.

- **R5, PF5, LT5** questions are all the same, which ask if the equipment, component, or asset has redundancy or backup in place. This will make the failure less critical since operations can continue. This will only be applicable if the two components are not switched at the same interval as previously discussed in this book; if not, then the possibility is that both running and standby components can fail at the same interval. This question can only apply if maintenance people are making the correct switching pattern for their redundant or standby components.

- **R6, PF6, LT6** questions are all the same and are asking if alternate or substitute parts can be used instead of the original part. This can apply to lesser critical parts. There are many cases where the storeroom contains several items, which are the same but have a different part number since they have different brands, or there is a very slight difference in the construction of the part. Still, it can be retrofitted in the equipment. An example of this is oil filters, valves, belts, bearing with different brands, bolts, nuts, and so on. This problem ranks fifth in the top problems on MRO spare parts. This means that if the part has a substitute or alternative part that can be used, then we do not even need to stock as these alternative parts can be used.

- **R7, PF7, LT7** questions are all the same and ask if the spare part belongs to any bottleneck equipment in the process. A bottleneck is a constraint or congestion in production operations that occur when the inventory of the products is built up quicker than the equipment can process. In manufacturing, a bottleneck is a piece of equipment that frequently fails in the process or has a lower UPH (units per hour) compared to other equipment in the process. If this would be the case, or if we answer yes to the question, we move on to the next question on the column for operational consequences to answer a series of questions.

Once we answer **no** to the first six questions and **yes** to the seventh question, then we move on and proceed with the column on operational consequences. If any of the questions from O1 to O6 is likely to occur, then the decision is to stock the parts in the storeroom. For operational consequences, if the question is irrelevant or not applicable, we just move on to the next question. If any of these questions are unlikely to happen, then the default is not to stock the parts.

Without an algorithm or MRO Decision Diagram, the decision to stock or not to stock is often based on guesses or OEM recommendations. If we ask maintenance if we need to stock this part, 90% of the time, they will agree and say yes, especially when the boss always gets mad about the failure. The MRO decision diagram is just one side of the story as we indicate in this chapter since improving the storeroom is not just merely the sole responsibility of the storekeeper, since all functions will be involved in this, especially the users, which are the maintenance people.

Whether we like it or not, the storeroom is an expense to any industry. It is merely a contingency plan for breakdowns and failures. Hence, we need to do every necessary measure so that we can reduce the cost of doing maintenance to a minimum. Maintenance is always hungry with repairs and reactive works, which give us (myself included, as I am also maintenance) a bad reputation. They are blinded by work that does not involve them. Try to look at the dictionary and look for the word "maintain." It means to preserve or to sustain and not to repair. As I have said many times in my previous books the job of maintenance is not to repair the equipment. The word maintenance connotes a negative word to most industries, which degrades the reputation of maintenance mankind worldwide. A mechanic uses the hands 100 % of the time to repair equipment. The maintenance must use a balance of both the hands and the brain. This means that if the failure keeps repeating, then it is time to use the brain and analyze the problem. So help me spread the word that we are not repair people, and we are far better than that. The distinction between maintenance and a mechanic is that a mechanic uses his hands all of the time.

It is not only important to place people in the storeroom just for the sake of withdrawing parts. It is important to place people who can control and manage the spare parts list because Spare Parts Management is about having the right part at the right time when maintenance needed it most. I believe that the best people to manage spare parts are the maintenance people. MRO Spare Parts Management plays an important role in any maintenance improvement strategy. Much can be saved if we can just manage our spare parts more intelligently. Each industry has its own storeroom for keeping the parts needed for maintenance work, but not every industry has control over its MRO spare parts.

In contrast, maintenance uses a balance of the hands and the brain. This means that if the failure simply keeps on repeating, then it's time to let go of the hands and use our final weapon, which is the brain, and start analyzing why the failure keeps on repeating. For as long as we adhere that we are just here to repair, then the storeroom will never run out of problems, and most of the parts will be unavailable when needed. The better our equipment runs, the more we can manage the storeroom. All I can say is that maintenance is not measured by how fast they repair, but by how they were able to analyze the failure so that the failures will stop recurring.

Figure 7.8 is the actual MRO to Stock or not to Stock form to be used together with the MRO Decision Diagram or algorithm. I would strongly recommend that this must be done by a team and not by a single individual. The team can compose of people from Purchasing, Storekeepers, people involved in Preventive, Predictive Maintenance, Maintenance Planner, operations, and those involved in the repair process. I would also recommend that the team could bring their laptop where they can access their data such as lead-time, vibration data, Preventive Maintenance records, Bill of Materials (BOM) for the planner, and so on before they can proceed with the algorithm. Only one person should read each question on the MRO Decision Diagram and allow time for the team to deliberate and exchange their points of view with the team members. Once the team agrees with the answer, only can they move to the next question on the algorithm. In case, the team has some difficulty answering a specific question or they cannot reach a decision, it is best to move to the next spare part and invite someone to your next meeting who can answer that question.

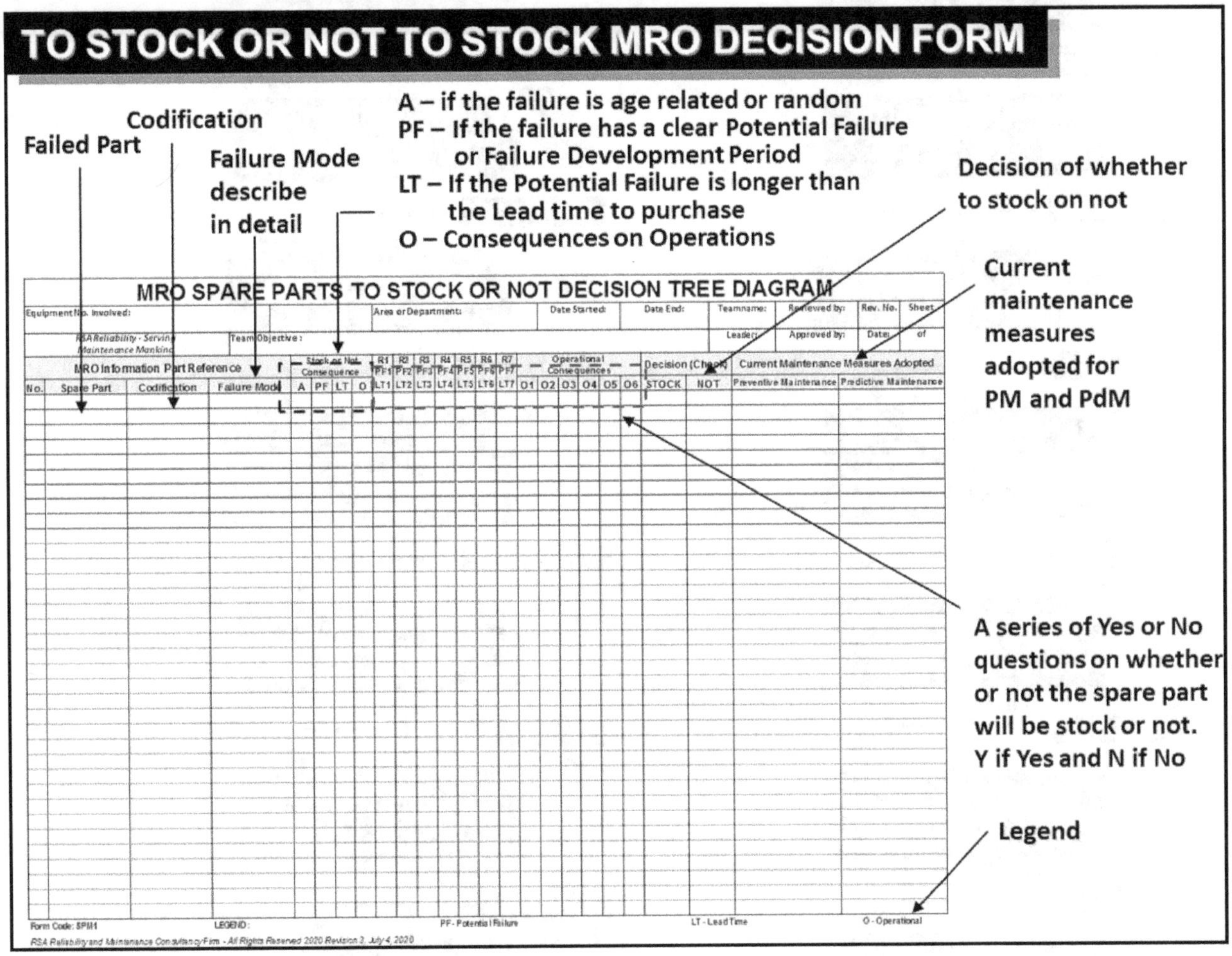

Figure 7.8: MRO to Stock or Not to Stock Form

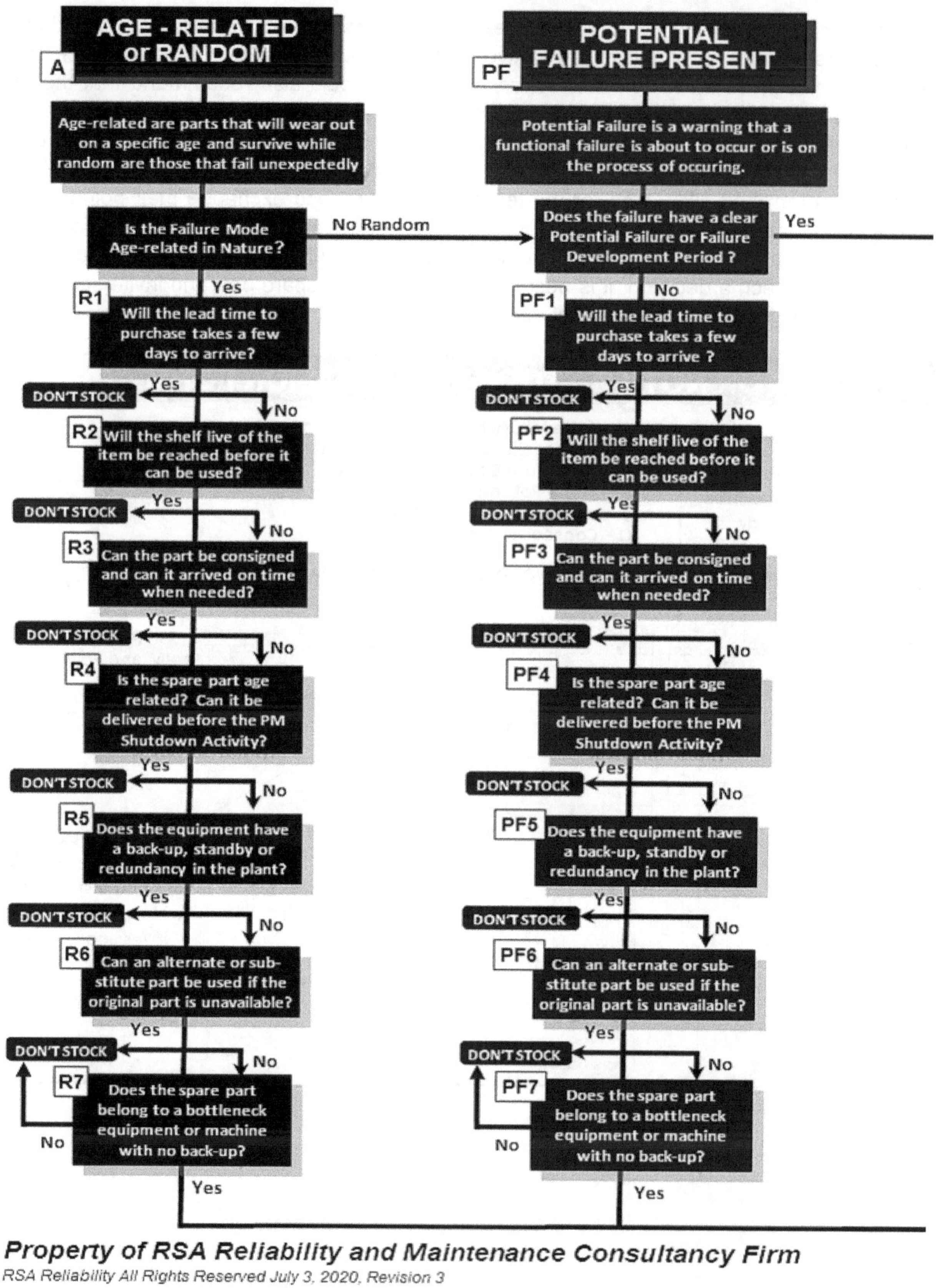

Figure 7.9: MRO Decision Diagram on Whether to Stock or Not to Stock a Part 1

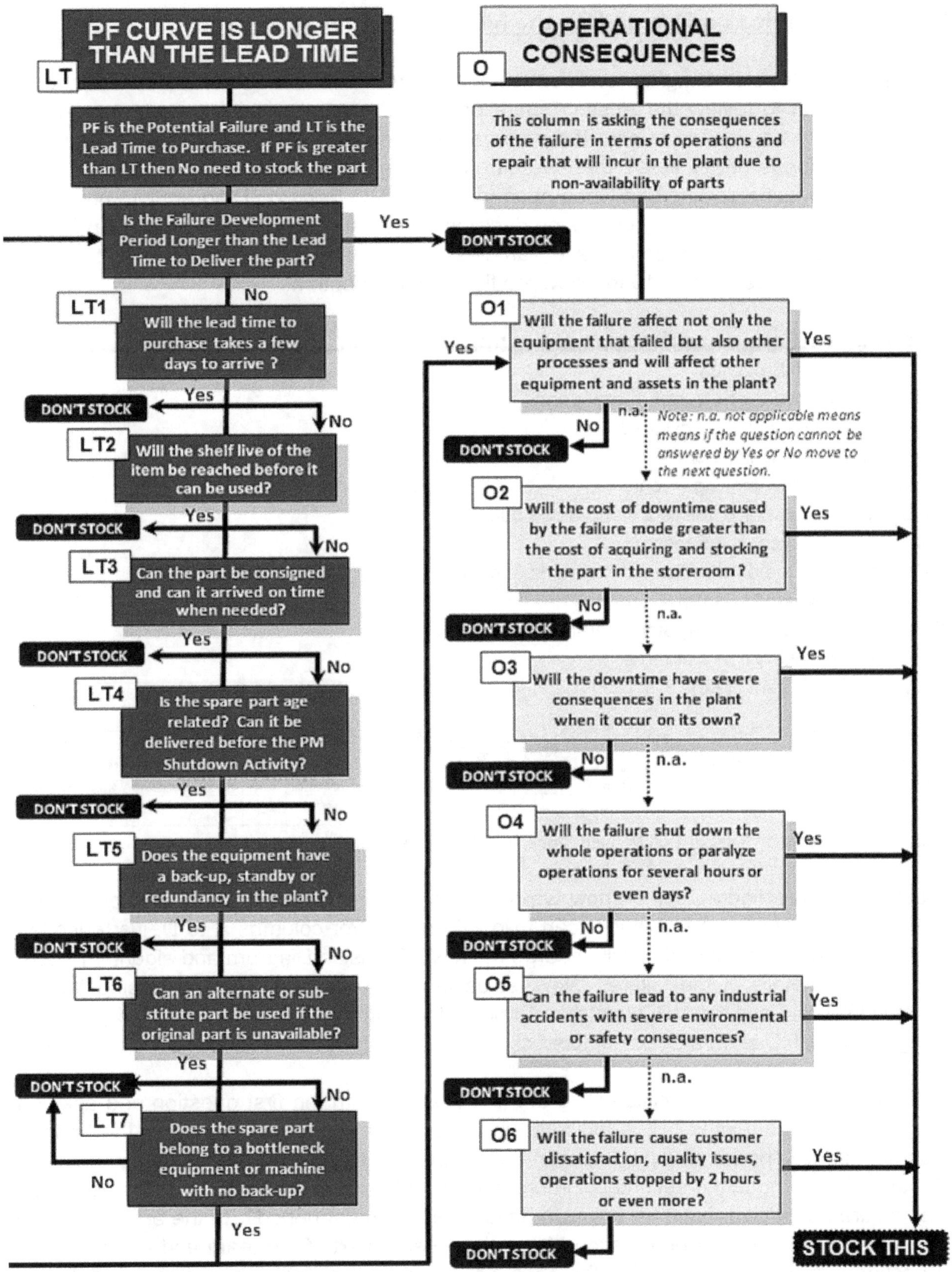

Note: If the team is uncertain on the question, move to the next question?

Figure 7.10: MRO Decision Diagram on Whether to Stock or Not to Stock a Part 2

7.8: Case Samples on How to Use the MRO Algorithm and Form

This MRO Decision Diagram and form that is provided in this book can be used for new spare parts and to review the existing spare parts held currently inside the storeroom if they actually need to be stocked or not. If the spare part is currently stocked in the main storeroom and the decision is not to stock, then we can just consume the remaining balance inside the storeroom and just order the part when needed. This MRO Decision Diagram can provide us a more rational and effective way of making decisions on what to stock or not to stock parts in the storeroom. Let us provide some case samples for the reader to understand better. Kindly revert back to figures 7.9 to 7.10 to follow the flow of the algorithm.

Case 1: Spare Part is a sealed type of bearing for a motor.

- Part Number: BRB-6204JM
- Failure Mode: Bearing Seizure/Fails randomly
- Note that item cannot be consigned
- Vibration Monitoring in place
- Classification: Slow-moving inventory
- No standby unit in place, this is a stand-alone pump
- Lead-time of delivery is 1.5 months
- P-F Curve detected on vibration monitoring is 3 months
- Failure is considered random
- Spare part is critical and may stop operations
- Downtime will take 8 hours or more.
- Do we stock this item or not?

A	B	C	D	E	F	G	H	I	J	K	L	M	N	O	P	Q	R	S	T	U	V	W	X	Y
MRO Information Part Reference				Stock or Not Consequence				R1 PF1	R2 PF2	R3 PF3	R4 PF4	R5 PF5	R6 PF6	R7 PF7	Operational Consequences						Decision (Check)		Current Maintenance Measures Adopted	
No.	Spare Part	Codification	Failure Mode	A	PF	LT	O	LT1	LT2	LT3	LT4	LT5	LT6	LT7	O1	O2	O3	O4	O5	O6	STOCK	NOT	Preventive Maintenance	Predictive Maintenance
1	Bearing	BRB-6204JM	Bearing Seizure	N	Y	Y	---	---	---	---	---	---	---	---	---	---	---	---	---	---		x	Perform lubrication every month by PM.	Perform regular Vibration monitoring.

Figure 7.11: Case Study Sample 1 on MRO to Stock or Not to Stock Form

Let us analyze figure 7.11 on how we arrived at the decision, not to stock (column W). I have placed a column so the reader can follow it easily. For columns A to D, just encode the spare part and part number that will undergo the MRO Decision Diagram and algorithm. Case 1 specifies the details of the failure. Now let us proceed here systematically using the MRO Decision Diagram and the MRO to stock or not to stock form. If the team answered yes, we just type **Y.** If the MRO team agreed not to stock, we just type **N**.

- Column E: In the MRO Decision Diagram (A), we start with the first question and ask, is the failure mode age-related? It is indicated in the case study that the failure of the bearing is random, so we answer no (N) on column E in figure 7.11; we place N for no.
- Column F: Moving to the second column of the MRO Decision Diagram (PF), it is asking does the failure has a clear potential failure or failure development period? In the actual meeting of the team, for example, the Vibration Monitoring user is part of the team and can answer this question easily. It is indicated in the case that vibration monitoring is in place; therefore, our answer, in this case, is yes, so we place (Y) on column F on the form.

- Column G: Moving on to the next column of the MRO Decision Diagram (LT), the question is, is the Failure Development Period Longer than the Lead Time to deliver the part? We asked the purchasing who is also present during the meeting, and based on his records, the lead-time to purchase this type of bearing is 1.5 months. This means that the Failure Development period is longer than the lead-time for the part to arrive at the storeroom by 1.5 months. Easily if we continue with the algorithm, **the decision is not to stock the bearing.** Finally, we write a brief summary of the current maintenance measures adopted to address the failure mode. For Preventive Maintenance, routine lubrication is done monthly, while for Predictive Maintenance regular monitoring of the bearing is conducted monthly for any changes in vibration on the FFT spectrum. Well, that was a bit easy. Now let us move on to the second case.

Case 2: Part is a sealed type of bearing for motor

- Part Number: BRB-6204JM
- Failure Mode: Bearing seizures and fails randomly
- Note that item cannot be consigned
- Vibration Monitoring is in place
- Classification is a slow-moving inventory
- With triple redundancy (3 motors and pumps in place)
- Lead-time of delivery is 3 months
- P-F Curve detected on vibration monitoring is 3 months
- Failure is considered random
- No downtime for the system, but the time to repair the pump is 8 hours
- When the duty pump fails, the standby pump will activate automatically.
- Do we stock this item or not?

A	B	C	D	E	F	G	H	I	J	K	L	M	N	O	P	Q	R	S	T	U	V	W	X	Y
MRO Information Part Reference				Stock or Not Consequence				R1 PF1	R2 PF2	R3 PF3	R4 PF4	R5 PF5	R6 PF6	R7 PF7	Operational Consequences						Decision (Check)		Current Maintenance Measures Adopted	
No.	Spare Part	Codification	Failure Mode	A	PF	LT	O	LT1	LT2	LT3	LT4	LT5	LT6	LT7	O1	O2	O3	O4	O5	O6	STOCK	NOT	Preventive Maintenance	Predictive Maintenance
2	Ball Bearing	BRB-6204JM	Bearing Seizure	N	Y	N	---	N	N	N	N	Y	---	---	---	---	---	---	---	---		x	Perform lubrication every month by PM.	Perform regular Vibration monitoring. Order the bearing at the earliest sign of Potential Failure.

Figure 7.12: Case Study Sample 2 on MRO to Stock or Not to Stock Form

- Columns E and F: This will be the same for Case 1, so we write the spare part and part number or codification. The first question in the MRO algorithm is the failure age-related? In this case, we answer no (E) since the failure is random. For column F, the question is there a clear potential failure? We answered yes (F) since the case indicates that it has a clear potential failure. We move on to the next column on lead-time (LT).
- Column G: In this column on Lead Time (LT) from the MRO Decision Diagram, the question is, is the failure development period longer than the lead time for the item to arrive? In this case, both the lead time and the failure development period are the same as 3 months, therefore if we ask the question, is the failure development period longer than the lead time? Our answer is no. We move to LT1.
- Column I: In this column, on the MRO Decision Diagram, we ask the question on LT1; will the lead-time to purchase take a few days to arrive? The team answered no, so we place N since the lead-time is given as 3 months. We move to LT2.

- Column J: In this column on LT2, the question is, will the shelf life of the item be reached before it can be used? Our answer is no, so we place N. We move to LT3.
- Column K: In this column on LT3, the question is, can the part be consigned and arrive when we need it? The answer here will come from the member from Purchasing. Let us assume that the part cannot be consigned at the moment since the purchasing said that no vendor will accept that. Hence, the team agreed that it cannot be consigned. So we agree to say no and place N under LT3. We now move to LT4.
- Column L: In this column on LT4, the question is whether the failure is age-related and part of any PM activity. The failure here is given as random, so it is not part of any Preventive Maintenance replacement or overhauling activities, so the team answered no. The MRO team now moves to LT5.
- Column M: In this column on LT5, the question is, does the equipment have a backup, standby, or redundancy in place? Our answer is yes since it has a triple redundancy. Since our answer to LT5 is yes, then the **default decision is not to stock this item.** We check or place an x mark on not to stock and write the current maintenance measures done to address the failure mode.

Case 3: Pump Impeller erodes due to wear and tear

- Part Number: PIMP-3256IM
- Failure mode: Erosion / Age-related / takes 2 years
- Note that item cannot be consigned
- Classification: Non-moving item for 2 years
- Lead-time of delivery (including fabrication) is 6 months
- No standby unit in place and this is a stand-alone pump
- There is no Vibration Monitoring in place
- Failure is considered age-related
- When the duty pump fails, operations will stop
- Time to replace the impeller will be 24 hours
- Do we stock this item or not?

A	B	C	D	E	F	G	H	I	J	K	L	M	N	O	P	Q	R	S	T	U	V	W	X	Y
MRO Information Part Reference				Stock or Not Consequence				R1 PF1	R2 PF2	R3 PF3	R4 PF4	R5 PF5	R6 PF6	R7 PF7	Operational Consequences						Decision (Check)		Current Maintenance Measures Adopted	
No.	Spare Part	Codification	Failure Mode	A	PF	LT	O	LT1	LT2	LT3	LT4	LT5	LT6	LT7	O1	O2	O3	O4	O5	O6	STOCK	NOT	Preventive Maintenance	Predictive Maintenance
3	Pump Impeller	PIMP-3256IM	Erosion and Wear Out	Y	---	---	---	N	N	N	Y	---	---	---	---	---	---	---	---	---		x	PM replacement will be done on the impeller after 24 months or 2 years.	Continuous Monitoring of the pumps flow rate.

Figure 7.13 Case Study Sample 3 on MRO to Stock or Not to Stock Form

- Column E: In this column, the given sample above, which is the impeller, is stated as age-related, which means that after a couple of years, the rate of flow will drop that will no longer be acceptable to operations, hence the impeller will be replaced. Therefore, in the MRO Decision Diagram, the first question we ask is whether the failure is age-related? Our answer is yes, and we move on to R1.
- Columns F and G: This will be skipped since the failure is age-related. We move to R1.
- Column I: In this column for R1, the question is will the lead time take a few days to arrive? In this case, the lead time, which includes fabricating the impeller, will take around six months; therefore, our answer here is no. We move to R2.

- Column J: In this column for R2, the question is will the shelf life of the item be reached before it can be used? The impeller does not have a shelf life except that it can corrode, so our answer here is no.
- Column K: In this column for R3, the question is, can the part be consigned? Let us assume that Purchasing gave us a reply of no, so we move on to the next question on R4.
- Column L: In this column for R4, is the spare part age-related? Can it be delivered before the PM shutdown activity? In the given sample, the lead time to deliver, which includes the fabrication, is 6 months. The PM shutdown will happen after 2 years, giving us a lot of lead-time for the impeller, so in this case, it is not advisable to order the impeller immediately. We can order the impeller 8 to 10 months before the actual Preventive Maintenance shutdown schedule since the fabrication time is 6 months. The Preventive Maintenance will happen in the next 2 years, so our default decision, in this case, is **not to stock the impeller.**

Case 4: Clogged Oil Filter

- Part Number: FIL-146578JT
- Failure Mode: Clogged oil filter / Age-Related / 3 months
- Note that item cannot be consigned
- Failure Pattern: Age-related
- Classification: Slow Moving item, used 4 times a year
- Lead-time of delivery: Can be delivered in less than 24 hours to a maximum of 2 days
- Clogged oil filter can be monitored by the drop of pressure in the pressure gauge
- Repair Time: Time to replace the oil filter can take 30 minutes
- Experience shows this failure mode happens every 3 months
- Do we stock this item or not?

A	B	C	D	E	F	G	H	I	J	K	L	M	N	O	P	Q	R	S	T	U	V	W	X	Y
MRO Information Part Reference				Stock or Not Consequence				R1 PF1	R2 PF2	R3 PF3	R4 PF4	R5 PF5	R6 PF6	R7 PF7	Operational Consequences						Decision (Check)		Current Maintenance Measures Adopted	
No.	Spare Part	Codification	Failure Mode	A	PF	LT	O	LT1	LT2	LT3	LT4	LT5	LT6	LT7	O1	O2	O3	O4	O5	O6	STOCK	NOT	Preventive Maintenance	Predictive Maintenance
4	Oil Filter	FIL-146578JT	Clogged Filter	Y	---	---	---	Y	---	---	---	---	---	---	---	---	---	---	---	---		x	PM replacement will be done evey 3 months.	Continuous Monitoring on the pressure drop.

Figure 7.14: Case Study Sample 4 on MRO to Stock or Not to Stock Form

- Column E: In this column, again, the question is the failure mode age-related in nature? We agree that this is age-related since the clogged oil filter happens every 3 months consistently where the pressure gauge will indicate a pressure drop. Our answer is yes, and we move to R1.
- Columns F and G: The question here will be skipped since the failure is age-related.
- Column I: In this column for R1, the question is, will the lead time take a few days to arrive? Our answer here is yes since it is clearly indicated in the case sample study. Therefore, since the filter will be clogged every 3 months and it takes 24 hours to a maximum of 2 days for the filter to be delivered to the storeroom, then we purchase the filter one week before the replacement. Hence, our default decision will be **not to stock the oil filter.**

Case 5: Back-up Motor recommended by OEM to be stocked in the storeroom

- Part Number: MOT-6790LO
- Failure Mode: Burnt motor/random
- Note that item cannot be consigned

- Classification: Non-Moving Item
- Lead-time of Delivery: 6 months
- Infrared thermography is in place: P-F curve is 1 month
- No redundancy in place and this is a standalone component
- Experience shows that this failure mode has not yet occurred before
- System is critical, and failure mode can shut down the whole operations
- Do we stock this item or not?

A	B	C	D	E	F	G	H	I	J	K	L	M	N	O	P	Q	R	S	T	U	V	W	X	Y
MRO Information Part Reference				Stock or Not Consequence				R1 PF1	R2 PF2	R3 PF3	R4 PF4	R5 PF5	R6 PF6	R7 PF7	Operational Consequences						Decision (Check)		Current Maintenance Measures Adopted	
No.	Spare Part	Codification	Failure Mode	A	PF	LT	O	LT1	LT2	LT3	LT4	LT5	LT6	LT7	O1	O2	O3	O4	O5	O6	STOCK	NOT	Preventive Maintenance	Predictive Maintenance
5	Oil Filter	MOT-679LO	Clogged Filter	N	Y	N	---	N	N	N	N	N	N	Y	Y	---	---	---	---	---	X		Follow OEM to have a back up of the motor where the MRO Algorithm also agrees	

Figure 7.15: Case Study Sample 5 on MRO to Stock or Not to Stock Form

- Column E: In this column, again, the question is the failure mode age-related in nature? We have a situation here in which the current operations do not have a standby motor. The OEM recommends buying another motor and having it installed as a backup. In this case, the failure of the motor will be random and not age-related, so we answer no.
- Column F: Does the failure have a clear potential failure or failure development period? Our answer here is yes. We move to the next column on LT.
- Column G: In this column on the lead-time, it will take 6 months for the new motor to arrive, and this will be longer than the Failure Development Period. The lead-time for the motor to arrive is 6 months, and the user of Infrared thermography can detect the failure 1 month before it fails. The MRO team answered no to this question, and so we move to LT1.
- Column I: For the question on LT1, will the lead time take a few days to arrive? Our response to this is no since the lead time will be 6 months. We move to LT2.
- Column J: For the question on LT2 about shelf life, our response is still no. The motor has no shelf life. We now move to LT3
- Column K: for the question on LT3 about consignment, we asked the purchasing, and the response the purchaser gave us was no. We move to LT4 on the algorithm.
- Column L: On LT4, is the spare part age-related, and can it be delivered before the PM Shutdown activity? The failure is clearly indicated as random, and will not be part of the PM activities so our response here is still no, and we move to LT5.
- Column M: On LT5, does the equipment have a backup, standby, or redundancy in the plant? Our answer is no since there is no backup in place. We move to LT6.
- Column N: On LT6, the question is, can an alternate or substitute part be used if the original part is unavailable? Our answer here is no. We move to LT7.
- Column O: On LT7, the question is, does the spare part belong to bottleneck equipment or machine with no backup? Our answer here is yes, and we move to O1 on the next column of the algorithm.
- Column P: Will the failure affects not only the equipment that failed but also other processes and will affect other equipment and assets in the plant? Our answer here is yes, and following the MRO algorithm, we have no option this time but to stock the motor, which means that we are in agreement with the OEM to install a backup motor in the future for our operations. Our decision is to **stock the motor in the storeroom.**

Case 6: Pump and Motor Coupling

- Part Number: COUP-9878SI
- Failure Mode: Broken Coupling / Random
- Note that item cannot be consigned
- Classification: Non-Moving Item
- Lead Time of Delivery: 2 months
- Vibration Monitoring is not in place
- With backup redundancy in place
- Experience shows that this failure mode occurred once a year. When coupling fails the standby pump, and the motor will be activated manually and take 5 minutes
- Do we stock this item or not?

A	B	C	D	E	F	G	H	I	J	K	L	M	N	O	P	Q	R	S	T	U	V	W	X	Y
MRO Information Part Reference				Stock or Not Consequence				R1 PF1	R2 PF2	R3 PF3	R4 PF4	R5 PF5	R6 PF6	R7 PF7	Operational Consequences						Decision (Check)		Current Maintenance Measures Adopted	
No.	Spare Part	Codification	Failure Mode	A	PF	LT	O	LT1	LT2	LT3	LT4	LT5	LT6	LT7	O1	O2	O3	O4	O5	O6	STOCK	NOT	Preventive Maintenance	Predictive Maintenance
6	Coupling	COUP-9878SI	Broken Coupling	N	N	---	---	N	N	N	N	Y	---	---	---	---	---	---	---	---		x	Perform regular inspection for any signs of cracks or fracture on the coupling - Conduct alignment yearly	Check for Vibration Monitoring

Figure 7.16: Case Study Sample 6 on MRO to Stock or Not to Stock Form

- Column E: In this column, the question is the failure mode age-related in nature? Our answer here is no. We move to the next question on PF in column F.
- Column F: Does the failure mode have a clear Potential Failure or Failure Development Period? When the maintenance was asked the question, based on their experience, the breakage of the coupling happens instantaneously so in this case, we answer no, and we move to PF1
- Column I: In column G, the question on PF1 is, will the lead-time to purchase take a few days to arrive? Our answer here is no since the lead time is 2 months. We move to PF2.
- Column J: In column J, the question on PF2 is about the shelf life; our answer here is no, so in this case, we move to PF3 of the algorithm.
- Column K: In column K, the question on PF3, can the part be consigned, and can it arrive on time from the vendor when needed? Our answer is no. We move to PF4.
- Column L: In column L, the question for PF4 is the spare part age-related, and can it be delivered before the PM Shutdown activity? Our answer is no. We move to PF5.
- Column M: In column M, the question for PF5 is, does the equipment have a backup, standby, or redundancy in the plant? Our answer here is yes since, in this case, there is a backup pump, and it takes five minutes to activate it manually. Therefore, the team recommends purchasing the coupling when the standby pump will be activated. In this case, the MRO team agreed that for the broken coupling, **we do not stock the coupling.**

Case 7: Pump and Motor Coupling

- Part Number: COUP-9878SI
- Failure Mode: Broken Coupling / Random
- Note that item cannot be consigned
- Classification: Non-Moving Item
- Lead-time of Delivery: 2 months
- Vibration Monitoring is not in place
- No backup in place

- Experience shows this failure mode occurred once a year
- When coupling breaks it will take 6 hours to replace if a part is available
- Equipment belongs to a bottleneck and will induce downtime in the process
- Do we stock this item or not?

A	B	C	D	E	F	G	H	I	J	K	L	M	N	O	P	Q	R	S	T	U	V	W	X	Y
MRO Information Part Reference				Stock or Not Consequence				R1 PF1	R2 PF2	R3 PF3	R4 PF4	R5 PF5	R6 PF6	R7 PF7	Operational Consequences						Decision (Check)		Current Maintenance Measures Adopted	
No.	Spare Part	Codification	Failure Mode	A	PF	LT	O	LT1	LT2	LT3	LT4	LT5	LT6	LT7	O1	O2	O3	O4	O5	O6	STOCK	NOT	Preventive Maintenance	Predictive Maintenance
7	Coupling	COUP-9878SI	Broken Coupling	N	N	---	---	N	N	N	N	N	N	Y	Y	---	---	---	---	---	x		Perform regular inspection for any signs of cracks - Conduct alignment yearly	Recommend to purchase Vibration Monitoring

Figure 7.17: Case Study Sample 7 on MRO to Stock or Not to Stock Form

- Column E: In this column, again, the question is the failure mode age-related in nature? Our answer here is no, since if the coupling breaks, it is random, and maintenance has experienced this in their operation several times.
- Column F: Does the failure have a clear potential failure or failure development period? Our answer here is no since the failure happens instantaneously. This means that the failure is random, but the failure development period is just too short as the breakage happens instantaneously. We move to PF1.
- Column I: In this column, since we answer no in column F, we move to PF1. The question is, will the lead-time to purchase take a few days to arrive? Our answer is no since the lead-time will take 2 months. We move to PF2.
- Column J: For the question on PF2, will the shelf life of the item be reached before it can be used? Our answer here is no; the coupling has no shelf life. We move to PF3.
- Column K: For the question on PF3, can the part be consigned, and can it arrive on time when it is needed? Our answer here is no. We move to PF4
- Column L: For the question on PF4, is the spare part age-related, and can it be delivered before the PM Shutdown activity? Our answer here is no since the failure is random. We now move to PF5 of the algorithm
- Column M: For the question on PF5, does the equipment have a backup, standby, or redundancy in the plant? Our answer here is no since it is stated in the given that there is no backup in place. We move to PF6 of the algorithm.
- Column N: For the question on PF6, can we use an alternate or substitute part if the original part is unavailable? Our answer here is no. We move to PF7.
- Column O: For the question on PF7, does the spare part belong to bottleneck equipment or machine with no backup? Our answer here is definitely yes, so we move to the next column on operational consequences and move to O1.
- Column P: The question in O1 states that can the failure affects not only the equipment that failed but also other processes? Our answer here is yes, and following the algorithm, our decision is **<u>to stock the coupling.</u>**

Figure 7.18 is the summary of the seven case study samples on how the MRO Decision Diagram is used. What we now have is a clear and concise decision-making process on whether to stock the parts in the storeroom or not. The team performing the MRO to Stock or Not to Stock Decision diagram can meet a minimum of once a week and spend at least a minimum of a couple of hours to discuss if the parts identified by the team actually need to be stocked or not in the storeroom. Just imagine if there are parts that are currently stock and the decision is not to stock, the storekeeper can just deplete the inventory, and the parts will no longer be replenished. The parts can just be ordered at the most appropriate time when it is actually needed. This algorithm will definitely save the company a large amount of savings in the future (by the way, it took me a lot of beer thinking about how to develop this algorithm). Just target 10 to 15 items at the most per meeting, and that will be a good start. Just make sure that this is done regularly by a team and not by a single individual. By consistently doing this, we can now free our storeroom of items that do not really need to be stocked.

MRO SPARE PARTS TO STOCK OR NOT DECISION TREE DIAGRAM

Equipment No. :	Equipment Type :	Area or Department :		Date Started			Date End :					Reviewed by:								Teamname :			Rev. No.	Sheet
RSA Reliability - Serving Maintenance Mankind	Team Objective :											Approved by:								Leader :			Date :	of
A	B	C	D	E	F	G	H	I	J	K	L	M	N	O	P	Q	R	S	T	U	V	W	X	Y
MRO Information Part Reference				Stock or Not Consequence				R1 PF1	R2 PF2	R3 PF3	R4 PF4	R5 PF5	R6 PF6	R7 PF7	Operational Consequences						Decision (Check)		Current Maintenance Measures Adopted	
No.	Spare Part	Codification	Failure Mode	A	PF	LT	O	LT1	LT2	LT3	LT4	LT5	LT6	LT7	O1	O2	O3	O4	O5	O6	STOCK	NOT	Preventive Maintenance	Predictive Maintenance
1	Bearing	BRB-6204JM	Bearing Seizure	N	Y	Y	---	---	---	---	---	---	---	---	---	---	---	---	---	---		x	Perform lubrication every month by PM.	Perform regular Vibration monitoring.
2	Ball Bearing	BRB-6204JM	Bearing Seizure	N	Y	N	---	N	N	N	N	Y	---	---	---	---	---	---	---	---		x	Perform lubrication every month by PM.	Perform regular Vibration monitoring. Order the bearing at the earliest sign of Potential Failure.
3	Pump Impeller	PIMP-3256IM	Erosion and Wear Out	Y	---	---	---	N	N	N	Y	---	---	---	---	---	---	---	---	---		x	PM replacement will be done on the impeller after 24 months or 2 years.	Continuous Monitoring of the pumps flow rate.
4	Oil Filter	FIL-146578JT	Clogged Filter	Y	---	---	---	Y	---	---	---	---	---	---	---	---	---	---	---	---		x	PM replacement will be done evey 3 months.	Continuous Monitoring on the pressure drop.
5	Oil Filter	Motor	MOT-679LO	N	Y	N	---	N	N	N	N	N	N	Y	Y	---	---	---	---	---	x		Follow OEM to have a back up of the motor where the MRO Algorithm also agrees	
6	Coupling	COUP-9878SI	Broken Coupling	N	N	---	---	N	N	N	N	Y	---	---	---	---	---	---	---	---		x	Perform regular inspection for any signs of cracks or fracture on the coupling - Conduct alignment yearly	Check for Vibration Monitoring
7	Coupling	COUP-9878SI	Broken Coupling	N	N	---	---	N	N	N	N	N	N	Y	Y	---	---	---	---	---	x		Perform regular inspection for any signs of cracks - Conduct alignment yearly	Recommend to purchase Vibration Monitoring

Figure 7.18: Summary of MRO to Stock or Not to Stock Form

Chapter 8

Involving Operators in Maintenance

Operators are important in any maintenance and reliability strategy since it is believed that operators are the first line of defense on any failure that occurs in our equipment. This means that they are the people closest to the asset during the time of failure. They will be the ones who will encounter the failure first before the maintenance.

8.1: Why Operators Should be Involved in Maintenance

Although in the majority of industries I served, operators are not involved in maintenance whatsoever. Some industries may have a set of inspection activities done during their shift which is generated by the maintenance function. There seems to be a division of tasks that originated during the time of their employment as it is clearly indicated that their role is to solely operate the equipment. In this case, we have the I operate, you fix syndrome. Much worst is the lack of communication between operators and maintenance. When a breakdown occurs, operators tend to disappear after calling maintenance. And when the concept of Operator and Maintenance partnership is bought forth to the table, some operation managers simply kill the idea immediately and tell maintenance that the idea simply won't work here. My stand on this is that operators are important in any maintenance and reliability strategy for two reasons. First, it is believed that operators are the first line of defense against any failure that occurs in our equipment since they are the people closest to the asset. They will be the ones who will encounter the failure first before the maintenance. Second, maintenance can only advance to any continuous improvement effort and advancement if operators accept the responsibility that they should play a major part in establishing Basic Equipment Conditions.

Operators must understand that maintenance is a shared responsibility for both operations and maintenance people. There are maintenance activities that must be done by the operators themselves. Maintenance must understand that involving operators in maintenance can only work if they teach operators about their equipment. Both operators and maintenance should understand that they cannot co-exist without each other and both should create a solid partnership because maintenance is simply a shared responsibility for both operators and maintenance. One of the major pillars of TPM is Autonomous Maintenance which is usually done in 7 steps excluding Step 0. The objective of implementing Autonomous Maintenance is first to work hand in hand with maintenance in establishing the basic equipment condition. The second objective is to develop their own standards, which will include cleaning standards, lubrication standards, and inspection standards on the equipment. The only problem with Autonomous Maintenance is that this is a slow process as it will take a minimum of 5 years or even more to reach Step 7, but one thing I experienced is that as the operators complete each step, their knowledge of their equipment increase as well as their attitude towards their equipment changes. What Japanese people believe is that if the operators know their equipment intimately, then they will start to care. But this will require continuous mentorship from the maintenance and maintenance must share what they know with operators which should be a continuous process.

Autonomous Maintenance Step 3 Standard - For Weekly Activities

SAMPLE STANDARD

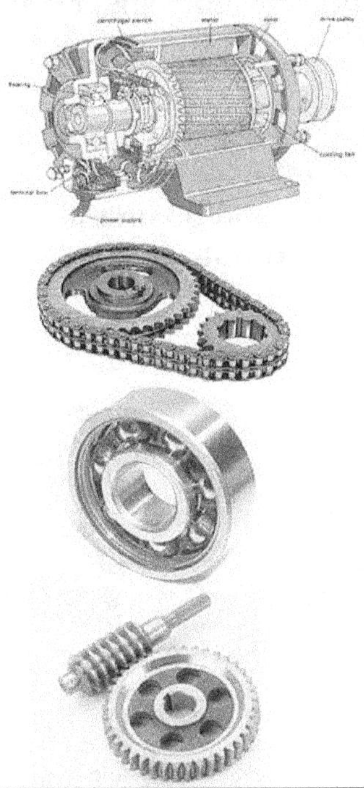

PART	STANDARD	METHOD	TOOL	TIME	WHO
A) Motor Section	- No dirt and spills	- Wipe	- Dry Cloth	- 5 minutes	
- Transmission	- No vibration, ab-normal nose and overheating		n.a.	- 30 sec	Operator
- Oil level Gauge	- Must be on the specified quantify		n.a.	- 30 sec	Operator
- Chain & sprocket	- No abnormal Noise - Adequate LUB		n.a.	- 30 sec	Operator
B) Outboard Bearing	- Clean, no dirt	- Wipe	- Dry Cloth	- 5 minutes	Operator
- Gland	- No leak			- 1 min	Operator
- Bearing	- No overheating, - Normal vibration			- 1 min	Operator
- Worm Bearing and Worm Wheel	- No unusual noise, thread deformation or overheating		n.a.	- 1 min	Operator

Figure 8.1: Sample of Autonomous Maintenance Checklist

When operators reached Step 4 of Autonomous Maintenance, they will already have a raw checklist that includes the three standards, but as the operators move into the next steps, these standards will be polished to come up with the final standards that include cleaning, lubrication, and inspection standards. One benefit of having Autonomous Maintenance in the plant is that those activities performed by operators can now be eliminated from the PM task list reducing the activities to be done by the PM group. Autonomous Maintenance activities will usually deal with routine activities performed on the equipment. The goal of establishing Autonomous Maintenance on the equipment is to enhance the senses of operators so that they can capture the problem when it is still small. Big problems are simply an accumulation of small problems that have been left neglected on the equipment. These small problems when not addressed during their early stage can lead to major catastrophic problems in the equipment which makes the operators the first line of defense against any equipment-related failures and breakdown.

Implementing Autonomous Maintenance or having an operator maintenance partnership brings production and maintenance to work together to accomplish a common goal which is to establish basic equipment conditions and to prevent accelerated deterioration where the life of a part or item did not reach its lifespan. Operators learn to carry out important daily tasks such as cleaning, proper lubrication, inspections, and other light maintenance tasks including simple repairs and replacements. Autonomous Maintenance is designed to help operators learn more about how their equipment functions, what common problems can occur, their impact, and how to prevent them by early detection and treatment of abnormal conditions. Autonomous Maintenance tasks prepare operators to be active partners with the maintenance and engineering in improving the overall performance and reliability of their equipment. What is important in an operator's inspection is knowing the limit or threshold and what to do in case it reaches the limit. This means that operators should understand what is the maximum measurement that can exist in which the equipment will still be allowed to operate. This is important so that operators can understand the maximum limit that the equipment can still operate safely before we can expect failure to occur in the equipment.

OPERATORS CHECKLISTS SHOULD BE QUANTITATIVE

Standards for cleaning, lubrication, inspection should be detailed and quantitative and not subjective. Operators should know what is acceptable and not

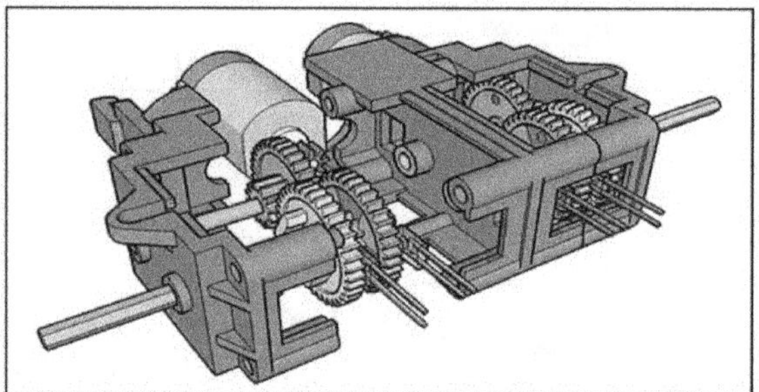

Subjective ✗	Quantitative ✓
• What: Check gearbox • When: Monthly • Who: AM Team	• What : - No loose or missing fasteners - Oil level within the min and max level - Oil temperature at 160 deg F +/- 10 If condition is not acceptable detail the things to be done

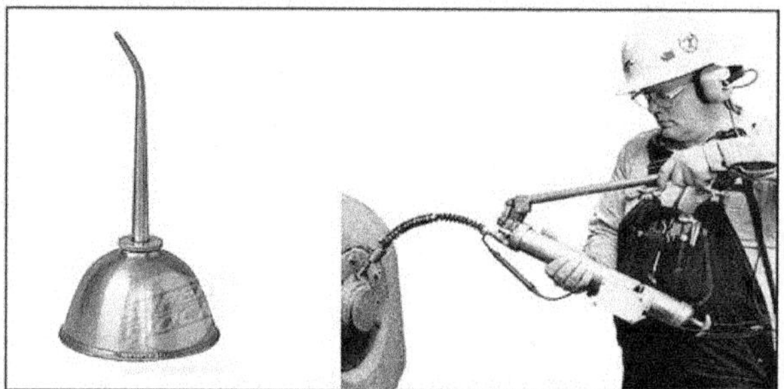

Subjective ✗	Quantitative ✓
• What: Lubricate motor • When: Quarterly • Who: AM Team	• What specific lubricant should be use oil/grease • If grease what specific NLGI number • What specific range of temperature of grease • What type of grease gun • How many grams of grease will be required

Figure 8.2: Operators' Checklists Should be Quantitative

One unique feature of establishing an Autonomous Maintenance activity is for the operators to develop inspections using Visual Controls. When these deviations are clearly visible and apparent to all, corrective action can be taken to immediately correct these problems and anomalies before they become bigger. Visual Control allows us to see the problems more easily. They are also meant to provide instructions and simply convey information. Visual Control, sometimes called Visual Communication, is a technique imported from Japan. In the early '70s, U.S. executives went on tours to Japanese industries to see why they were beating them so badly in terms of quality. They noted that the workplaces were filled with pictures and diagrams of what to do and what not to do. With the way things were displayed, anyone passing by whether an executive or a new employee could actually sit down and do the assembly without knowing the Japanese language. It is one unique feature of Autonomous Maintenance that tells you what to do and what not to do.

8.2: The Most Important Thing About Operator's Involvement

While we usually hear the sentiment from maintenance that we prepare the checks, operators are just too stubborn, or too naive to follow them ending up in maintenance doing the inspections themselves. First, let me explain that this is not how Autonomous Maintenance works. In implementing Autonomous Maintenance, the maintenance will play a very important role since they will act as their mentors. Maintenance will be responsible for having a continuous partnership with operators. Their role is to teach the operators about their equipment until the operators know their equipment intimately. Operators need to understand why this inspection is important and what happens if it will be neglected. What will be the effect? What will be the consequences of the failure?

Here's the thin line, in Autonomous Maintenance. The operators are responsible for developing their own checklist and standards. This means that the task and activities listed will not come from maintenance but from the operators themselves. This could only happen if the operators know their equipment intimately and understand the effects and consequences of each failure that is possible to occur in their equipment. The key to Autonomous Maintenance is if the operators themselves wrote the checklist, then it will be followed by them religiously. The role of maintenance is to feed them with the knowledge they need to know regarding their equipment. That is why implementing Autonomous Maintenance takes time as we need to change the mindset of operators before performing any activity on the equipment they operate. Nurturing operators with knowledge is not a one-time event, but this is a continuous process. For as long as maintenance prepares their checklist, then the chances of operators performing them will be 50-50.

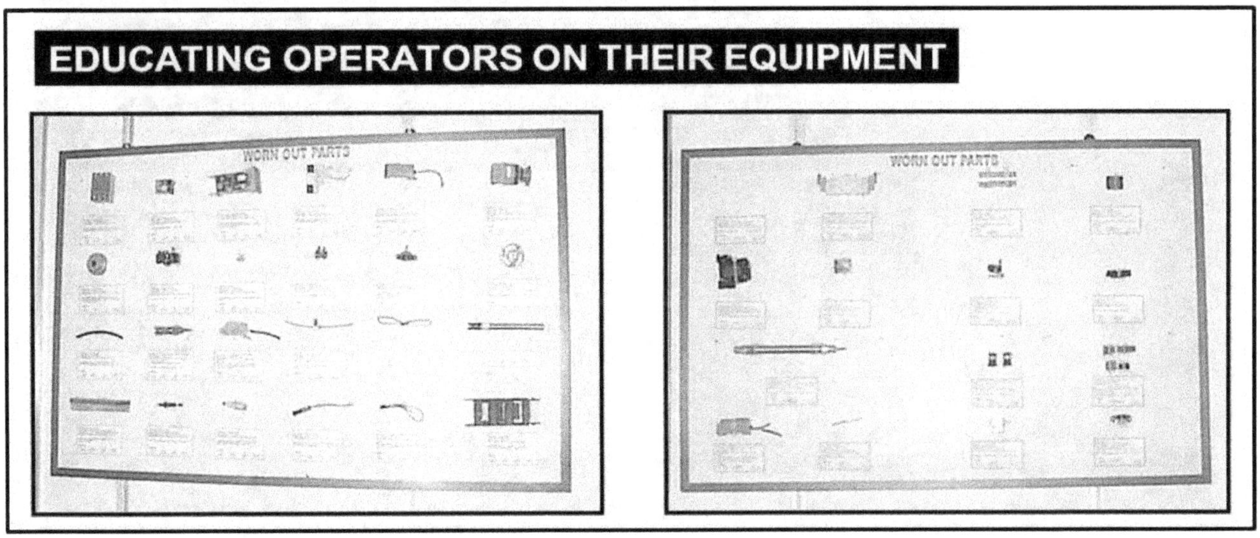

Figure 8.3: Educating Operators About Their Equipment

During my employment days, I asked the maintenance leaders that if something failed on the equipment, do not throw them and use them as a teaching aid to operators. These worn-out items are being mounted on a 4 x 8 feet whiteboard which is used as a teaching aid to operators. The maintenance spends 10 to 15 minutes discussing one part per shift, its name, function, and the impact if this part fails on the equipment. The next day another part or item will be taught to operators. This is just one of the many initiatives we generated to educate operators on their equipment. The thing is, once the operator understood their equipment intimately, then they will generate the checks on their equipment.

8.3: Standards to be Developed by Operators

Many industries try to pass some maintenance activities to operators instantly and operators just ignored them. In the end, these industries complain that having an operator-maintenance partnership is not possible at all. Believe me when I tell you that this is not how it is done. One of the key success factors in implementing Autonomous Maintenance is the support from the Planned Maintenance team. As I have written in my third book on Reliability – A Shared Responsibility for Operators and Maintenance, if Planned Maintenance is weak, then implementing Autonomous Maintenance will be a complete disaster. For those implementing TPM, the strongest pillar should be Planned Maintenance as their role is two folds, first, perform the Phases of Planned Maintenance phases, and second, understand their role in implementing Autonomous Maintenance. They will be responsible for mentoring operators about their equipment and assets.

Figure 8.4: Autonomous Maintenance 7 Steps

Without the mentorship of Planned Maintenance, Autonomous Maintenance will stall. It is as simple as that. Three standards will be prepared by operators, first, we have the cleaning standards which started from Step 1 Initial Cleaning activities. This standard is aimed at understanding which parts of the equipment should be maintained and cleaned at all times. The operators will challenge themselves by performing Kaizen activities when the equipment gets dirty once again after a short period by addressing sources of contamination. When the operators reached Step 3, they will be adding another standard which is about lubrication. But before operators can perform these tasks, they need to be trained on lubrication. What happens when the equipment lacks or applies excessive lubrication? What contamination can do to the lubricating oil or grease and how to select the correct lubricant for their equipment. Once the Autonomous Maintenance reached Step 4, one more standard will be established which is the Inspection Standards. Visual Control will be added to make the inspection more precise and easy. The challenge for the Autonomous Maintenance team is to generate these 3 standards in which the time of performing these standards is just sufficient as they also need to produce the needed and required production.

While Step 0 is the Preparatory Phase where the team prepares and plans their activities for Step 1 which is the initial cleaning activities. Steps 1 to 3 will include the activities to prevent deterioration. This is a parallel activity for Planned Maintenance restoration activities. AM teams will place priority on abolishing environments that caused accelerated deterioration and establishing basic equipment conditions on the machine. Steps 4 to 5 will be the activities to measure deterioration. AM Team leaders teach inspection procedures to their team members and perform a general inspection which expands their activities from individual equipment to the whole process itself. The last 2 steps 6 to 7 activities are aimed to restore deterioration. These steps are designed to upgrade AM and improvement activities by standardizing systems as well as empowering operators by changing the culture and upgrading the skills of each operator. In the last step of Autonomous Maintenance, operators achieve Full-Self Management which means that the role of the operator is no longer routine. The three standards that Autonomous Maintenance will be developing will include;

Cleaning Standards: During the Step 1 Initial Cleaning process, the operators exerted their best efforts to remove all the dirt, grime, and excess grease and keep their equipment clean. After the initial cleaning has been completed, the machine should be sustained so that it will not revert back to its old dirty condition. The Autonomous Maintenance team will identify all parts of the equipment that needs to be maintained clean always. The cleaning will either be performed 5 to 10 minutes before their shift ends. The operators will also design special cleaning tools for those hard to clean or hard to reach areas of their equipment that they experienced during the Step 1 Initial Cleaning activities. If operators will be involved in these tasks, they should have their own tools and lubrication kit as in figure 8.5.

Lubrication Standards: Once the operators are trained on lubrication and know the fundamentals and principles of lubrication and oil contamination control, they will try to develop visual controls mostly on areas on the equipment that needs to be lubricated and if the oil level is sufficient. They will also be responsible for performing greasing. If there are 2 or more types of grease that need to be applied, a color-coding will be done so as not to mix the grease as this will create incompatibility issues with the grease.

Inspection Standards: These are those standards that operators need to inspect in their equipment. Visual Control is added to ease up the inspection process. Once the operator has developed the three standards, these standards will be combined into a single universal standard that contains the cleaning, lubrication, and inspection activities. If such activities as cleaning, lubrication, and tightening activities are included in the current PM lists, the CMMS is revised and the responsible person is changed from maintenance to operator since this will be finalized in the system. The checklists operators generated at this stage are very different from the checklist provided by maintenance in a traditional industry since the operator understands now the value of doing it

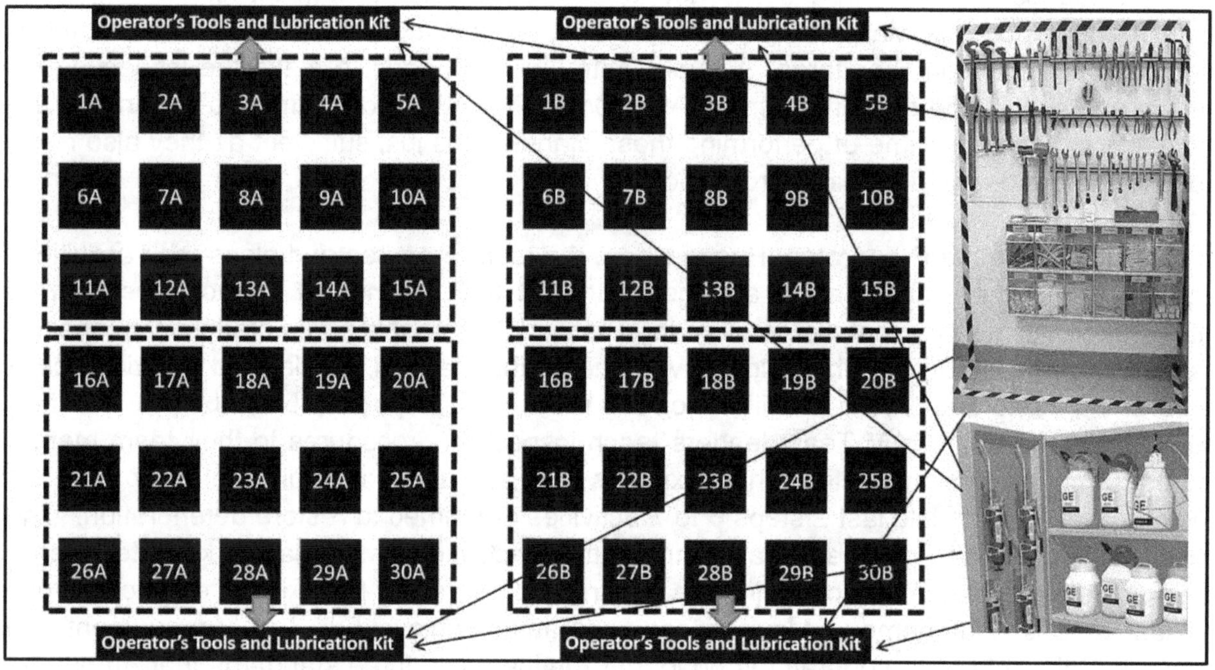

Figure 8.5: Operator's Tools and Lubricating Kit Set-Up

By thoroughly reviewing the standards for duplicated or omitted inspection items, the allocation of routine tasks between operators and maintenance is finally determined. It is important to consolidate the activities performed by both maintenance and operators on the equipment in this step to avoid duplication and redundancy on the maintenance side.

8.4: Consolidating Operators and Maintenance Activities

The general inspection process is organized so that the correct frequency and interval can be derived from the operator's checklist. Operators prepare an inspection process needed to be done by them every shift, daily, weekly, monthly, quarterly, and as so required. The activities of Step 5 are done to generate a sustaining process in the previous steps of Autonomous Maintenance. The inspection is carried out regularly and equipment failures and breakdowns are minimized at this stage since they can now be detected by operators. As the operators become full partners in this shared responsibility, maintenance can focus on carrying out their Planned Maintenance activities on their equipment. Because of the training and mentoring from Planned Maintenance gained from the previous steps of Autonomous Maintenance, operators' skills and knowledge are truly enhanced. At this step, operators are capable of detecting abnormalities and irregularities in their equipment and what to do in case they spot one.

For industries implementing Autonomous Maintenance, when the operators reach Step 5, they will have a sit down with the Planned Maintenance just like the La Cosa Nostra of Mafia and consolidate the activities performed by both maintenance and operators on the equipment on this step to avoid duplication and redundancy on the maintenance side. If such activities such as cleaning, lubrication, and tightening activities are included in the current PM lists CMMS is revised and the responsible person is changed from maintenance to operator since this will be carried out by the AM teams. Inspection finalized and generated at this stage is very different from the checklist provided by maintenance in a traditional industry since the operator understands now the value of doing it. Surprisingly, the standards set forth by operators are sometimes superior to those set by the maintenance department and passed on to operators. By thoroughly reviewing the standards for duplicated or omitted inspection items, the allocation of routing tasks between operations and maintenance is finally determined. Also in this Step operators combine inspection standards prepared for each category into a single routine inspection standard for each piece of equipment. Operators are once again finally challenged to reduce the time to perform these standards. The standards developed by the operators will help create a sustaining activity in establishing the basic equipment condition. A piece of clean equipment will also reduce the chances of minor stoppages occurring in their equipment.

Information Reference			Consequence Evaluation				H1 S1 O1	H2 S2 O2	H3 S3 O3	Proposed Tasks	Maintenance Classification (Check Classification)					Initial Interval	Can be done by
F	FF	FM	H	S	E	O	N1	N2	N3		PM	PdM	NSM	FFT	RED		
Shiftly Activities																	
1	L	2	Y	N	N	Y	Y			Check oil level every shift.		do				Every shift	Lubeman change to Operator
3	A	1	Y	N	N	Y	Y			Monitor cyclone feedpot pressure regularly and adjust pump speed accordingly if variable speed or run a		do				Every shift	Operator
1	C	1	Y	N	N	Y	Y			Monitor mill power draw.		do				Every shift	Operator
3	B	1	Y	N	N	Y	Y			Monitor the overflow of individual cyclones regularly and reduce the number of operating units or switch fe		do				Every shift	Operator
3	C	1	Y	N	N	Y	Y			Observe cyclone feedpot pressure gauge needle for any fluctuation.		do				Every shift	Operator
3	D	1	Y	N	N	Y	Y			Observe cyclone feedpot pressure gauge needle for any surging.		do				Every shift	Operator
Daily Activities																	
1	L	6	Y	N	N	Y	Y			Monitor drive motor power draw and temperature.		do				Daily	Elect. - change to Operator
1	A	2	Y	N	N	Y	N	Y		Restore ball levels regularly.	do					Daily	Services - change to Operator
Activities Every 2 Days																	
1	A	1	Y	N	N	Y	N	Y		Restore rod levels regularly.	do					2 days	Services change to Operator
Weekly Activities																	
1	D	17	Y	N	N	Y	N	Y		Adjust tertiary cone crusher discharge setting.		do				Weekly	Operator
1	L	1	Y	N	N	Y	Y			Monitor pump bearing housing temperature and vibration.		do				Weekly	PM - change to Operator
1	K	2	Y	N	N	Y	N	Y		Monitor pump running time and cyclone feedpot pressure.	do					Weekly	PM - change to Operator
Activities Every Two Weeks																	
1	A	9	Y	N	N	Y	N	Y		Adjust impeller to suction liner clearance. Standard 1/8" gap.	do					2 weeks	Maintenance
1	A	8	N			N	N	Y		Change pump FPL insert.	do					2 weeks	Maintenance
1	A	5	Y	N	N	Y	N	Y		Change pump suction liner assembly.	do					2 weeks	Maintenance
1	A	6	N				N	Y		Change pump throatbush.	do					2 weeks	Maintenance
Activities Every Three Weeks																	
1	A	4	Y	N	N	Y	N	Y		Change pump casing.	do					3 weeks	Maintenance
1	A	7	Y	N	N	Y	N	Y		Change pump volute liner.	do					3 weeks	Maintenance
Monthly Activities																	
1	A	3	Y	N	N	Y	N	Y		Change pump impeller.	do					Monthly	Maintenance
1	D	18	N				N	N	Y	Change pump impeller.	do					Monthly	Maintenance
1	L	4	Y	N	N	Y	Y			Monitor drive motor bearing temperature and vibration.		do				Monthly	PM - change to Operator
1	K	1	N				N	Y		Monitor rod and ball mill liners as per liner inspection schedule.	do					Monthly	PM - change to Operator
Quarterly Activities																	
1	A	14	Y	N	N	Y	N	Y		Change packings.	do					3 months	Maintenance
1	D	10	N				N	N	Y	Change vortex finder.	do					3 months	Maintenance
Activities Every Four Months																	
1	D	19	N				N	N	Y	Change cyclone inlet head liner.	do					4 months	Maintenance
1	A	13	Y	N	N	Y	N	Y		Change shaft sleeve.	do					4 months	Maintenance
Semi-Annual Activities																	
1	A	11	Y	N	N	Y	N	Y		Change worn-out sump screen.	do					6 months	PM - change to Operator
Yearly Activities																	
1	D	3	N				N	N	Y	Change upper and lower cone liners.	do					Yearly	Maintenance

LEGEND: PM - Preventive Maintenance NSM - No Scheduled Maintenance FFT - Failure Finding Tasks RED - Redesign

Figure 8.6: Consolidating Operators and Maintenance Activities

8.5: Tips on How to Involve your Operators in Maintenance

First, when the equipment fails, ask the operator to stay while maintenance is repairing the equipment. When you have experienced a failure before, tell the operator stories about your experiences with that failure. Create bonding with the operator. I recalled that in our previous Planned Maintenance Journey, I asked the members, what support have we given to our operators? One maintenance told me that when the equipment fails, I ask the operator to stay and I told her my experiences with the part that failed. I said that's good. He told me that during lunch break, they eat together and tell her more stories and things she should know about their equipment. As I was about to speak, he said that he was not finished yet, since, after ending their shift, I invited her to have a couple of drinks and as we drink, I told her stories about my experiences with the equipment which she greatly appreciates. Today they are now married and have three loving children. That maintenance guy took my advice literally.

Second, always keep the part that failed. Make it a habit to teach the operator the name of the part that failed and what is the function of this part in the equipment. Discuss why this part is important. By doing this, the operator can learn a great deal about the part that failed unless the operators are engineers themselves. Most of the time, operators are non-technical people. They find it difficult to communicate with maintenance regarding what went wrong with their equipment because they do not even know the name of the part that failed. They can only describe the failure to a limited extent. Teaching operators just one part a day and spending 10 to 15 minutes each day per part, can facilitate better communication between the two. During my time, we mount these failed parts on a 4 x 8 whiteboard and discuss one part or spare each day and teach operators as in figure 8.3. It created better communication between them.

Third, most people complain to me that maintenance prepares some checklist for operators to check on the equipment, yet operators do not actually perform the checking themselves. Remember that operators will only perform these checks religiously if they are the ones who made the checklists themselves. That is the main difference in doing Autonomous Maintenance; operators will be the ones who will develop their own checklists. Maintenance must understand that operators are capable of developing their own checklists if they will be trained by maintenance both formally and informally. The role of maintenance is to coach and guide operators in allowing them to make decisions on what they think are the critical things to inspect in their equipment. If this will be done, then operators will perform inspections themselves since they were the ones who made their own checklists and they understand the consequences of the failure if they ignore these checks.

Whether operators learn from maintenance or maintenance itself learning from operators, the involvement of operators plays a vital role in any reliability improvement strategy. The equipment is always a shared responsibility for both operators and maintenance. They are partners for life. They should be. Maintenance can only start doing their true tasks if operators themselves will be involved in maintenance. Operators will play a major role in establishing this very basic equipment condition on their equipment and asset. The truth is simple, maintenance will always remain in a fire-fighting and reactive mode if operators will not accept their responsibility for maintenance. Doing maintenance is not only for the maintenance function. It is both an operators and maintenance function.

When we first hired our JIPM consultant, the first thing he checked was the membership of Autonomous Maintenance. Our lineup of Autonomous Maintenance includes a supervisor, a couple of operators, and a cross-selection of maintenance. The consultant changed the line-up to 100% operators and remove all non-operators from the team. He told us that if this will be your set-up then the operators will just lean-to maintenance and will not become empowered. This was where I truly realized what Autonomous Maintenance is all about. It is all about empowerment.

8.6: Detailed 7 Steps Implementation of Autonomous Maintenance

The key to a successful Autonomous Maintenance implementation lies in Planned Maintenance. This means that if PM is weak, then Autonomous Maintenance will collapse or stall. It is as simple as that. Why? Because it will be the maintenance who will mentor the operators in their Planned Maintenance journey. This means that Planned Maintenance activities and steps should be ahead before implementing Autonomous Maintenance. As mentioned in the previous chapter, one of the key Preparatory Phase for Planned Maintenance is conducting a machine ranking so we can determine how many Rank A (worse), Rank B (getting worse), and Rank C (good) machines we have so far. The pilot machine for Planned and Autonomous Maintenance should belong to the Rank A category and if possible they should be the same machine so that Planned Maintenance can monitor the activities of Autonomous Maintenance. Below are the detailed seven steps of Autonomous Maintenance.

Figure 8.7: Planned and Autonomous Maintenance Parallel Activities

Step 0: Preparatory Phase: This Preparatory Step is the beginning of Autonomous Maintenance activities. The membership of Autonomous Maintenance should be composed of 100% operators who can either belong to different shifts or a common shift. The pilot machine to be selected should be identical to the pilot machine of Planned Maintenance. Setting up the Autonomous Maintenance operator team can be done horizontally or vertically. This means that if the team is a horizontal set-up, then the members will be from the same shift. For plants that operate 24 hours a day, a vertical set-up means that the members will be from different shifts. This must be discussed during the beginning since if the members will be in a vertical setup, then the 2 members need to be paid overtime so they can join the meeting. This will be more difficult but more effective. There will be a weekly or twice-a-week meeting among the members of the AM team to complete the preparatory phase and succeeding steps of Autonomous Maintenance. The main highlight of these activities in Step 0 is to identify all the cleaning materials needed for Step 1 Initial Cleaning and to map out the positions of the members on the areas on the equipment they will clean and what cleaning materials they need to have with them. Safety is also being discussed before the initial cleaning process such as ensuring that the machine will not run accidentally when the process of initial cleaning is taking place. Remember that if an operator gets injured during the initial cleaning process, then this will end all initiatives for Autonomous Maintenance. A date will be set for the initial cleaning of the pilot machine selected. One more thing to consider is not to use an air gun to blow away the dust as they will just suspend and return back to the equipment. This might also cause eye injuries to the members of the team.

Figure 8.8: Step 1 Initial Cleaning Activities on AM

Step 1: Initial Cleaning: Pilot equipment selected by the AM team will be cleaned as thoroughly as possible without an initial time limit depending on the industry, equipment selected, and boundaries for both manufacturing and non-manufacturing plant. The purpose of conducting this initial cleaning is to expose abnormalities that will be found during step 1 activities and tag them. The team will also expose all hidden problems and hard to access and hard to clean areas they experience on the equipment. Cleaning is the starting point of Autonomous Maintenance. Cleaning is inspection. The physical act of touching the equipment reveals problems and abnormalities Equipment is restored by establishing Basic Equipment Condition which means eliminating causes of accelerated deterioration. Cleaning consists of removing all dust, dirt, grime, oil, grease, and other contaminants that adhere to the equipment and parts. The purpose of this step is to expose hidden defects and abnormalities in the machine. Cleaning also reveals big surprises such as cracked frames or shafts that had been covered with dirt and grime throughout the year. Abnormalities uncovered are being tagged by the operators which can be done simultaneously with Planned Maintenance. All those easy-to-address abnormalities will be corrected. Clean equipment as thoroughly as possible. Clean parts with layers of grime and scale built up over many years. Open previously ignored covers, guards to expose the dirt. Make sure the machine is shut down and remove the fuse to prevent accidents. Do not give up when parts get dirty again after the initial cleaning. The focus of Step 1 is to eliminate dirt, dust, and grime, expose all abnormalities and start correcting minor flaws and establish Basic Equipment Conditions. At the end of the initial cleaning activities, the team needs to identify how many abnormalities they tagged and how many have been corrected together with their equivalent percentage. In Step 1 there is no time limit on conducting Initial Cleaning and depends on the number of people cleaning and how big the equipment or machine is.

WHAT AUTONOMOUS MAINTENANCE TEAMS SHOULD TRACK

WHAT TO TRACK	STEP 1 :	STEP 2 :	TOTAL	PERCENTAGE
1. Abnormalities Tagged	145	+ 25	170	Percent = 90 %
2. Team Suggestions	25	+ 20	45	Percent = 67.0 %
3. One Point Lesson	20	+ 25	45	Percent = 100.0 %
4. Questions Generated	20	+ 25	45	Percent = 100.0 %
5. Difficult to Clean Areas	5	+ 6	11	Percent = 100.0 %
6. Contamination Sources	4	+ 0	4	Percent = 75.0 %
7. Kaizen Activities	0	+ 20	20	Percent = 75.0 %

Figure 8.9: What Autonomous Maintenance will be Tracking

Step 2: Address and Eliminate Sources of Contamination: After completing Step 1, the AM team realizes that dirt tends to accumulate after some time on their equipment. The team discusses problems with areas and parts of the equipment which are difficult to access and hard to clean areas. The AM team brainstormed on ways to eliminate or reduce sources of contamination and if specialized cleaning tools are required for hard-to-reach or to clean areas. The Autonomous Maintenance team shares their experiences in performing Step 1 Initial Cleaning with the group and initiates simple Kaizen activities to address their problems with cleaning and contamination. They likewise continue to address the tags that had been left from Step 1. The team begins with the goal of eliminating or reducing sources of contamination that makes the equipment or critical parts of the equipment dirty once again. They also become aware of hard-to-reach or hard-to-clean areas and feel obliged to think about improving their accessibility. Reduce cleaning time by preparing cleaning charts and reducing contamination sources, making hard-to-clean places more accessible, and devising more efficient cleaning tools. The operators begin to think of ways to prevent sources of leaks, oil spills, dirt, and other contamination sources. Operators must understand what contamination can do to the equipment and how to prevent or eliminate its source. The goal of Step 2 is to reduce the time it takes for cleaning, inspection, checking, and lubricate the machine. The AM team thinks of simple yet innovative ways to clean hard-to-reach areas that are inaccessible to them during the initial cleaning period. Teams continue to correct abnormalities found, raise questions to further enhance knowledge, develop One Point Lesson based on most of the questions and suggestions, and start to perform their Kaizen activities. In this step, the operators take note of how long it takes the part to become contaminated and where the contamination is coming from. The AM team will think of ways how to reduce the cleaning time to a minimum as they can go on with the activities of Autonomous Maintenance. The team develops specialized cleaning tools and kits for hard-to-access and reach areas on their machine. At this step, Autonomous Maintenance will also start preparing visual controls on their equipment.

Step 3: Establish Initial Cleaning and Lubrication Standards: The goal of this step is to lock in place the gains of Steps 1 and 2 to ensure maintenance of basic equipment conditions and to keep the equipment in peak operating condition. During Step 1 of Autonomous Maintenance activities, the team put tremendous effort into cleaning, correcting minor flaws, and establishing Basic Equipment Conditions. In Step 2 they reduce the time required for these tasks by addressing sources of contamination making cleaning, and lubrication easier to perform. During Step 3, AM team prepares tentative standards and establishes checkpoints for cleaning and lubrication. Operators with the help of maintenance try to understand why these tasks are important and what harm they can do to their equipment if it is neglected. AM will seek guidance from Planned Maintenance on which activities will be performed when the machine is running and those which will be performed when the machine is stopped. At this step, the operators prepare tentative cleaning and lubrication standards and note the time or duration to conduct the activity on the equipment. These standards specify what needs to be done. What parts of the equipment need to be clean and lubricated? Many industries have these standards but very few follow them thoroughly because the people who set the standards are seldom the ones who apply them. These standards and inspection checklists will be followed if the people who do them truly understand what these standards are all about and most importantly if the operators are the ones who set the standards. Before operators perform these standards maintenance must fully train operators on the importance of doing them. Remember that it will be the operators who will witness the failure first and not the maintenance people. Let operators identify what needs to be inspected on the equipment. The role of the maintenance is merely to guide them.

Step 4: Develop General Equipment Inspection Standard and Training: In Steps 1 to 3, the operator had experience performing initial cleaning, detecting abnormalities, identifying hard clean areas, generating lubrication, and cleaning standards as well as basic knowledge about safety and machine function. In Step 4, maintenance conducts training for the operators to understand the fundamental theory as well as its principles. Training needs identified in the previous steps will be conducted in this step so that the operator can develop and perform the standards they generate. Next, decide what the operators should learn to be able to check these items, pass-fail criteria, inspection procedures, and actions to take when abnormalities and deviations occur. Timing of Step 4 is very critical which means that when the AM teams reach this stage, training modules and cut-out models to be used for simulation should already be available which means that when AM is in Step 1 to 3, PM teams or SMEs (Subject Matter Experts) should already start preparing the training modules for operators in advance. In Step 3, a Tentative Standard is developed for cleaning and lubrication. In Step 4, this is reviewed and analyzed to further shorten the time that is performed on these activities and a final standard is added which is the inspection standards. Since the operators have experience with the actual cleaning, and performing lubrication on their equipment and had been taught the basic principles of lubrication, bolting, and other relevant topics they should know. AM teams with the guide of the PM group finalize the best frequency and interval to perform these activities on their equipment. The purpose of this step is for the team to understand its structures, functions as well as principles to learn their optimal conditions to achieve maximum efficiency of their equipment. In this step, depending on their equipment, fundamentals of machine elements, hydraulics, pneumatics, lubrication, bolts, nuts, fasteners, electrical, power transmission, and rotating machine elements are learned by operators so that operators can find confidence in themselves in performing the standards generated on lubrication, cleaning, and inspection. Unlike the first three steps, Step 4 will be the longest step in Autonomous Maintenance to complete so what is important is not to rush this activity since we are building the skills of the operators. If done correctly, at the end of this step, the company should realize around a 60 to 80% reduction in the breakdown and other forms of downtime such as minor stoppages. The final standard which is the inspection standards will complete the standard maintenance activities for operators to be performed on their machines

Step 5: Autonomous Maintenance Standards and Consolidation of PM Tasks: In the cleaning and lubrication standards set in Step 3 and the tentative inspection standards prepared for each given category in Step 4 are now combined in Step 5 where they are combined into a unified Autonomous Maintenance Standard which includes the three standards cleaning, inspection, and lubrication. They are now finalized to accomplish a higher check efficiency and eliminate the possibility of human errors. If problems occur, operators work with maintenance to develop inspection points that can prevent the problem from recurring. At this step, operators continue to think of ways to shorten the time to perform cleaning, lubrication, and inspection of their assets. The general inspection process is organized so that the correct frequency and interval can be derived. Operators prepare an inspection process needed to be done by them every shift, daily, weekly, monthly, quarterly, and as so required. The activities of Step 5 are done to generate a sustaining process in the previous steps of AM. The inspection is carried out regularly and equipment failures and breakdowns are minimized at this stage since they can now be detected by operators. As the operators become full partners in this shared responsibility, maintenance focus on carrying out their Planned Maintenance activities on their equipment. It is important to consolidate the activities performed by both maintenance and operators on the equipment in this step to avoid duplication and redundancy on the maintenance side. If such activities such as cleaning, lubrication, and tightening activities are included in the current PM lists CMMS is revised and the responsible person is changed from maintenance to operator since this will be carried out by the AM teams. Inspection finalized and generated at this stage is very different from the checklist provided by maintenance in a traditional industry since the operator understands now the value of doing them. Surprisingly, the standards set forth by operators are sometimes superior to those set by the maintenance department and passed on to operators. By thoroughly reviewing the standards for duplicated or omitted inspection items, the allocation of routine tasks between operations and maintenance is finally determined. Also in this Step operators combine inspection standards prepared for each category into a single routine inspecting standard for each piece of equipment. Operators are once again finally challenged to reduce the time to perform these standards.

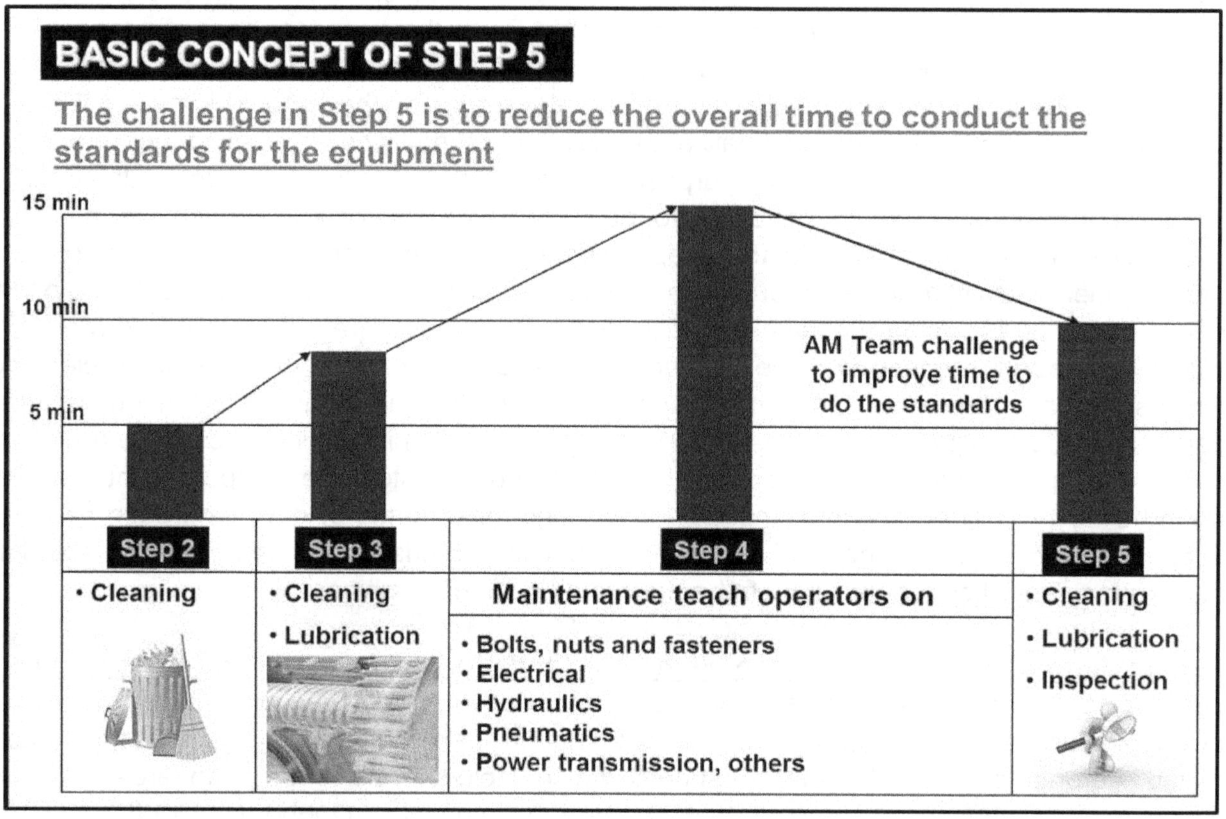

Figure 8.10: Finalizing the Standards for Autonomous Maintenance

Step 6: Systematic AM and Manage the Workplace: In Step 6, the AM activities expand beyond the equipment to other aspects of the work environment. In layman's terms, let us say that the first 5 steps refer to cleaning and inspecting your house. In this step, the owner likewise would like to maintain his backyard and garage and clean up unnecessary things that are not needed. Using simple principles of housekeeping, visual control, and 5s, teams begin eliminating clutter and unnecessary items within their work areas and then organize the things that should be stored in place. One of the most important activities in Step 6, is to clearly specify the details of process quality at the exit point of the equipment to prevent the outflow of defective products. The 5-Criteria for Quality Assurance will be used in which the quality conditions should be clearly defined and understood, easy to set, must resist any variation, easy to detect, and easy to restore. All activities geared toward completing Steps 1 to 5 focused on reducing equipment breakdowns and minor stoppages. In Step 6, AM activities are focused on aiming for Zero Waste and Zero Defects. Hence, operators, maintenance, engineers, and quality people must have a clear identification of process quality and strict adherence to quality conditions. This can be achieved by not only improving equipment but as well by work methods, operators' human senses, application of visual controls, and Poka-yoke or Mistake Proofing devices are taught to operators to eliminate defects. Operators should not only be keen on determining abnormalities in their equipment but also should learn to distinguish materials that can contribute to waste and quality defects. Step 6 aims to foster the development of operators knowledgeable in matters of quality, equipment, and procedures for fully implementing autonomous supervision without the need for full detailed instruction by supervisors and managers. Although most industries want equipment with minimum breakdowns more industries are interested in matters of quality since achieving Zero Defects in quality is much more difficult to achieve than Zero Unplanned Breakdowns.

CAUSES THAT CAN AFFECT QUALITY

Causes	Major Activities
1) Raw Materials used	• Detailed classification of raw materials • Incoming quality for raw materials • Adhere to 5 Criteria of Quality Assurance • Set raw material control standards 
2) Measuring Apparatus Control	• Detailed and classify measuring apparatus • Calibration of measuring apparatus • Totally inspect measuring apparatus • Assess measuring apparatus in terms of 5 Criteria of Quality Assurance 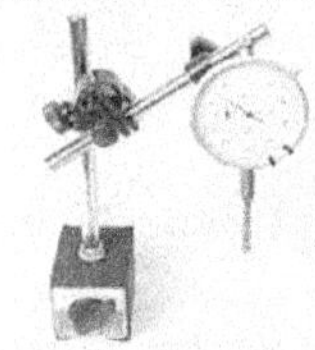
3) Machining Condition	• Detailed and classify machine conditions • Totally inspect machine conditions • Assess machine conditions in terms of 5 Criteria of Quality Assurance 
4) Jig and Die Control	• Set jig and die control standards • Method of storage for jigs and die 

Figure 8.11: AM Step 6 will be Focused on Quality

Step 7: Practice Full Self-Management and Empowered Workforce: Step 7 is the last step of Autonomous Maintenance and takes around 5 years to achieve. Not only have we improved the equipment in terms of reducing breakdowns and defects but having a highly skilled and self-directed operator is now achieved at this stage. At this last step, operators become independent, skilled, confident, and well-motivated and can perform simple repairs and equipment restoration by training them in repair skills. Operators' improvement skills are raised as they join Focused Improvement initiatives and projects in eliminating The 6 Big Losses of the equipment. In Step 7 knowledgeable operators conduct autonomous supervision and follow the standards set by themselves on the orderly shop floor, where any deviation from the normal or optimal operating condition is detected at a glance by anyone. It is a very long journey to reach this stage, breakdowns and quality defects will tend to revert back as frontline people, engineers, maintenance including plant managers are being replaced every year. What is important is to sustain the enthusiasm of the team on every step that the team will undertake.

Empowered teams provide a vehicle for employees to take on the responsibility typically reserved for managers or supervisors in the workplace. Organizations that acquire a capable workforce will offer a culture that matches a new workforce. Teams will offer greater participation, and feelings of accomplishment. Industries with self-directed teams will retain the best people, while most will not. But perhaps the biggest reason for the movement towards an empowered workforce is the fact that these teams really work. In the long run, they will reduce costs, boost morale and improve productivity.

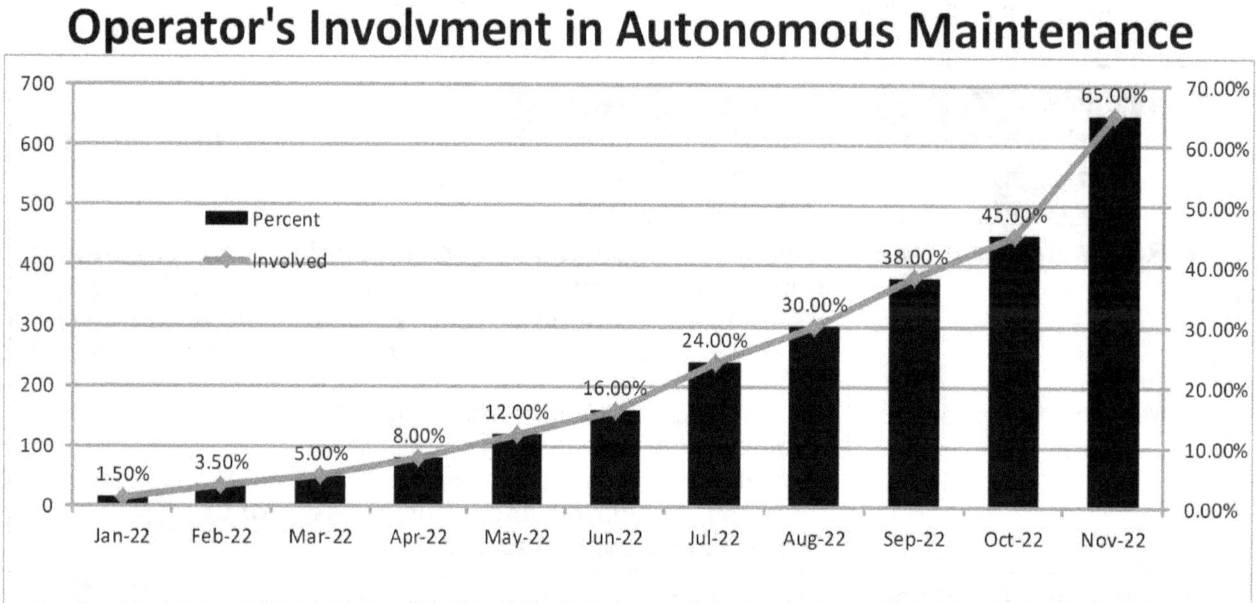

Date	Jan-22	Feb-22	Mar-22	Apr-22	May-22	Jun-22	Jul-22	Aug-22	Sep-22	Oct-22	Nov-22
Involved	15	35	50	80	120	160	240	300	380	450	650
Percent	1.50%	3.50%	5.00%	8.00%	12.00%	16.00%	24.00%	30.00%	38.00%	45.00%	65.00%

Figure 8.12: Operators' Involvement Monitoring

One of the main roles of the TPM Office is to monitor the number of people involved in each of the pillars of TPM. Figure 8.12, shows an example regarding the number of operators involving themselves in the TPM pillar of Autonomous Maintenance. Assuming that there are 1000 operators in the plant, we monitor the involvement monthly. Just imagine the power of this operator's workforce carrying out routine and simple maintenance tasks on their equipment such as cleaning, lubrication, and inspection which serves as a big time-saver for the maintenance function. Yet again, this will not happen overnight and what is important is continuous coaching and mentorship from the maintenance side of the house. This is simply the power of the Japanese workforce in industries and explains the difference in how they operated from the west. The key lies in their operators and that is the difference between having an empowered workforce.

Chapter 9

The Role of Predictive Maintenance on PM

Predictive Maintenance is never an isolated group. It is not a replacement for Preventive Maintenance. Both Preventive and Predictive Maintenance will have their roles in the overall maintenance function. Communication, and collaboration between the two should always remain open and is the key to sustaining the equipment.

9.1: Predictive Maintenance Explained

Before we discuss the role of Predictive Maintenance on PM, let us define what Predictive Maintenance is in the simplest way possible. A human being is gifted with five senses. These are the sense of smell, touch, taste, hearing, and sight. Except for the sense of taste, the rest of these senses can be used to detect early problems with the equipment. Similarly, in maintenance, many available instruments can check the actual condition of the equipment to detect any signs of early problems with precision accuracy. The only difference between our human senses and these instruments used in Predictive Maintenance is that these instruments are just a higher form of our human senses. Our human senses are simply limited. Predictive Maintenance instruments can sense problems in the equipment far more than our human senses can detect them.

According to Merriam-Webster Collegiate Dictionary, Predictive simply means to declare or indicate something in advance, especially foretell based on observation, experience, or scientific reason. Predictive comes from the Latin word "pre," meaning before, and "diction," which is also from the Latin word "dicare," meaning to proclaim. Although we are not using a set of cards, crystal ball, or any psychic person to conduct some prediction to predict when equipment will fail, we are using some proven non-destructive diagnostic instruments with precision accuracy to provide us signs that the equipment is in the process of failing. When it comes to our equipment and assets, these Predictive Maintenance instruments are simply an extension of our human senses.

Proven Predictive Maintenance techniques can capture failures and decrease the costs of unexpected breakdowns. Predictive Maintenance does not extend the life of the equipment. It will just tell us if there is a problem with the equipment or not. The decision and action of the maintenance team will be the key factor that will extend the life of the part or asset being monitored through Predictive Maintenance. For example, doing an infrared scan does not extend life. The scan may show a hot connection, cleaning, and retightening of the hot electrical connection done will extend the asset's life. Predictive Maintenance allows maintenance to make decisions by being able to peek inside the equipment and components and replace them just before they actually fail without dismantling them. We are not basing our maintenance decision on the scheduled time interval but rather on the actual condition of the equipment. It will be the decision of the maintenance and reliability team that will extend the life of the part or item being monitored. These non-destructive instruments will just tell us if there is something wrong with them or not. While doing Preventive Maintenance will be based on the schedule of the equipment, Predictive Maintenance will be based on the actual condition of the equipment. This also allows the maintenance to save cost because unlike Preventive Maintenance only parts confirmed with problems will be replaced since we are basing maintenance on its condition rather than going on time-based or running hours.

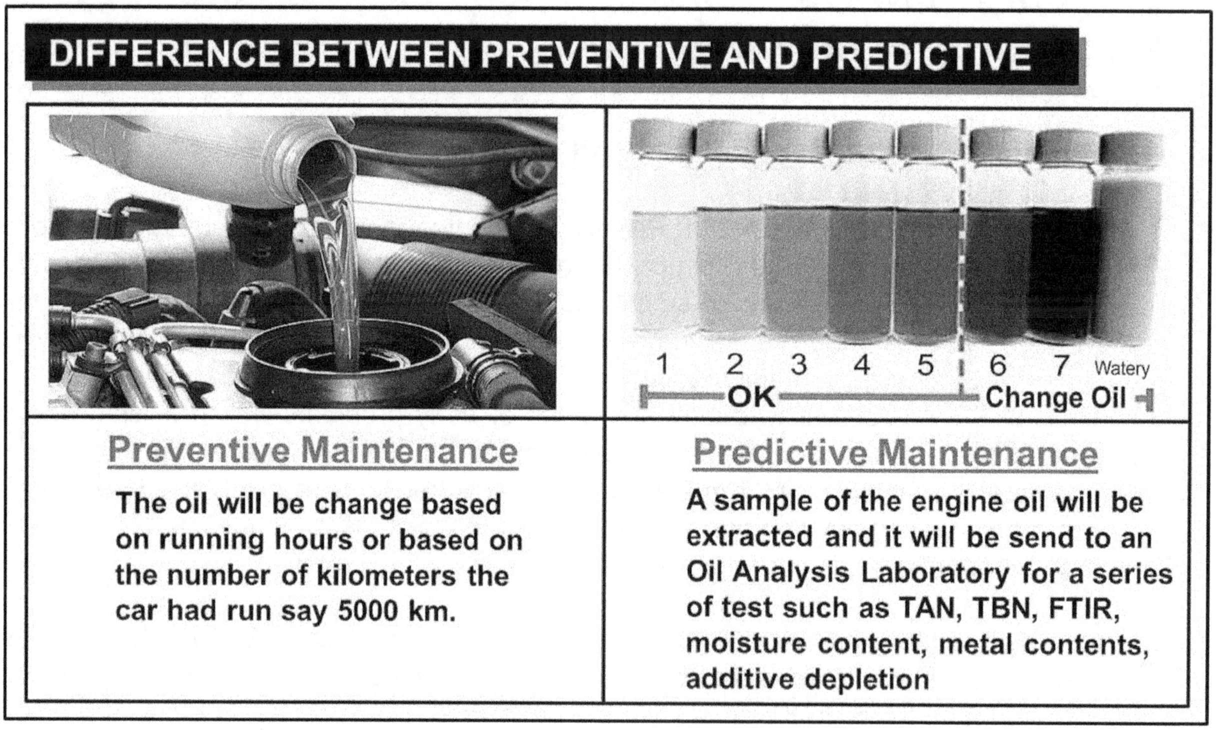

Figure 9.1: Difference Between Preventive and Predictive Maintenance

Predictive Maintenance is never a replacement for Preventive Maintenance. PM will always have its place in the overall maintenance strategy of World Class Maintenance Management. What we are saying here is that there are instances where checking the actual condition of the equipment is a much better strategy than overhauling the equipment itself. Besides, there will be times when overhauling the equipment will induce infant mortality or start-up failures. A good Predictive Maintenance strategy is not just about monitoring failures using these instruments. Data and information gained from the instrument can aid us in determining the root cause of the problem.

9.2: Potential and Functional Failures

There are failure modes that provide us signs or symptoms that they are in the process of failing. These failure modes are the best candidates for Predictive Maintenance. An ailing bearing will provide signs such as an increase in both temperature and noise or an increase in vibration. Predictive Maintenance instruments such as Vibration Monitoring and Infrared Thermography can be used to capture these symptoms. Predictive Maintenance tasks include checking the equipment for any signs of potential failures, so that sound decisions and actions can be taken to prevent the functional failure or to avoid the consequences of a functional failure from occurring. A potential failure is defined as an identifiable physical condition that indicates that a functional failure is about to occur or is in the process of occurring. Functional failure is defined as the inability of an item to meet a specific performance standard. It means that a failure has occurred and the equipment already failed. The P-F curve or others called this the P-F interval is the interval between the emanation of potential failure and its decomposition to its final stage, which is a functional failure. The P-F interval is actually a warning period or lead-time to failure, better known as the failure development period. What is important is that the P-F interval is determined so that a maintenance task can be performed before the functional failure takes place. This means that if the P-F interval is one week, a maintenance activity such as replacement or overhauling should be performed on the equipment before reaching the functional failure stage. If we perform maintenance after the failure, then again, this makes our maintenance tasks reactive instead of proactive. A Predictive Maintenance task aids us in determining the potential failure or symptoms that the equipment is in the process or verge of failing and allows us to make a decision before the failure happens.

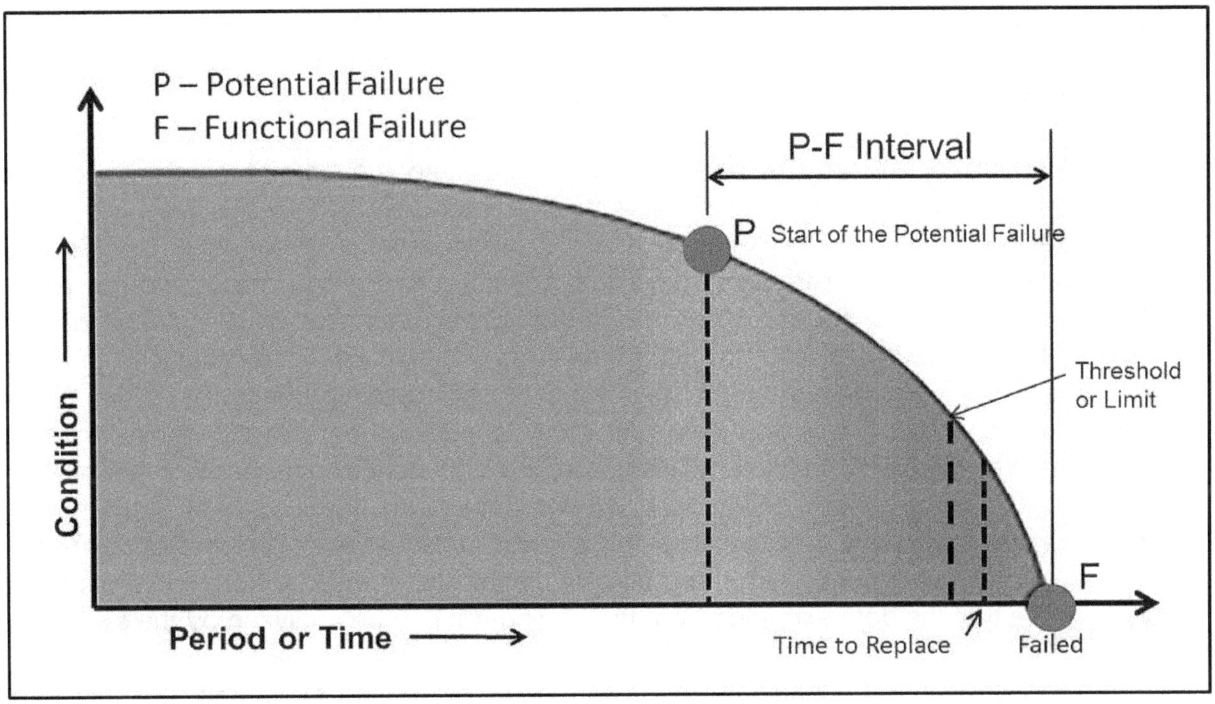

Figure 9.2: Potential and Functional Failure Interval

Although figure 9.2 is called the P-F Curve or Interval, sometimes I wonder why they called this the P-F Curve. The truth is that these Predictive Maintenance instruments will never know when an item or part will exactly fail. What these instruments will provide is a threshold or limit so that the user can recommend maintenance to perform a corrective action. I have been thinking about this P-F Curve or P-F Interval. And been asking myself why did they name it the P-F Interval or P-F Curve? By definition, the P-F Interval is the interval between the emergence of a potential failure until decay into its functional failure. Its recommendation is to use CBM tasks. But does the instrument really knows when the item, spare, or part will exactly fail? What these PdM instruments will provide is the threshold or limit on when to conduct corrective maintenance so that it will not reach the failure point. Hence, if a limit is provided by these instruments, how long will it take before it fails, will it be 2 minutes, 2 hours, 2 days, or 2 weeks. I guess nobody knows unless we run it to fail. And even if we run it to fail, these parts will not fail at the same time since the failure is random. So my thought is that the P-F Interval is really unknown and what is only known is the time these instruments will recommend us to perform corrective maintenance. I believe that this curve should be called the P-CM and not the P-F Curve or Interval which is just the time to perform Corrective Maintenance (CM) on our assets.

There are specialized diagnostic instruments that can aid us in detecting the following potential failures such as an increase in heat or temperature, vibration, increase in noise, change in pressure, increase in lubricant contamination, change in flow rate, leak detection, wall thickness decrement, increase in the rate of corrosion, crack detection, fluid flow, motor resistance and more. In addition, there are failure modes that will give us some sort of warning that they are about to occur. For example, the P-F curve indicates how a failure starts, deteriorates to a point where it can be detected, and if the failure is undetected, it will deteriorate at an accelerating rate where the end result will be a functional failure.

9.3: Difference between CBM and Predictive Maintenance

Although they are almost the same, both will base their maintenance on the actual condition of the equipment. Unlike Preventive Maintenance, where the tasks are based on schedule, for Predictive Maintenance and Condition-Based Maintenance, the decision to perform maintenance is based on the actual condition of the equipment and not based on the schedule. There are two types of Predictive Maintenance, which can be done both offline and inline (others refer to this as online). For Offline Predictive Maintenance, an instrument will be required. The user will have to place probes and sensors on the equipment to monitor data and then plot them to see if they are still within limits. With Maintenance 4.0, there are now wireless sensors that can be placed permanently on the equipment. These sensors will directly provide data to a CMMS or EAM without manually going to the equipment. In this case, an analyst is required to interpret and monitor the readings. But still, we are just on the birth of digitalization and developments have already been made that the data extracted from the equipment can be analyzed by the system itself.

On the other hand, Condition-Based Maintenance is also the same, but this is much wider in scope since this is about monitoring the condition of the equipment either with the use of Predictive Maintenance instruments or by using our human senses. For example, checking a drop of pressure on a pressure gauge will indicate that the strainer or filter is starting to clog, or inspecting the equipment with the human senses to check the vibration, temperature, or noise is a form of Condition-Based Maintenance. Besides, if we consider the instruments, monitoring, analysis, system, and software, we speak about Condition-Based Maintenance. Predictive Maintenance tends to be equipment-oriented while, on the other hand, CBM is system's oriented. Therefore, CBM deals with the entire system as a whole.

9.4: Vibration Monitoring

All equipment emits some form of vibration when loaded. Vibration can be defined by these common terms, such as swinging back and forth, oscillating, unbalanced, and shaking. The vibration occurs when a machine or machine component moves from its neutral or normal position of rest to a lower and upper extreme travel limit. This happens because the force is applied to the rotating equipment and its components. Vibration can also be cyclic or can be a pulsating motion of a machine or machine component from its point of rest. It is also a motion that repeats itself after some time.

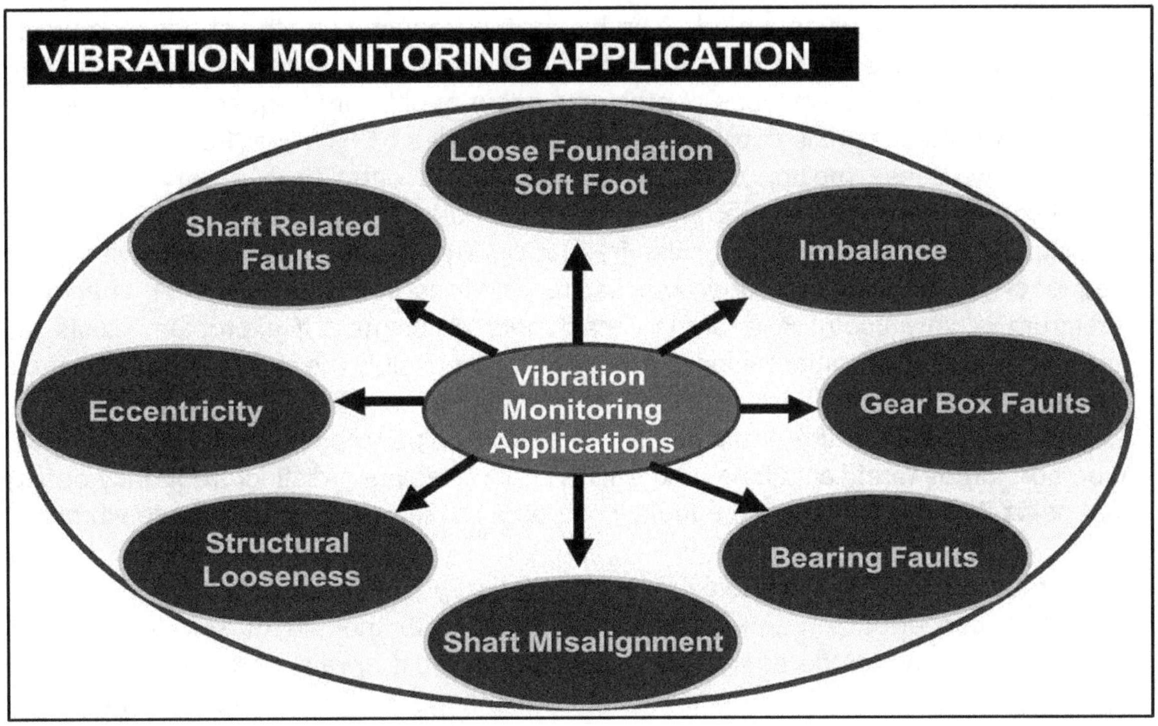

Figure 9.3: Vibration Monitoring Applications

All machines vibrate when loaded. Excessive vibration provides a destructive force that leads to premature failure of mechanical and rotating parts of the equipment. The vibration will accelerate the weakest mechanical part in the equipment since vibration will not only be confined to just one part of the equipment. It will transmit this destructive force to other parts of the equipment such as shafts, couplings, keyways, bearings, and the other parts where the end result will be total destruction or damage to the equipment. Vibration increases as the condition of the equipment worsen. When we touch the equipment, we feel the vibration from the machine. Inside the equipment are different mechanical parts that provide movement in a specific direction. Each of these mechanical parts will emit some form of vibration. Therefore, when we touch the equipment, we feel the summation of all vibrations produced from the different mechanical parts inside the equipment. Each of these mechanical parts will have its own unique set of vibration signatures or frequencies. A change in vibration can tell us an actual change in the condition of the equipment. Vibration monitoring can provide the user with useful information regarding the condition of mechanical and rotating parts inside the equipment since each part provides a unique set of vibration characteristics or frequency. Vibration analysis is one of the most common types of Predictive Maintenance techniques that analyze these plotted signals to diagnose the cause of abnormal vibration. This is the most used technique for analyzing the condition of rotating machinery. Vibration monitoring and analysis include trending vibration data, time waveform, and the FFT or Fast Fourier Transformation. Vibration values are compared to previous measurements and set alarm points to determine if the equipment is vibrating more than normal. ISO 10816 Standards provide guidance for evaluating vibration severity in machines operating in the 10 to 200Hz (600 to 12,000 RPM) frequency range.

Frequency is the number of oscillations per second. It is also the number of times an event occurs over a given period. The higher the frequency is, the shorter the wavelength. Frequency is measured in Hertz (Hz), which is the same as oscillations per second or cycles per second since different types of problems occur at different frequencies. It is important to measure vibration signals concerning their frequencies. FFT (Fast Fourier Transformation) breaks down these multiple vibration signals, which often form a very complicated and complex time waveform, and arranges them according to their frequencies. The amount of amplitude occurring at any particular frequency is called the amplitude of vibration of that frequency. A plot of amplitude over its frequency is known as the FFT spectrum. FFT spectrum is very useful in analyzing machinery problems. If a problem exists, the FFT spectrum helps determine the location, cause, and possibly the duration of the P-F curve. A thorough and careful analysis of this spectrum can provide the user with the cause of excessive vibration in the equipment. Frequency helps to pinpoint the source of vibration. Vibration forces rarely act alone, and each vibration leaves a distinct frequency or fingerprint that helps the user to identify the faulty component that is causing excessive vibration.

Vibration analysis is used to extract the vibration pattern from the machine. The vibration pattern is converted through an electrical signal with the aid of sensors. Mounting of sensors or probes on the equipment is critical and must always be done in the same exact location; otherwise, the instrument will record a change in signal. To surmount this problem, the mounting of sensors is done via magnet, and the most common sensor used is the accelerometer. Vibration monitoring must always be done in the same location on the equipment, having the same speed to get a consistent reading.

The amount of vibration is known as the amplitude, which can be expressed by either displacement, velocity, or acceleration. Displacement probes range from 10 to 1,000 hertz and will be used for low frequency. Velocity is the most common measurement of vibration, which covers mid-frequency responses from 5 to 2000 hertz. Acceleration usually relates to high frequency, such as bearings and gears. This is being measured through the use of accelerometers. They can operate from 1 to around 100 kHz. Accelerometer transducers available in the market have a range from around 1 to 10,000 Hertz

9.5: Infrared Thermography

Thermography is a Predictive Maintenance technique that uses the emission of infrared energy to determine its operating condition. Infrared technology is predicated on the fact that all objects with a temperature above absolute zero emit some form of energy or radiation. Absolute zero is the temperature where the molecules stop moving, which is at 273 ° C. For as long as molecules are moving, it emits some form of heat. For hot objects, the molecules are moving in a fast direction, while for cold objects, the molecules are moving slowly. Infrared radiation is one form of this emitted energy, which is a function of its surface temperature. For Infrared Thermography users, only the heat and energy that is being emitted by the object will be important to the user. Reflected and transmitted energy must be filtered out from the reading to provide an accurate measurement of data. Since heat can be transmitted, emitted, absorbed, reflected, or refracted, an infrared camera might be detecting heat that is not actually being emitted from the object itself but from other sources. Users of this instrument must have knowledge regarding the emissivity of objects to minimize errors in readings. The cameras that are used in Infrared Thermography simply see the radiated energy that is emitted from the surface of the object that it is viewing. These types of infrared cameras can be used for routine applications for monitoring electrical and mechanical inspections where the thermal image of the equipment and temperature measurements can warn us of an impending failure.

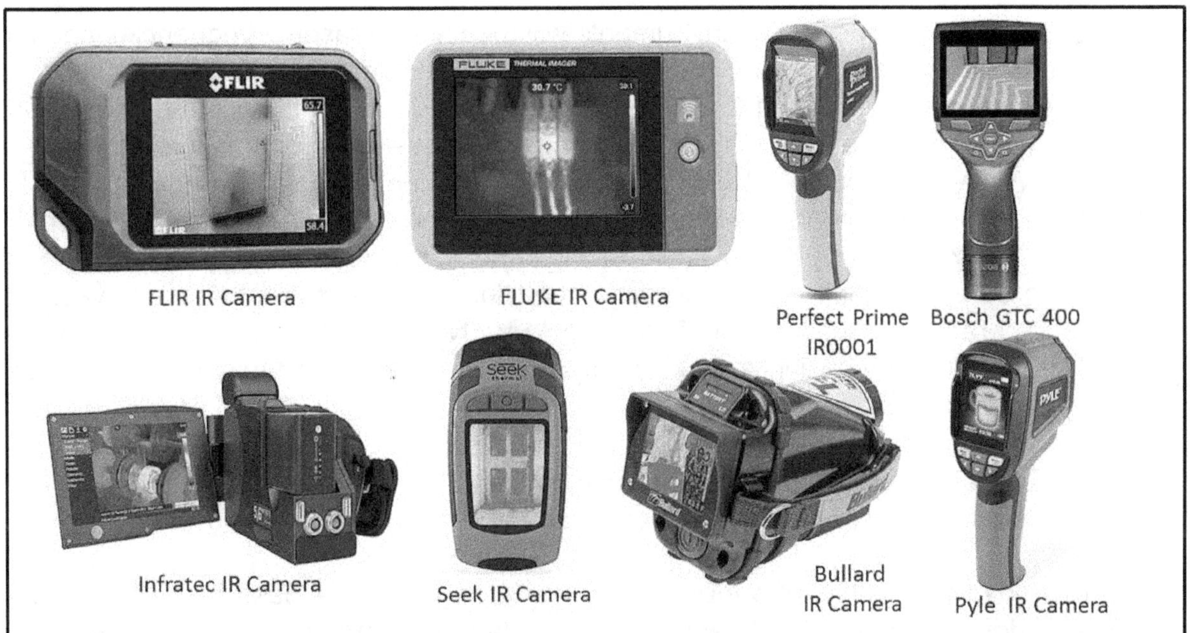

Figure 9.4: Different Brands of Infrared Thermography Camera

For industry applications for infrared cameras, they can be used for electrical systems. They can detect poor electrical connections, short circuits, overloads, load imbalances, faulty, mismatched, or improperly installed components since poor connections will induce more heat. Heat raises its resistance, and more resistance produces more heat. Infrared thermography pinpoints the hottest part of an object in exact precision, known as hot spots. For mechanical systems, it can detect problems in rotating equipment, power transmission, fluid flow, and insulation systems. Mechanical systems produce heat due to insufficient lubrication resulting in friction.

9.6: Ultrasonic Monitoring

Frequencies below the sonic range are called infrasonics, few animals in nature can generate or respond to these sound ranges. For example, elephants can hear infrasonic sounds and they use this ability to communicate at very large distances. Elephants have big ears since they need to hear infrasonic sounds which have large wavelengths. They can hear up to 15 Hz. Frequencies above the sonic range are called Ultrasonics. Bats use a very high-frequency burst of sound to locate and classify insects in the dark. High-frequency sound waves do not bend around objects and reflect in straight lines. This allows the bat to navigate and locate objects without the benefit of sound. Dogs can hear ultrasonic frequencies, they can hear up to 40-kilo hertz and they can respond to dog whistles. Ultrasonic monitoring will allow us to translate high-frequency sound to the sonic range which can be heard by the human ear.

- Sonic/Audible Range: 20 Hz to 20 kHz
- Infrasonic Range: Below 20 Hz
- Ultrasonic Range: Above 20 kHz

These Ultrasonic Monitoring instruments will measure ultrasonic frequencies and through a process called heterodyning, translate these frequencies down into the audible range. Heterodyning is the mixing of 2 waves. This produces the sum and the difference of their original waves which allows the shifting of a high-frequency sound to the audible or sonic range. While Vibration Monitoring can monitor frequencies in the range of 1 to 20,000 Hertz, Ultrasonic Monitoring can monitor a much higher frequency range between 20,000 Hertz to 100 Kilo Hertz. It is used to monitor the noise generated by the equipment and assets to determine their actual operating condition. Let's assume that we have a bearing that is generating a sound signal of 31 to 33 kHz. Note that the human ear can hear up to 20 kHz, therefore, by mixing a 30 kHz constant frequency sound wave with this signal we will get the following:

- Adding up the frequency: 61 to 63 kHz (cannot be heard)
- Subtracting up the frequency: 1 to 3 kHz (can be heard)

Heterodyning consists of a mixer, local oscillator, and a low pass filter. For example, a signal is detected at 21.50 kHz at the local oscillator (LO) which is at 30.30 kHz. The sum and difference are shown at 51.80 kHz (not detected by sonic) and 8.80 kHz (detectable by sonic). When this composite signal passes through the low pass filter only the difference of 8.80 kHz will appear at the output since this is the one detected and audible by the human ear. Mechanical movements produce a wide spectrum of sound. By focusing on a narrow band of high frequencies, this ultrasonic monitoring instrument can detect changes in amplitude and sound quality. It then heterodynes these undetectable sounds into the audible range and can be heard through headphones.

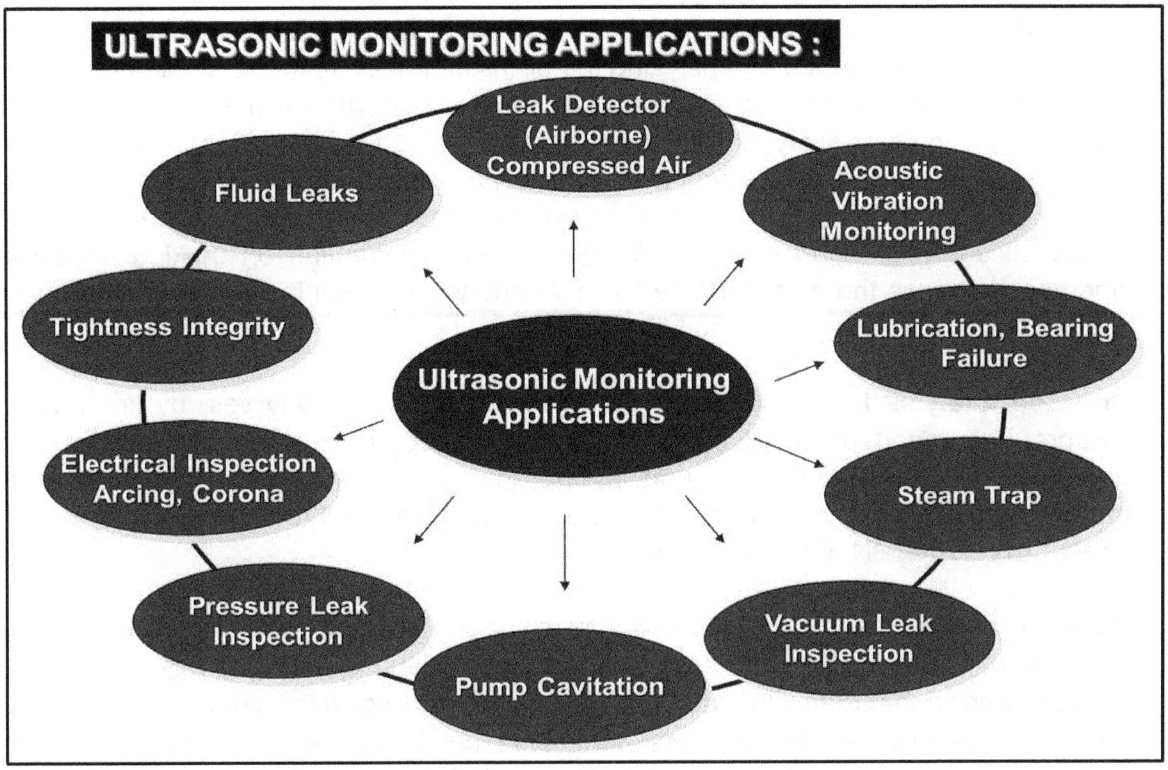

Figure 9.5: Ultrasonic Monitoring Applications

There are many applications of using Ultrasonic Monitoring. Compressed air is one of the most costly utilities in plants today. The Department of Energy has estimated that around 30% of all compressed air produced in the US is lost to leaks. An ultrasonic compressed-air leak survey can offer fast payback. One plant averaged an electric cost of $1,760,000 to generate compressed air. The survey using Ultrasonic Monitoring identified 3,561 leaks totaling 6,340 CFM at an annualized cost of $597,000.00 of energy waste. A program of leak detection and repair can have a dramatic effect on the profitability of a company such as this. In addition to being a source of wasted energy, leaks can also contribute to other operating losses. Leaks cause a drop in system pressure, which can make pneumatic equipment function less efficiently, adversely affecting production. It is important to develop a scheduled monitoring program for detecting, tagging, and repairing leaks. Inexpensive ultrasonic leak detectors are now widely available and easy to use, even in noisy plant environments. In-house resources generally can handle this task, and several service companies specialize in this service.

9.7: Oil Analysis Explained

When we are sick, we go and schedule an appointment with our doctor, and the doctor will try to check and diagnose us. Upon physically checking us, the doctor may recommend drawing a blood sample from us and sending it to a medical laboratory to have it checked and analyzed. The blood will be sent to a medical laboratory for diagnosis. Upon receiving the blood test results from the medical lab, the doctor renders a medical opinion or advisory that we are free to accept or reject. Similarly, the oil inside our equipment is like our blood. It can dictate its condition through what we call an "Oil Analysis Program." Therefore, like blood, the oil contains a great deal of information about the equipment itself. Useful information can be learned by analyzing the oil inside our equipment. While many industries are still changing their oil on a time-based dominated pattern, which is in the form of running hours, some industries are shifting from running hours to a condition-based approach by sending oil samples to an independent oil analysis laboratory, while some industries even have their own oil analysis laboratory, which gives them an added advantage. The results of the oil analysis report can provide them useful information on making a decision on whether to change the oil or still used it. By conducting a regular oil analysis sample, it will answer the following important questions.

Does oil analysis tell us when to change our oil? Definitely yes, by knowing how to interpret the report provided by the oil analysis laboratory, maintenance, and reliability people can make sound decisions on whether to continue running the equipment or change the oil completely. Several instruments can tell us if we need to change the oil or not such as RULER, TAN/TBN, FTIR, and others.

Does oil analysis tell us when and where machine parts fail? Yes, the test on wear metal debris analysis such as analytical ferrography and spectroanalysis can tell us what metallic and non-metallic elements keep on increasing over time. By trending and monitoring these elements every time a sample is sent and understanding the limits of each element, maintenance can make decisions on what part is starting to wear out inside their equipment. By knowing this information, maintenance can advise operations, order the spare part affected, and schedule the equipment for repair instead of waiting for it to fail and perform emergency work. In short, oil analysis, like any other condition-monitoring program, is a strategy that allows the user to look inside the equipment without actually dismantling it. The report will provide crucial and critical information on what is going on inside our equipment. In this case, what is important for the maintenance function is to know the metallurgy or materials in which the mechanical components inside their equipment are made of.

Does oil analysis tell us what particles are currently present in our oil? Yes, for example, if we send an engine oil and the report indicates large traces of silicon present, and it is increasing over time, then most probably there is dirt ingression happening in the equipment. If the element of chromium is increasing in ppm (parts per million) in a diesel engine, then this means that the piston ring is starting to wear out since most piston rings are made from the element of chromium.

Different types of tests can be conducted through oil analysis and each test will try to look at a different aspect of the oil. What is important for maintenance is to determine the test they need on the oil as there are more or less than 200 to 220 different tests that can be conducted on both new and used oil. Unlike other Predictive Maintenance where you just need one instrument to measure such as vibration, or thermography, oil analysis is composed of many test instruments. Therefore, what is important is to determine the test that best suits your application because this is where the industry can derive the most benefit from its application. Oil Analysis makes it possible to look inside an engine, transmission, gearboxes, or hydraulic system by monitoring the actual condition of the oil and determining the presence of contaminants inside. It can tell us if the additives in the oil are still present or had already been depleted, or why is the oil burning so fast? Some tests can detect how much moisture is present in the oil. Moisture also promotes oxidation of the oil's base stock as well as inhibits rust and corrosion. For industries that consume a large amount of lubrication consumption, I would recommend having their own oil analysis laboratory. Prepare a Feasibility Study on having an in-house oil analysis laboratory and include in the report the most important part, which is the Return on Investment (ROI) and payback.

An oil analysis program can save us not only in lubricant costs but also in spare parts costs, energy costs, labor costs, breakdown costs, downtime costs, and fuel costs. In fleet applications, oil analysis has been applied to determine when additives had been depleted so we can decide whether to change the oil or not. Some tests can determine if the wrong oil had been added to the system. There are so many benefits to having an oil analysis program for the maintenance function, such as the following.

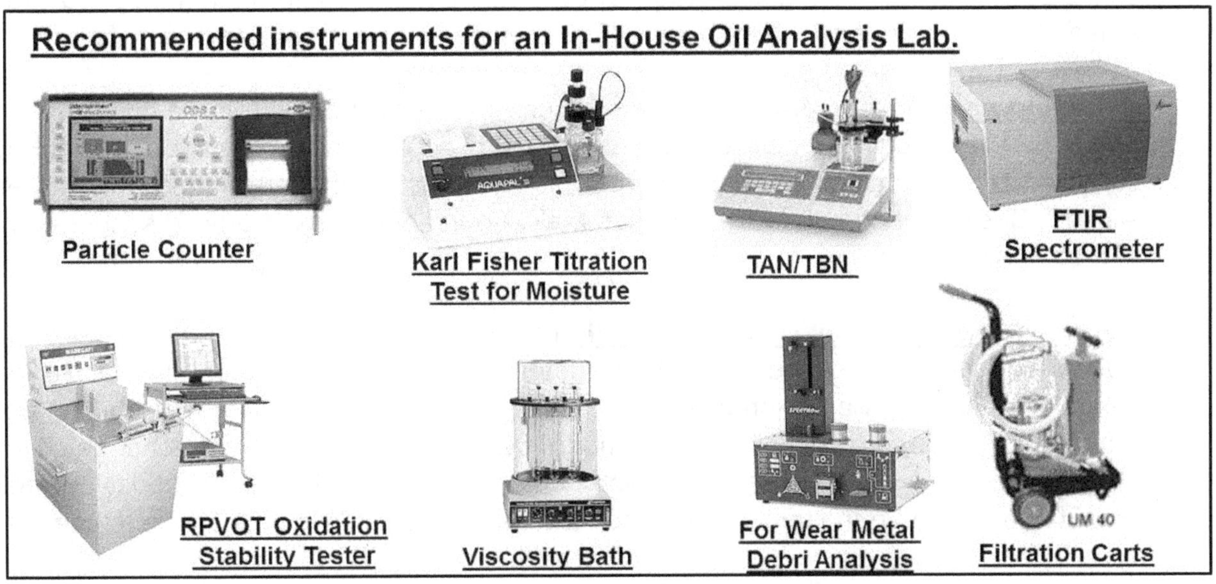

Figure 9.6: Recommended Oil Analysis Instruments for In-house Laboratory

Prolong the change interval of oil: Oil contains two things, the base, and additives. When the additives deplete, then the oil is no longer fit to use and needs to be changed. Hence, the best decision on when to change the oil is based not on the number of hours or miles the equipment has run but rather on the number of contaminants present in the oil. An oil analysis program will also suggest methods to reduce accelerated wear and contamination on mechanical parts. Thus, oil now becomes a working history of the equipment itself.

Reduction in maintenance cost: Oil analysis plays a major role in reducing maintenance costs and increasing the life of the equipment. Improving oil contamination helps to improve the combustion efficiency of engines by monitoring and optimizing fuel system efficiency, thereby decreasing harmful emissions into the atmosphere. The most traditional way of maintaining our equipment is going on a time-based schedule mode or Preventive Maintenance in which assets, systems, components, and equipment are overhauled on a time-dominated pattern. Parts are replaced even when there is nothing wrong with them just to comply with the scheduled PM. With Oil Analysis, there are actually two ways on how the cost is being saved, first on the amount of lubricant consumed and the amount of reactive maintenance or breakdown maintenance that needs to be attended. My experience tells me that the cost of doing repair work is doubled or even tripled compared to the number of lubricants consumed by the plant. Let us say that if the cost of your lubricant monthly in the mining industry is around $100,000.00, multiply this by a minimum of two, and that is the rough estimated cost of all the breakdowns and failures attributed to lubrication. The cost of parts that fail, labor, and downtime cost is always higher than the costs of lubricant consumed.

Early Detection of Problems: An oil analysis program can provide early detection of problems, mostly on mechanical parts. It can anticipate problems in advance and provide maintenance adequate time to inform operations people to schedule repair work to avoid long downtime during its critical time of use. By establishing trends from oil analysis tests, it can indicate a change in engine or lubricant condition, and the information they provide can be used to correct abnormal conditions before they can cause more damage or failure.

Evaluating used equipment: A complete record of oil analyses performed can prove to be a great tool when selling or buying a piece of used equipment. It shows potential buyers how you have maintained the equipment and any adjustments you made during its entire lifespan. Besides, it can allow the buyer to make sound decisions in buying second-hand or used equipment.

Determine the frequency of changing oil: Through oil analysis, it can allow maintenance to decide on when to change their oil and determine the correct interval needed for an oil change to determine if an oil change is necessary or not. Since we are now shifting from a running hour mode to a condition-based mode, the decision on whether to change oil or not will depend on the oil analysis report and not on its running hours. If additives are still present and are within the target limits, then the default decision is to continue to use the oil.

So how does oil analysis works? The oil that has been inside any moving equipment or asset for a specific period reflects the exact condition of the equipment. Oil is in contact with mechanical components and parts. Small metallic debris mixed with the oil. These metal fragments are so small, usually measured in microns that they remain suspended in the oil. Many products of the combustion process also become trapped in the circulating oil. Thus, the oil becomes a working history of the equipment. Since these particles will mix with the oil, through oil analysis, we can establish trends that indicate a change in engine or lubricant condition. The information they provide can be used to correct abnormal conditions before they cause more serious damage. This contamination present in the oil is measured using different tests; hence, maintenance has an indication of the rate of wear and contamination present inside their engine or equipment. An oil analysis will recommend methods to reduce accelerated wear and contamination, as well as what parts are starting to wear out. Many benefits can be reaped off when implementing an Oil Analysis Program correctly. However, the question most maintenance managers have in mind is, how come I still have many breakdowns and emergency repairs work in operations even if I send oil samples outside to an independent oil analysis laboratory?

There may be more than a hundred reasons why industries do not benefit from their oil analysis program, first, it will all boil down to training. Before implementing or initiating an oil analysis program in the plant, whether you will be sending samples outside to an independent oil analysis laboratory or planning to build your own oil analysis laboratory, all maintenance should be trained in oil analysis and lubrication. I do not recommend vendor training since, at the end of the training, there will always be a sales pitch from them. This is very critical, most especially in sampling oil. If the oil samples are to be sent outside, kindly note their response time. If the report is provided to you in a couple of weeks, then look for another oil analysis lab. There are cases where the failure already took place before the report had been provided to you. This is one of the many benefits of having an oil analysis in-house rather than sending them outside since you can have an oil analysis report in a matter of minutes. You see, when you send samples outside, your bottle needs to queue or fall in line with the other bottles since most outside laboratories follow a strict rule on a first-come, first-serve basis. Just like when you go to the airport, you need to queue. Remember that you are not the only industry on the planet that they are serving, and the oil analysis lab needs to earn a profit. In oil analysis, there are sensitive tests and less sensitive tests, meaning there are tests that need to be prioritized while the oil is still warm, and in my experience, the most sensitive test is the particle counter. If the test is conducted a couple of days after the sample had been sent, then the data might not be conclusive since it no longer represents the actual sample since some of the contaminants had already stuck on the bottom part of the bottle. Hence, when you have your own oil analysis laboratory inside your plant, there is what you call a tier of importance on which type of test needs to be prioritized and not on a first-come, first-served basis, just like an outside oil analysis laboratory does.

Another main reason I see for the failure of the oil analysis program is setting up the sampling points. The golden rule to follow is that the sampling points should be consistent and done on the same spot or location. A no-no in oil analysis sampling is trying to get the sample from used oil or acquiring samples from the drain plug. Remember that this is where all the dirt and contaminant are settled. The actual oil sample may not be the actual representation of the oil itself and a greater chance that the oil analysis report will end up failing with the laboratory technician's recommendation to change the oil. Note that the laboratory technician will only provide recommendations on what to do with the oil. These technicians might not know exactly how your equipment actually operates, but we, on the maintenance side, are the ones that know the equipment well. Always use the same sampling ports since variation in the contaminants can be seen at different locations inside the reservoir. It is best to place a minimess (faucet-like) on every sampling port if necessary so that location of the sample is consistent. If you have a lubrication system with several components, the sampling points should be located after the component or part and not before the part so that we can have a more accurate way of monitoring the part itself. The sampling ports should also be located before and not after the oil filter.

Oil samples must be drawn while the unit is in operation or the equipment is running. If that is not possible, the sample must be drawn right after shutdown, so that separation of any particles or water in the reservoir does not occur yet, or the contaminants in the oil are still in suspension. If you have around 2000 pieces of equipment, it is unlikely that the laboratory technician alone will be the one to extract all the samples here. So once again, training will play a key role in extracting samples correctly from your equipment.

Some industries might be hesitant to have their own oil analysis laboratory because of the investment, but my advice is that if your lubrication cost is very high, I think you need to make some thinking and reconsider having one. Remember that oil analysis instruments require two investments. The first will be the oil analysis instruments, and the second is training. Laboratory technicians should be certified in Oil Analysis. To justify the cost, you need to provide a feasibility study, present that to your decision-makers, and answer the most important question, which is the payback or rate of return, so that you can convince your top management to have an oil analysis approved onsite. Make a feasibility study and try to include the following, cost of failures due to lubrication for the past 2 to 3 years, your plant's oil consumption costs that you can easily get from your purchasing department. Try to explain in layman's terms what the benefits of oil analysis are. What tests and instruments are required and why you need them, estimated the cost of each instrument will be needed, where the lab will be located inside your plant, rate of return, estimated savings, success stories from those that have adopted onsite oil analysis. By doing this, then there is a greater possibility that management can rethink their decision on the importance of having an oil analysis inside your plant than sending samples to a third-party laboratory especially if your industry is a heavy consumer of lubricants just like a mining industry.

9.8: The Role of Predictive Maintenance on PM

Although the key to using Predictive Maintenance is detecting the potential failure or the Failure Development Period. The failure development period is the time where the potential failure is spotted by the instrument until the time the equipment finally reached its functional failure or this is when the equipment is actually declared fail.

Although originally, the P-F curve is designed for random failures. This means that some parts and items before they fail will emit a signal or symptom that they are on the verge of failing. An example of this is a bearing in which the temperature is increasing which can be captured through an Infrared Thermography instrument or a misaligned shaft in which the vibration is increasing which can be captured by a Vibration Analyzer. The P-F curve or P-F interval is what the users are monitoring so that they can make a decision on when to recommend the equipment to be stopped for corrective maintenance before the equipment finally gets into a failed state. This means that these instruments will allow us to perform maintenance tasks before the failure happens.

So the question is, how can Preventive Maintenance benefit from the use of Predictive Maintenance. First, Predictive Maintenance can remove tasks that include human inspections as these instruments will sense the problem much further than the human senses can. If a bearing starts to become noisy in which the noise can be audible, then this will just be a matter of days before it fails. A Vibration Analysis instrument can detect the problem 2 to 3 months before the failure happens. An Ultrasonic Monitoring instrument can even detect the problem much further than the vibration analysis. This provides maintenance a lot of time on making plans for replacement.

Another important role for Predictive Maintenance is conducting a Pre-Post monitoring test before and after executing Preventive Maintenance. By having a Pre-test Predictive Maintenance inspection, the thermography, vibration, and oil analysis users can be used to inspect the different parts of the equipment with their instrument to detect anomalies, and problems, which can be discussed with the planner and the people who will execute the PM tasks. If a part or item is subject to a replacement, Predictive Maintenance can inspect the item if it really needs to be replaced or not. This means that if a certain sub-assembly or auxiliary equipment needs to be overhauled or a part needs to be replaced but the Predictive Maintenance instrument does not detect a problem, then there is really no need to perform a task on a particular sub-assembly. The PM work will be focused on where the problem really exists. Likewise, having a post-monitoring will provide the PM group if the Preventive Maintenance execution was successful, thereby minimizing the chances of infant mortality failure from occurring. For example, if the overall vibration inspected before the PM was 8.4 mm/sec and after completing the Preventive Maintenance, the vibration was reduced from 8.4 to 4.5 mm/sec, then we know for a fact that the PM execution was indeed successful. Having a Post Inspection Test such as this can lessen the chances of Infant Mortality occurring right after an extensive Preventive Maintenance Shutdown

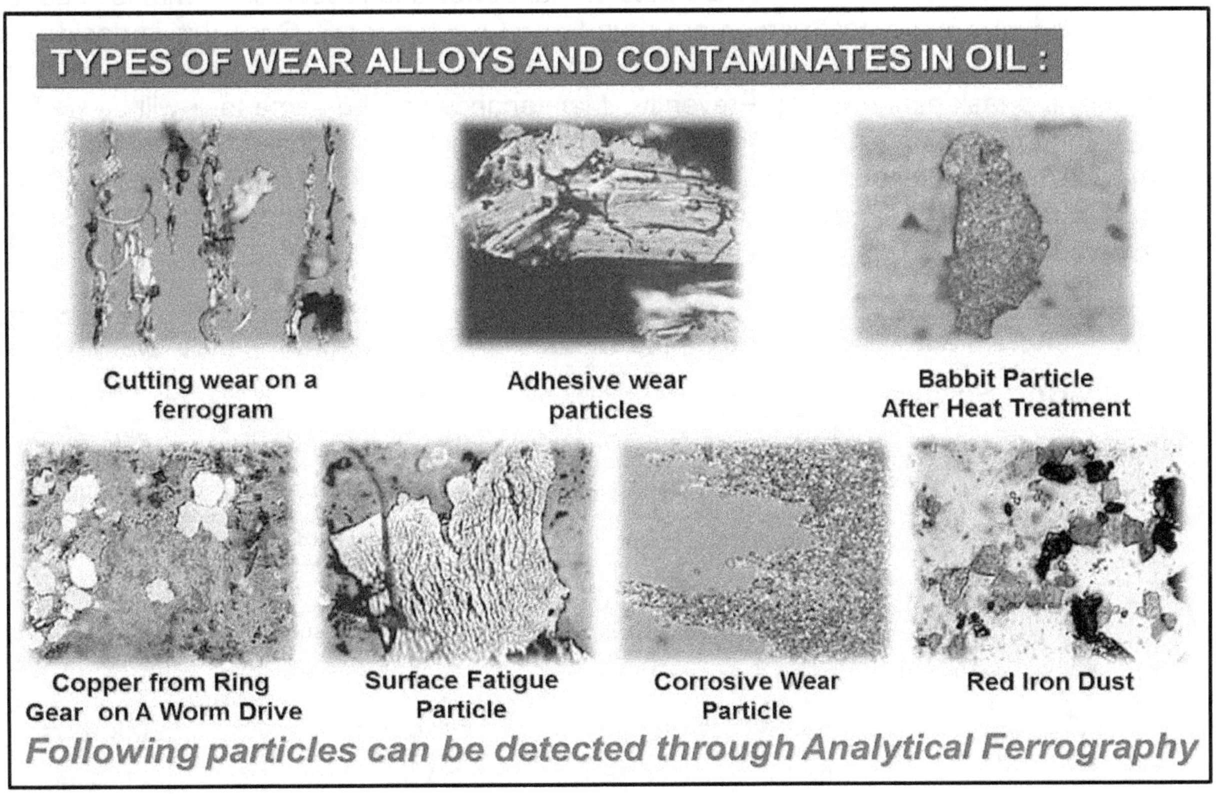

Figure 9.7: Wear Metal Debris Captured through Ferrography Test

Although as explained previously that what Predictive Maintenance is after is the P-F curve or P-F interval and Preventive Maintenance is feasible for wear-out or age-related failures. The question is how do we know if a mechanical part inside the equipment is already in the process of wearing out since we cannot see them as it is inside the equipment. Gages and charts can provide us a hint. If the flow rate of a pump is dropping from a rated capacity of 1000 GPM to 960 GPM, then this means that the pump is starting to erode. If a pressure drop is indicated in the pressure gauge, perhaps the strainer is starting to clog which can signal the maintenance to schedule a cleaning on the strainer. Predictive Maintenance can also spot wear out since this is also an indication of a potential failure. Metal elements detected from analytical ferrography and spectroanalysis can indicate what parts or items are starting to wear out and can be scheduled for a Preventive Maintenance replacement.

9.9: Combination of Tasks

There are cases when a particular failure mode will require a combination of different maintenance tasks to be addressed. A Combination of Tasks is the use of two or more types of maintenance to address a particular failure mode such as Preventive and Predictive Maintenance which can be done at different intervals by different people. This should be well differentiated from a redundant task. Redundant tasks are the same task done by different groups of people which can either be the PM Crew or a third party contractor. These are considered intrusive and should be deleted from the Preventive Maintenance lists since this will just end up in a total waste of manpower and resources. What is important is to decide who is the best person or group to perform the task. Once this is decided, then the same task should be deleted from the final Preventive Maintenance tasks. An example of this is a task performed on Preventive Maintenance yet the same task will be performed by a third-party contractor. A Combination of Tasks includes the use of different tasks which may or may not compose of different people to address a particular failure mode. This may be the use of both Predictive and Preventive Maintenance which can be done at different intervals yet the failure mode being addressed is the same.

Failure Mode	Maintenance Tasks	Type of Maintenance	Responsible
Bearing Seizure	• Perform Vibration Monitoring	Predictive Maintenance	PdM Group
	• Perform Ultrasonic Monitoring	Predictive Maintenance	PdM Group
	• Perform Infrared Thermography	Predictive Maintenance	PdM Group
	• Perform Oil Analysis	Predictive Maintenance	PdM Group
	• Thermal Gun	Predictive Maintenance	Operators
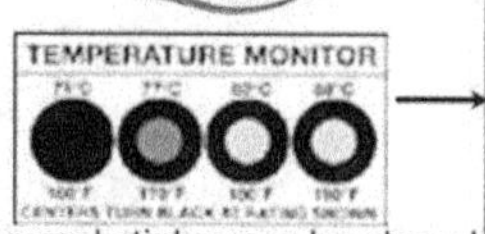	• Inspect Thermal Sticker	Preventive Maintenance	Operators
	• Perform Regular Greasing	Preventive Maintenance	Operators
Thermal sticker can be placed on the motor to indicate an in increase temperature on the bearing	• Inspect for Audible Noise	Preventive Maintenance	Operators
	• Replace when necessary	Preventive Maintenance	PM Crew

Figure 9.8: Combination of Tasks

Figure 9.8 is an example of a combination of tasks that will be an activity to address a bearing failure. In this case, perhaps a combination of vibration monitoring, ultrasonic monitoring, and thermography is used to monitor the condition of the bearing in the motor. Routine lubrication is also performed. The operator may also place a thermal sticker or may use a thermal gun to detect the temperature of the motor. The operator may also be using their senses to detect an increase in noise in the bearing. The activities performed in this case are different and can be done by different people at different intervals. This combination of tasks will ensure that maintenance will be one step ahead of failure.

Chapter 10

Deriving the Maintenance Tasks

We must understand that the goal of any Preventive Maintenance initiative is not about eliminating failures. This is an impossible feat to accomplish. Preventive Maintenance aims to anticipate, prevent, prolong, or delay the process of failure from occurring in our equipment. All activities being performed on PM must be geared towards anticipating and planning for the failure itself. It will always be less expensive if maintenance is ahead of failure.

10.1: The Reason RCM Was Developed by the Airline Industry

The Airline industries have a scary KPI which is called the accidents per million take-offs. In the United States, this is being monitored by the FAA or Federal Aviation Administration. The Federal Aviation Administration (FAA) is the largest transportation agency of the U.S. government and regulates all aspects of civil aviation in the country as well as over surrounding international waters. This means that from all the available airports, for every 1 million planes that will take off, how many will fail. We all know that more than the majority of failures when the plane is airborne will lead to the destruction of the plane and the death of the passengers. Another problem the airline industry faced in the 1950s was the high cost of doing maintenance. The Federal Aviation Administration (FAA), formerly the Federal Aviation Agency, started in 1958. The mission of the FAA is to regulate all civil aviation and U.S. commercial space transportation, maintain and operate air traffic control, provide navigation systems for both civil and military aircraft, and administer programs relating to aviation safety and the National Airspace System. In 1958, upon reviewing how an aircraft is maintained, the FAA found out that it was not possible to control the failure rate content or frequency of scheduled unreliable types of engines by any feasible changes either in the content or frequency of scheduled overhauls. This means that even if the interval of performing scheduled overhaul tasks was reduced, the amount of failure possible to occur still remains the same. This was a critical situation for the FAA since the accidents per million take-offs are high and second was their initial plan of developing a much larger aircraft which is the BOEING 747. This means that if the maintenance tasks for smaller aircrafts such as DC8 are high, then it would not be feasible to fly this new aircraft as the revenue it will generate will just be eaten up by the maintenance expenses.

In the 1960s, a tasks force was formed consisting of representatives from the FAA and reliability people from the commercial aviation industry to investigate the capabilities of Preventive Maintenance, which led to the establishment of the FAA/Industry Reliability Program and the development of MSG-1. This MSG1 includes a rudimentary decision-diagram technique which was devised in 1965 and in June 1967. A paper on its use was presented at the AIAA Commercial Aircraft Design and Operations Meeting. This maintenance technique was applied by the maintenance steering group on BOEING 747 airplane. The document was termed MSG-1 and the key person involved here was Thomas Matteson. Further developments led to MSG2 and the latest which was the MSG3 document which is the basis for maintaining all types of civil aircraft. The development of MSG3 by Stanley Nowlan and Howard Heap was based on comprehensive research over 30 years, to make airlines safer while making their maintenance costs more effective. This document was first used commercially in the 1970s to determine the maintenance program of BOEING 747. This document emerged from the most important piece of research ever done on an aircraft and a report was finally submitted by Nowlan and Heap from the United Airlines to the Pentagon in 1978.

Thomas Matteson resigned from United Airlines and started his consulting works and met Anthony Smith (Rest in Peace to this great legend who just recently passed away last February 2022). While on the other end, Stanley Nowlan I believed mentored a bright person by the name of John Moubray who was the author of RCM II. Although it can be said that Matteson, Mentzer, Nowlan, and Heap were responsible and considered the pioneers of RCM which started in the airline industry. It was Anthony Smith and John Moubrey who can be said as the pioneers of RCM for the land industry as these are the very people who first studied how RCM can be adapted by the land industries. According to John Moubray, the most common problems with maintenance programs that if not correctly designed, will lead to around 40 to 60% of the Preventive Maintenance tasks serving very little purpose whatsoever. Therefore, evaluating our current Preventive Maintenance System should lead us to the following conclusions:

- Many PM tasks duplicate other tasks, which means that this is being done by different groups.
- Some tasks are done too often while others are not enough.
- Some tasks serve no purpose whatsoever.
- Many tasks will be intrusive (forced) and overhauled based whereas they should be condition-based
- Some tasks cost more to do than the failure it is meant to prevent.
- Some tasks are costly by replacing perfectly good parts since we are basing the PM tasks on time instead of their actual condition.

Although the story does not end here. John Moubray created the Aladon Network in which he expanded his reach by having trained and certified consultants on RCM on the different continents of the planet which includes Europe, Southeast Asia, America, and Australasia. These consultants from different continents start spreading the concept and how RCM can be implemented in their assets and equipment. After a few years, there was a group of RCM consultants that complained to John Moubray that other industries are complaining about the RCM process as it is cumbersome and slow. So they introduce a much faster approach to complete the RCM process. Moubray insisted on them to stick to the process and this group break away from the original Aladon group and form their own group which is a Streamlined RCM process. While some reviewed the RCM process of Moubray and these break-away groups come up with their own idea. That is why when you search the internet today for the keyword Streamlined RCM, there are a lot of versions. That is why John Moubray developed the standard of SAE JA 1011 and SAE JA1012 so that there can be a standard classification between what is the original or classical and those streamlined versions of RCM.

Although these two approaches can be used to derive and refine the plant's current Preventive Maintenance tasks. Let me explain briefly how it is done. If the reader wants more comprehensive reading materials for RCM, I have also written a book exclusively for RCM which is Volume 7 of this WCM series which is titled Decoding Reliability-Centered Maintenance Process for Manufacturing Industries.

10.2: Deriving the PM Tasks Through RCM

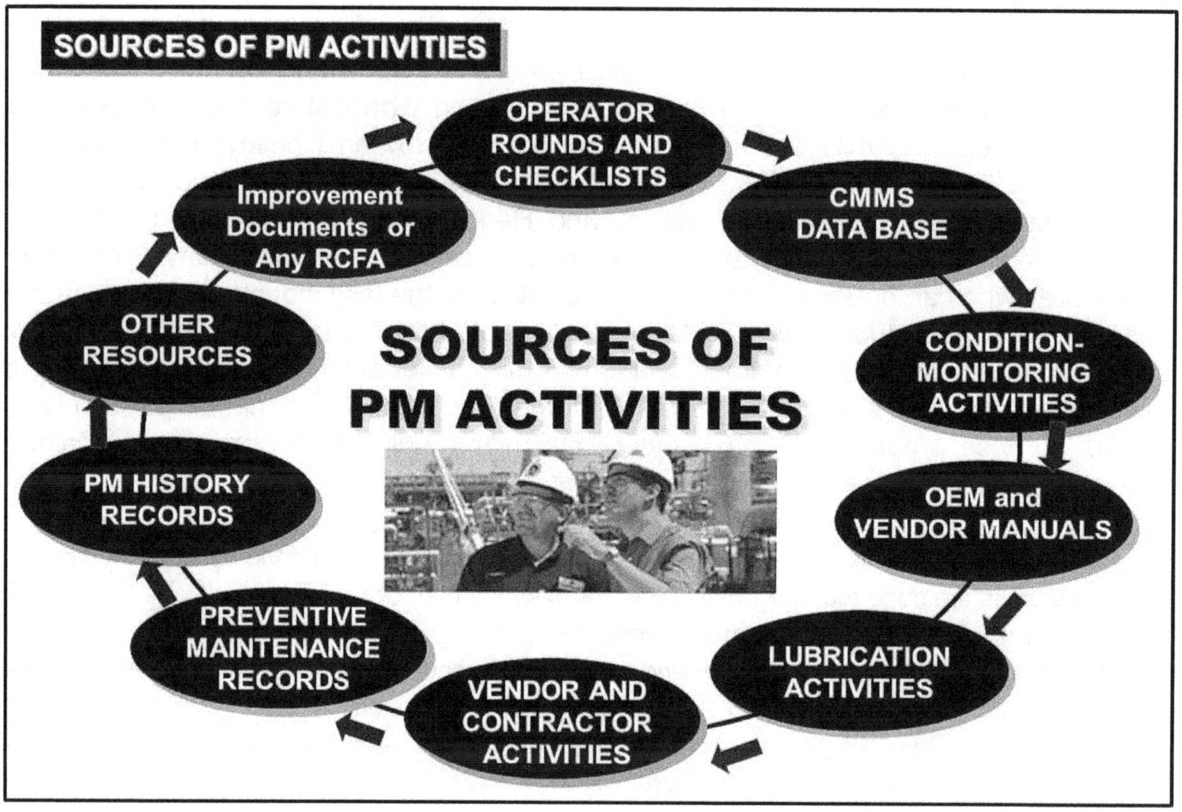

Figure 10.1: Sources of PM Activities

There are many sources for deriving the Preventive Maintenance tasks for the equipment as in figure 10.1. Some industries already have their existing Preventive Maintenance tasks, but there are also perhaps a few industries that do not have any PM tasks performed on their equipment and assets. Two methodologies can provide us a guide on how to develop the PM Tasks. These are Reliability-Centered Maintenance (RCM) and RCM Streamlined Version which is a shortcut process of the RCM version. The main difference between these two tasks is that RCM will start with zero-based and assume that the plant has no form of Preventive Maintenance, while the RCM Streamline Version will use the plant's current Preventive Maintenance tasks and refine them. Both these two methodologies will use a common algorithm or decision diagram. Another main difference between the two is that RCM will start by deriving the functions of the asset which is question number 1 in the classical RCM Analysis. Streamlined RCM will disregard both the functions and functional failures and start with the current Preventive Maintenance tasks the plant is using and identify the failure modes it is meant to prevent or predict.

Both RCM and other streamlined versions will not only limit all the tasks to Preventive Maintenance but will also see the bigger picture and consider all the available tasks or even a combination of the different maintenance tasks. Again so as not to confuse the reader, and as discussed in the previous chapters when using RCM or any streamlined version, Preventive Maintenance tasks will only include scheduled overhauls (restoration) and scheduled discard tasks which include parts replacement. The classical RCM approach will be done by answering the seven classical questions which are as follows:

1. **What are the functions and associated performance standards of the asset in its present operating context?** Equipment is used by operators to provide a product or service to its customer. Every piece of equipment has several functions, but the primary function of any equipment is to produce some product or service, just like a car where the main function is to travel from one distance to another. But just like the equipment, a car has several functions which are used by the operator or the driver. A side mirror is placed both on the left and right side of the car for the driver to see cars behind him. A brake light has its function, which is to signal the driver behind you not to come too close or provide a safe distance, most especially when you are on the expressway. From the brakes, clutch, transmission, steering, fuel lines, and so forth, each of them has its own unique functions, and we must understand that these functions are important to the operator. Therefore, the first step in any RCM process is to define the functions of the equipment in its present operating condition or the condition in which the asset is being operated.

2. **In what ways does it fail to fulfill its functions?** Once the function has been derived, when do we declare that the function of that part or system has actually failed? For the car, this simply means that if you have not covered the distance you are supposed to travel, or if you are going from point A to point B and your car stops midway from point B, the car has encountered a functional failure, or simply it stops because it fails. Similarly, if a pump whose function is supposed to discharge a rate of 350 liters per minute stops and there is no discharge at all, then we can declare that this pump is in its functional failure mode. Likewise, if the pump discharges less than 350 liters per minute, say at 300 liters per minute, we can also declare this to be in a failed state, and maintenance must look into it to verify what seems to be the problem.

3. **What causes each functional failure?** The third question asks us what causes this failure to occur or simply what are the probable causes of this functional failure. In this

question, we are seeking all the possible probable causes of failures or failure modes that can cause this failure to occur. For example, when a pump discontinues discharging, one possibility is that the valve was closed, another possibility is the motor burned out; still, another possibility is that the bearing seizes or the filter screen is blocked by contaminants. What we need to do is to list down all the possible failure modes or probable causes of failure at this point. When we are speaking about failure modes, we are referring to all the probability of failures and not the Root Cause of the failure itself.

4. **What happens when each functional failure occurs?** This question asks us that in the event this failure happens, what will be the effect of this failure. Will some alarm sound as a result of this, or stop production? What is the estimated time the machine will be down? Usually, how long will the maintenance be working on the equipment, or are spares needed to place the equipment back in operation available? Knowing the effects of this failure will make us understand what will take place after the failure had occurred on its own and how maintenance is going to respond to this situation.

5. **In what ways does each failure matter?** While question number 4 is asking about the failure effects, this question is asking about the consequences of the failure when it occurs on its own. Every failure has its own unique set of consequences. This question asks us that in the event of a failure, what will be the consequences of the failure itself? Will it affect the environment? Does it have safety consequences? Will the failure affect the operations, or will it not matter at all since the equipment is equipped with a backup or standby unit? RCM simply tells us that if the consequence of failure is not acceptable, then maintenance should exhaust all efforts to reduce or eliminate the risks of failure and when the failure has minimal consequences, then RCM may allow failure to occur on its own. RCM tells us that it is more important to focus on understanding the consequences of failure rather than completely eliminating the failure itself.

6. **What can be done to predict or prevent each failure?** The sixth question asks us if the activities performed by Preventive or Predictive maintenance can address the failure when it occurs on its own. If the failure mode occurs in wear-out patterns, then this can be anticipated using Preventive Maintenance replacements and overhauls, while if the failure mode provides some signs or symptoms that it is on the verge of failing, then these failures can be captured using Predictive Maintenance instruments. For example, infrared thermography can provide its users that an increase in heat or temperature in one of the motors they are monitoring can signal them that the motor has already reached its potential failure limit, these instruments can alert maintenance that something must be done and maintenance must intervene to avoid further damage.

7. **What should be done if a suitable proactive task cannot be found?** This is the last question in the classical RCM process, which states that if both Predictive and Preventive Maintenance will be unlikely to address the problem, then what remaining options in the maintenance tasks are left for the maintenance to address the failure mode itself. The remaining maintenance tasks left according to RCM will be failure finding tasks or functionality inspection for hidden protective and standby equipment, a run to fail situation, or allowing the part to fail as long as the only consequences of failure will be just the direct

costs of repair and the last option we have is to redesign and modify which means that if doing maintenance cannot help solve the problem then the final maintenance option will be to modify or redesign the system. Redesign and modification are applicable if the consequences of failure would simply be not acceptable at all in the plant.

Figure 10.2 is an extract from my book on Decoding Reliability-Centered Maintenance Process for Manufacturing Industries which is a complete maintenance task derived using the RCM Methodology. Note that Figure 10.3 represents the functions (F), functional failure (FF), and the failure modes on how figure 10.2 was derived. The RCM will compose of two parts which include deriving the first part which is the RCM Information Worksheet and subjecting them to an RCM Decision Diagram to finally derive the correct maintenance tasks to address each failure mode.

RCM Finalized Maintenance Tasks

N	F-FF-FM	Proposed Tasks	Interval	Who
		Weekly Maintenance Tasks		
1	1-A-1 3-A-2 8-A-1	a) Visual Inspection of pressure differential gauge at >2.5 in. H20.	Weekly	Facilities Engineer
		b) Replace primary or secondary filter when necessary. Note that the visual inspection can be done weekly, but the replacement will be done if the differential pressure gauge is >2.5 inches of H20	As Needed	Facilities Engineer
		Bi-Weekly Maintenance Tasks		
2	1-B-1	a) Visual inspection of filters to confirm the pressure of detached or disconnected units.	2 Weeks	Facilities Engineer
		b) Replace if there are missing or damaged filters.	As Needed	Facilities Engineer
		Monthly Maintenance Tasks		
3	4-A-4 4-A-5 6-A-1 6-A-2	a) Inspect the thermal paint on the outer motor bearing and blower pillow block bearing. The red color indicates an alarm that the temperature exceeds 85 ° C. This signifies that the bearing needs to be lubricated (pillow block) or replaced in the motor.	Monthly	Facilities Engineer
		b) Temperature measurement using a thermal gun is also permissible. The temperature operating range must be below 85 ° Centigrade.	Monthly	Facilities Engineer
		c) Perform routine grease lubrication when the temperature is nearing 85 ° C. Indicate the correct NGLI number and grease brand.	As Needed	Facilities Engineer
		e) Conduct Vibration Monitoring. If the vibration frequency is increasing, shorten the monitoring frequency.	Monthly	Facilities Engineer
		f) If the vibration limit is near and reaching its functional failure, replace the bearing.	As Needed	Facilities Engineer
4	13-A-1	a) Apply cleaning to the mechanical room and the entire unit when necessary. Perform 5s and housekeeping	Monthly	Facilities Engineer
		Bi-Monthly Maintenance Tasks (Every 2 months)		
5	1-A-2 12-A-1	a) Verify the functionality of the pressure differential gauge by comparison. Install the calibrated Magnehelic gauge and compare the reading. A 0.5 difference in reading signifies a defective gauge.	Bi-Monthly	Facilities Engineer
		b) Replace the pressure gauge when found defective.	As Needed	Facilities Engineer
6	2-A-1 2-A-2 2-A-3	a) Visual inspection of temperature and pressure gauge. The temperature range of supply should be </= to 10 ° C. Return =18 ° C. The acceptable temperature differential is 10 °C. The supply pressure is >/= 22 psi and return	Bi-Monthly	Facilities Engineer

	2-A-6 2-A-7 7-A-4	>/= 18 psi. The acceptable pressure differential is 5psi.		
		b) Conduct a walk-around check to inspect if cooling valves are still functioning, presence of cavitation, cooling coil leaks or clogged, and chilled water restriction. Details of walk-around inspection to be provided.	Bi-Monthly	Facilities Engineer
7	3-A-1 4-A-3 5-A-1 6-A-3 6-A-4	a) Inspect the belt condition and correct the tension if the belt is loose.	Bi-Monthly	Facilities Engineer
		b) Replace the belt when worn out. Record the RPM of the drive and driven pulley before and after replacing the belt. Compute for the belt slippage. 5% slippage is the acceptable limit. Beyond or >5% slippage yields belt tension for correction. Apply belt dresser if necessary.	As Needed	Facilities Engineer
		c) If any belt has slipped, correct the problem immediately.	As Needed	Facilities Engineer
		d) If any belt is missing on the pulley, replace it with a new one.	As Needed	Facilities Engineer
8	4-A-1 9-A-1 9-A-4	a) Periodic check-up and perform the restoration. If anomalies exist, execute retightening activities.	Bi-Monthly	Facilities Engineer
		b) Check the motor control circuit component and replace it if necessary or as needed.	Bi-Monthly	Facilities Engineer
		c) Perform infrared thermography inspection.	Bi-Monthly	Facilities Engineer
Quarterly Maintenance Tasks (Every 3 months)				
9	3-A-5 4-A-6 6-A-5	a) Check alignment of motor and pulley. Inspect the drive and driven pulley. Look for a loosening or wiggling effect.	Quarterly	Facilities Engineer
		b) If the motor and pulley are misaligned, correct the anomaly and retightened it whenever necessary. Perform alignment inspection.	As Needed	Facilities Engineer
10	7-A-1 7-A-2	a) Conduct regular test simulations for drain functionality. Pour approximately 1 pail of water into the drain and check whether it is disposed of easily.	Quarterly	Facilities Engineer
		b) If disposed of easily, a clogged drain line requires pressure washing or vacuuming to flush trap dirt restrictions.	Quarterly	Facilities Engineer
11	8-A-2	a) Regular inspection of filter frame condition. No sagging of filter required. The filter should be fitted into the filter frame easily. Inspect for weak support and welded joints. No presence of leak for the contaminant in filtration.	Quarterly	Facilities Engineer
Semi-Annual Maintenance Tasks (Every 6 months)				
12	1-B-2	a) Visual inspection of damper position to confirm if it is close. If open, fix the manual closer to retain a close position.	6 months	Facilities Engineer
13	2-A-4 3-A-3	a) Visual inspection of the by-pass or back-draft damper.	6 months	Facilities Engineer
		b) If the damper is fully or half-open, retain the normally closed position if found.	As Needed	Facilities Engineer
14	4-A-7	a) Check for ground fault and insulation condition of the feeder wire, motor stator, and winding. <5 megaohms require replacement of the feeder wire or reconditioning of the stator windings.	6 Months	Facilities Engineer
Annual Maintenance Tasks (Yearly)				
15	10-A-1	a) Inspect the condition of the access door. Look for possible leaks from defective door gasket. Check the hinges and replace them whenever necessary.	Yearly	Facilities Engineer
16	11-A-1	a) Inspect the condition of air ducting. Look for leakages from worn-out seal joints and the presence of punctured holes in the air ducting.	Yearly	Facilities Engineer

Figure 10.2: Final RCM Maintenance Tasks for AHU 401 Pilot

Function		Functional Failure		Failure Modes	
1	To condition the air by filtering, cooling, and recirculating at a filter differential pressure less than or equal to 2.5 inches of water.	A	When the filter differential pressure is greater than 2.5 inches of water.	1	The air filter is clogged with dirt, contamination, dust, and particles.
				2	Defective or clogged pressure differential gauge.
		B	If the air filter differential pressure is 0 inches of water, there is no flow.	1	The filters are either disconnected or detached
				2	The supply or return damper is closed
2	To provide a chilled water temperature differential of -12.2 °C.	A	If the Chilled water temperature differential is greater than -12.2 °C.	1	The Chilled water strainer is restricted
				2	Dirty or clogged cooling coils
				3	Presence of cavitation at chilled supply and return piping.
				4	Open by-pass damper
				5	Increase in heat load.
				6	Malfunction of cooling valves.
				7	Unbalance throttling of supply and return valves.
3	To provide a flow rate equal to or greater than 22,400 cfm.	A	The flow rate is less than 22,400 cfm.	1	The belt is loose or incomplete
				2	Clogged Filter
				3	Partial opening of backdrop damper.
				4	Incorrect pulley installed.
				5	Misaligned motor and blower
4	To provide a smooth, efficient motor with current drew within the rated range @ 62 amperes (50 hp. motor)	A	The motor will not run; there is no load, under-loaded or overloaded at 100% rated output.	1	Defective control circuitry assembly component.
				2	Out of specs power supply, 480 volts, 3 phase, and 60 hertz.
				3	Incomplete belt installed.
				4	Defective motor end bearing.
				5	Defective blower pillow block bearing.
				6	Defective puller drivers.
				7	The feeder wire is grounded.
				8	Defective supply duct auto damper
5	To maintain the belt tension to yield 1600 to 1700 rpm output.	A	Loose or over-tightened belt tension that results in uncontrolled RPM below 1600.	1	An incomplete belt was installed in the pulley
6	To maintain the blower running condition.	A	The blower rpm is less than 1000, which leads to a noisy operation.	1	Defective blower pillow block bearings
				2	Defective motor end bearing
				3	Belt Slippage
				4	Incomplete belt installed.
				5	Defective pulley drivers.

7	To maintain an efficient and functional drain line.	A	The drain line is unable to discharge the condensate.	1	Accumulation of excessive dirt, and algae on the sump results in drain line clogging.
				2	Defective or clog ged P-trap.
				3	Undersize drain pipeline installed
				4	Cooling coil leakage.
				5	Incorrect elevation of the drain line
8	To provide a rigid filter frame properly aligned with the locks provided.	A	Filter frame sags or deforms due to missing locks.	1	Clogged air filter units.
				2	Weak filter from support braces.
9	To provide a controlled circuit with an auto-manual selector, overload protection, and the pilot indicator light.	A	Malfunction of the motor control unit and its circuit components.	1	Defective motor control accessories, circuit breaker, delta contactors, start-stop switch buttons, control transformers, fuse, and transition timers.
				2	Undersized protective devices such as circuit breakers, fuses, and contactors.
				3	Out of range power supply at 460 volts with +/-15, 3 phase, 60 hertz as operating range.
				4	Loose terminal and connections on the motor and the circuit component.
10	To access the door free of leaks and should be properly insulated.	A	Outside door with leakage or outside air inflow.	1	Defective door hinge, lock, or gasket.
11	Air ducting insulation should be leak-free.	A	Leaks on Ducting.	1	Worn out seal of joints and presence of punctured holes on the insulation.
12	To have functional sensors and gauges.	A	Defective and unreliable or malfunctioned gauges and sensors for pressure, temperature, and Magnehelic gauges.	1	Malfunction unit dial gauges causing excessive rattling. Damage and unreadable gauges with broken glass and corrosion.
13	To maintain a clean unit's bottom side and paneling?	A	Dirty and untidy surrounding and unit paneling	1	Uncontrolled activities within the unit area.

Figure 10.3: Reference for Function, Functional Failure, and Failure Mode

10.3: Streamlined RCM Version

There are also many RCM streamline versions which is a shortcut process for the classical RCM version for improving the effectiveness of current maintenance programs and strategies. The difference is that the function and functional failure will no longer be included. Instead, it will start with the existing Preventive Maintenance program used within the plant. Working in cross-functional teams from the shop floor, the team identifies what type of duplication exists within their own environment and what elements of the current maintenance program are useful, and what tasks are inappropriate and intrusive. The following steps details how to conduct a streamlined RCM approach for their current existing PM program.

Step 1: Determine all existing activities and Preventive Maintenance tasks for the asset: In this step, we list all the activities being performed by all people in the equipment. Determine all recorded and unrecorded formal, and informal activities and tasks being performed on the equipment or asset. It is believed that maintenance is being performed by a wide cross-selection of people including the operators. Compilation of Maintenance Tasks and activities is simply a matter of writing down all the activities different people are performing in their equipment. It is uncommon for organizations to have an informal PM system in operation and rare for an organization to have none. In this step, we list all the maintenance task activities performed by the different people involved.

- Operators of the equipment
- Preventive Maintenance group
- Electrical and Mechanical group
- Sustaining and Repair crew
- Third Party Contractor or OEM
- Calibration group
- Condition-Monitoring or PdM Group
- Facilities group

TASKS	INTERVAL	YEARLY	RESPONSIBLE	ACTIVITIES	TOOLS	FAILURE MODE
1	Shift	365	Operator	- Inspect drop in Pressure Gage	Visual	Clogged Strainer
2	Shift	365	Operator	- Inspect Motor Temperature	Thermal Gun	Insulation Failure
3	Shift	365	Operator	- Inspect Shaft for Alignment	Visual	Misalignment
4	Shift	365	Operator	- Inspect Motor for Unusual Noise	Hearing	Bearing Failure
5	Shift	365	Operator	- Clean Motor Housing	Dry Rags / Solvent	Short Lifespan
6	Shift	365	Operator	- Monitor Drop in Flow Rate	Visual	Impeller Erosion
7	Weekly	52	PM Group	- Inspect base bolts for looseness	Torque Wrench	Excess Vibration
8	Weekly	52	PM Group	- Apply Lubrication	Grease Gun	Bearing Failure
9	Weekly	52	PM Group	- Inspect the coupling for looseness	Visual	Coupling Failure
10	Monthly	12	PdM Group	- Perform Vibration Analysis	Vibration Analysis	Bearing Failure
11	Monthly	12	Contractor	- Perform Monthly Greasing	Grease Gun	Bearing Failure
12	Monthly	12	PM Group	- Inspect Pressure Gage	Visual	Clogged Strainer
13	Quarterly	4	Contractor	- Replace motor bearing	------------	------------
14	Quarterly	4	PdM Group	- Thermography Scanning	Infrared	Overheating
15	Quarterly	4	Electrical	- Loose Rotor Bars	Vibration Analysis	High Vibration
16	Annual	1	Contractor	- Perform Shaft Alignment	Laser Alignment	Misalignment
17	Annual	1	Contractor	- Replace Impeller	New Impeller	Low Flow Rate
18	Annual	1	Contractor	- Check for Pipe Leaks	Ultrasonic	Leaks

Figure 10.4: Case Study on Overall Tasks Performed on Motor and Pump

Step 2: Determine all failure modes to justify each task: Once the list of all Maintenance tasks had been completed, create another column and determine the possible failure modes the tasks are meant to prevent. This will give us an indication if the maintenance task written really needs to be done or not on the equipment. All activities listed should address a particular failure mode. Streamline RCM method starts at the opposite where the PM tasks are reviewed to identify the failure for which it is meant to prevent. This revelation will tell us that activities that serve no purpose whatsoever will not benefit the current maintenance rather doing it is not only a waste of time but a time-consuming one. The tasks lists will also be checked if there are additional failure modes that must be added which is not currently present in the tasks list.

• Does this failure mode really need to be addressed by more than one person?
• Why was this activity performed? What failure mode does it specifically address?
• Are there similar maintenance tasks that are duplicated and done by different people?
• Who is the best person to perform this task?

TASKS	INTERVAL	YEARLY	RESPONSIBLE	ACTIVITIES	TOOLS	FAILURE MODE
1	Shift	365	Operator	- Inspect drop in Pressure Gage	Visual	Clogged Strainer
2	Shift	365	Operator	- Inspect Motor Temperature	Thermal Gun	Insulation Failure
~~3~~	~~Shift~~	~~365~~	~~Operator~~	~~- Inspect Shaft for Alignment~~	~~Visual~~	~~Misalignment~~
4	Shift	365	Operator	- Inspect Motor for Unusual Noise	Hearing	Bearing Failure
5	Shift	365	Operator	- Clean Motor Housing	Dry Rags / Solvent	Shirt Lifespan
6	Shift	365	Operator	- Monitor Drop in Flow Rate	Visual	Impeller Erosion
7	Weekly	52	PM Group	- Inspect base bolts for looseness	Torque Wrench	Excess Vibration
8	Weekly	52	PM Group	- Apply Lubrication	Grease Gun	Bearing Failure
9	Weekly	52	PM Group	- Inspect the coupling for looseness	Visual	Coupling Failure
10	Monthly	12	PdM Group	- Perform Vibration Analysis	Vibration Analysis	Bearing Failure
~~11~~	~~Monthly~~	~~12~~	~~Contractor~~	~~- Perform Monthly Greasing~~	~~Grease Gun~~	~~Bearing Failure~~
~~12~~	~~Monthly~~	~~12~~	~~PM Group~~	~~- Inspect Pressure Gage~~	~~Visual~~	~~Clogged Strainer~~
13	Quarterly	4	Contractor	- Replace motor bearing	------------	------------
14	Quarterly	4	PdM Group	- Thermography Scanning	Infrared	Overheating
15	Quarterly	4	Electrical	- Loose Rotor Bars	Vibration Analysis	High Vibration
16	Annual	1	Contractor	- Perform Shaft Alignment	Laser Alignment	Misalignment
17	Annual	1	Contractor	- Replace Impeller	New Impeller	Low Flow Rate
18	Annual	1	Contractor	- Check for Pipe Leaks	Ultrasonic	Leaks

Figure 10.5: Removing Duplicated Tasks

Step 3: Determine duplicated task: Duplicated tasks are similar maintenance activities that are done by different people. This will create redundant work and a waste of manpower and other resources. By sorting the data by failure mode, task duplication and redundancy can be easily identified. Tasks duplication is when the same tasks are being done by more than one person. We ask ourselves do these maintenance tasks need to be carried out by more than two trades or can be performed by a single group and who is the best trade to perform these tasks? In this step, the team reviews the different tasks generated and will try to add additional failure modes that are likely to occur on the asset. The team must decide who is the best trade to perform the tasks on each failure mode identified. Determine if duplicated tasks really need to be performed by different people involved in the equipment. Thoroughly reviewing each task will reveal that there are maintenance activities that do not address any particular failure modes and should be removed from the lists. Again determine what are the possible failure modes that can occur which is not currently reflected in the tasks.

In figure 10.5, task 13 is removed as this is the same as task 1. Task 3 is removed as this will be similar to task 16 and task 11 is removed as this is the same as task 8. If the plant is performing Autonomous Maintenance, in Step 5, both Autonomous and Planned Maintenance will sit down and consolidate the activities that will be done by both operators and maintenance so that duplicated activities can be removed so that only one group should perform the task.

Step 4: Check for Additional Failure Modes: Once these redundant or duplicated tasks are removed, the team will brainstorm if any other failure modes have been missed and will be included together with the task that needs to be done to finally address these failures modes. Once an additional failure mode has been added, the interval, responsible and maintenance tasks will be added to the list.

TASKS	INTERVAL	YEARLY	RESPONSIBLE	ACTIVITIES	TOOLS	FAILURE MODE
1	Shift	365	Operator	- Inspect drop in Pressure Gage	Visual	Clogged Strainer
2	Shift	365	Operator	- Inspect Motor Temperature	Thermal Gun	Insulation Failure
3	Shift	365	Operator	- Inspect Motor for Unusual Noise	Hearing	Bearing Failure
4	Shift	365	Operator	- Clean Motor Housing	Dry Rags / Solvent	Shirt Lifespan
5	Shift	365	Operator	- Monitor Drop in Flow Rate	Visual	Impeller Erosion
6	Weekly	52	PM Group	- Inspect base bolts for looseness	Torque Wrench	Excess Vibration
7	Weekly	52	PM Group	- Apply Lubrication	Grease Gun	Bearing Failure
8	Weekly	52	PM Group	- Inspect the coupling for looseness	Visual	Coupling Failure
9	Monthly	12	PdM Group	- Perform Vibration Analysis	Vibration Analysis	Bearing Failure
10	Quarterly	4	Contractor	- Replace motor bearing	------------	------------
11	Quarterly	4	PdM Group	- Thermography Scanning	Infrared	Overheating
12	Quarterly	4	Electrical	- Loose Rotor Bars	Vibration Analysis	High Vibration
13	Annual	1	Contractor	- Perform Shaft Alignment	Laser Alignment	Misalignment
14	Annual	1	Contractor	- Replace Impeller	New Impeller	Low Flow Rate
15	Annual	1	Contractor	- Check for Pipe Leaks	Ultrasonic	Leaks
16					ADD	Relief Valve Open
17					ADD	Excessive Corrosion
18					ADD	Wrong Direction

Figure 10.6: Checking for Additional Failure Modes

Step 5: Failure Modes to Undergo the Decision Diagram: Once the duplicated tasks have been removed and consolidated, these failure modes can be subject to an RCM

Algorithm or Decision Diagram to determine the correct task and can be used as a baseline for comparing the current maintenance task lists. The RCM Decision Diagram will determine the consequences of the failure mode and the best suitable maintenance tasks to be adapted. The tasks to be selected can either include Preventive Maintenance, Run to Fail, Predictive Maintenance, Failure Finding Tasks, Modification, or a combination of any of these tasks. If redundancy is in place, a switching interval will be decided on how long should the duty pump run and when to switch to the standby.

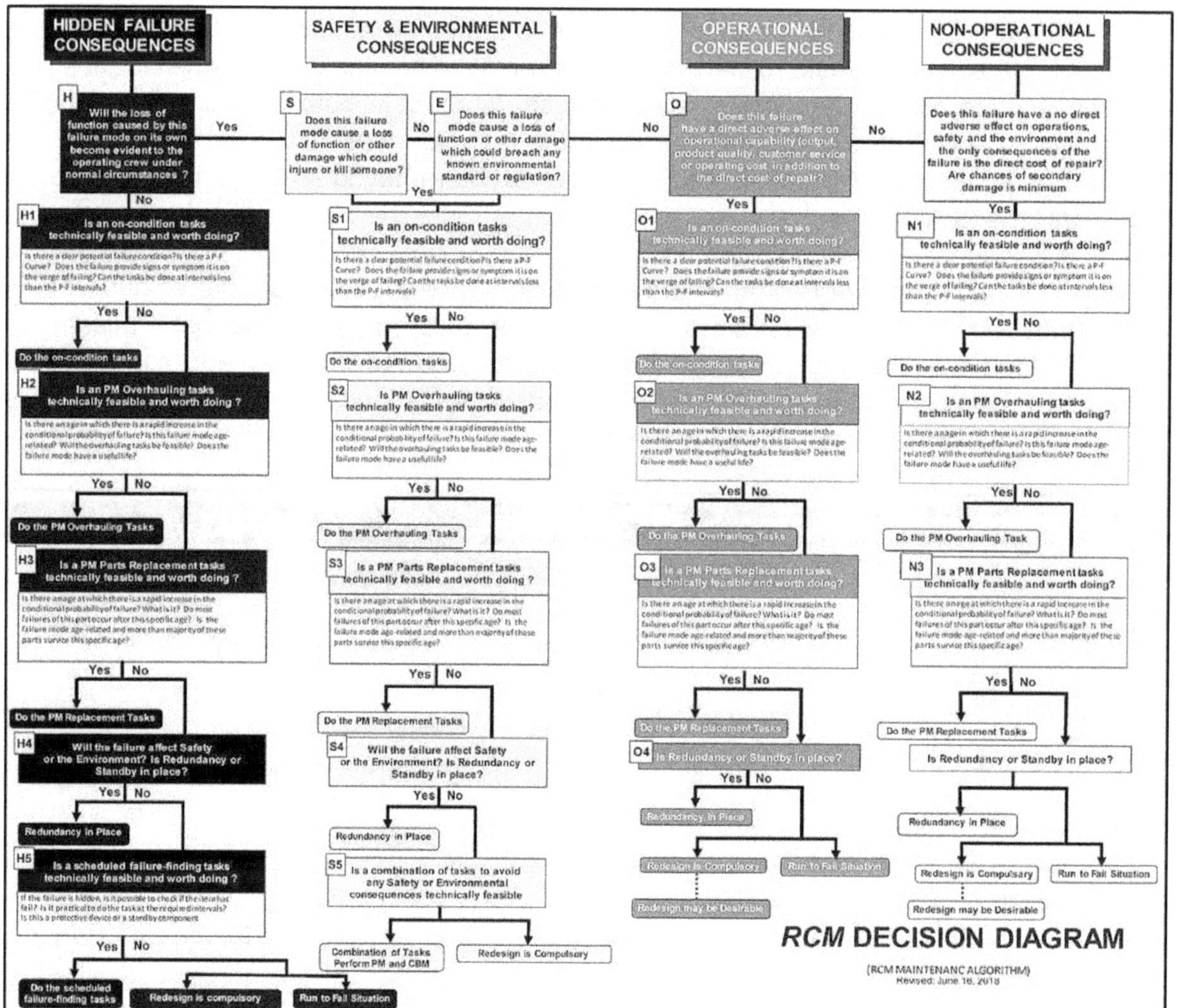

Figure 10.7: RCM Algorithm or Decision Diagram

Once this is completed, the revised PM task will be finally encoded in the system. The new tasks whether this is derived through RCM or streamlined RCM version should supersede their current PM tasks. This means that their current PM tasks should no longer be done on their equipment and assets.

10.4: Before Deriving the PM Tasks Consider the Operating Context

While many industries rely on their OEM or Vendor to derive the maintenance tasks or rely on the OEM manual and use the recommended maintenance tasks from the vendor.

What is important is that we should also consider how the equipment or asset will be operating in our plant. The problem with OEM manuals and recommendations is that their task is generic and conservative as well. If I have two similar Cummins Diesel Engine equipment here in the Philippines in which one is located in Manila and the other in Benguet where the temperature can reach 7 ° C, then the engine oil I would be using for Manila and Benguet would not be the same. For Manila where the temperature can be from 31 to 39 °C especially when it hit summer, I will be using SAE 40, with a NOACK Volatility Rate of 7 to 8%, but for Benguet in which the temperature is cold, then I will be using as 5W - 40 engine oil, where W stands for winter. This means that 5W represents the viscosity of the oil at cold temperature. Because the oil's viscosity changes concerning temperature, a multigrade oil or oil with 2 numbers were introduced to provide protection across a range of temperatures. This means that the lower the W rating, the thinner the oil and the faster it will flow when the generator will be started. The number after the W describes how thick the oil is at the engine's normal operating temperature. Here are the following operating conditions to consider before deriving the maintenance tasks.

1. **Batch and Flow Processes:** Are machines independent of each other, or will it affect the entire process if one piece of equipment failed? This means if a breakdown happens, will the failure be confined only to the equipment that failed, or will it affect the whole process or even the whole plant? Let us provide an example of this. In the semiconductor industry, we have a station called the wirebond. Wire bonding is the process of connecting the chip and the legs using a gold wire before it moves to the next process called molding, whose function is to encapsulate the chip with a plastic molding material to protect the chip inside the integrated circuit. Since this is a large semiconductor plant that can produce different packages (meaning integrated circuits), it has around 20 stations called wirebond. Each wirebond station has around 100 to 200 wirebond machines. Hence, if one machine failed, the rest will still operate at a reduced capacity. All wirebond machines are pneumatic, which means that compressed air will be supplied from the facility, connected through a pipe. If the compressor from the facilities failed, every single wirebond connected with that compressor would stop, and downtime occurs. In this case, the failure of the wirebond will only affect itself, while the compressor's failure from the facilities will affect several if not all the wirebond machines.

2. **Redundancy of Standby:** Is the equipment or system equipped with a standby just in case the equipment on duty fails. Note that even if the standby is the same type and model as the duty equipment, the maintenance tasks for both stand-alone and standby equipment will differ even if they are of the same type or make. This means that for stand-alone equipment with no standby in place, a run to fail option is not feasible since it will affect the operations. While for a plant with duplicated or standby equipment, the components can fail, and a better alternative is to create a switching pattern on both standby and duty components. Note that having standby equipment will make the system less critical compared to a stand-alone component.

3. **Quality Standards:** What will be the impact any quality issue will have on the customer's requirements? Is there a possibility of losing a valued customer or client because of any deviations from quality? Has this happened in the past? In my previous

books, I explained that breakdown is a broad topic; in this case, quality is a much broader topic than breakdown. This means that quality is more difficult to address than breakdowns since most quality problems are chronic by nature. Quality defects and problems not only come from the equipment but also from other sources, such as raw materials, human errors, processes, or even the environment to which the equipment is exposed.

4. **Environmental Standards:** Were there any previous failure modes in the past that could have affected any breach in any environmental issues that affected society? For powerplants, most of the failure modes may result in environmental consequences, which might be given priority in the RCM analysis. Remember that failure modes with safety or environmental consequences cannot be allowed to fail. Run to fail will not be valid for failures with environmental consequences.

5. **Safety Hazards:** Are there any failure modes that risk the possibility of hurting, injuring, or even killing someone? Is there any previous history of failure modes in the past that have any safety consequences? Again, in this case, run to fail should not be used for failures that have safety consequences.

6. **Shift Arrangements:** Will the equipment be running for 24 hours a day, 7 days a week straight, consecutively for a month or more, or will it only run for only one or two shifts depending on the load production demand?

7. **Repair Time:** What is the average speed of response and repair time of this failure? What are the most common failure modes experienced on this equipment, and how long does it take maintenance to repair the equipment on average?

8. **Spares:** Are critical spares held in stock in the storeroom in case of a breakdown? If not, what is the average lead time to acquire the part? How long will the equipment be down in case the part is unavailable in the storeroom? Will the operation be affected? Can the part be easily be obtained from the storeroom? In case the equipment or component has a standby, this might not be an issue but not with stand-alone assets or equipment.

9. **Market Demand:** Are there any specific months of the year that the equipment will run at its peak operating season in which the products or services have a very high demand with its customers in which the plant cannot afford any one or two of their operators or maintenance to use their vacation leave? If yes, specify these months in writing the operating context.

10. **Is the Machine Dedicated:** For manufacturing plants, equipment will not be dedicated to consistently producing the same product. This means that some equipment can manufacture different products using the same machine. In this case, a downtime called conversion, set-up, or change-over will be experienced on this equipment to change the parameters, settings, tooling, dies, and jigs when running the equipment to produce a different product. In this case, the team must understand if the same failure modes occur on the equipment when manufacturing different products or if a different failure mode is evident to occur only on this type of product.

11. **Environment:** Is the equipment running in a clean environment or not. Specify in the operating context if the machine is running in a clean, dusty, acidic, humid, or corrosive environment. Equipment with electronic components may fail prematurely in a humid and corrosive environment compared to a cleanroom environment.

12. **Specify Parts that Fail Prematurely:** Try to specify the spare parts that frequently fail in the equipment undergoing the RCM analysis that both operators and maintenance have always experienced.

13. **Failure Occurrence:** Determine the failure occurrence or how frequently this type of failure mode occurs in the equipment. Past history records, logbooks, and CMMS can provide this data for reference. If there is no record of its occurrence, the only thing left is for the team to make their judgment on their experience.

14. **Operators Competency:** Are operators certified and well versed in their equipment? Can they address minor issues and problems with their equipment such as minor stoppages? Are the skills of the operator sufficient in operating their equipment?

These operating contexts will provide maintenance an insight into the degree of maintenance requirements needed for the equipment. By understanding the operating context or conditions in which the equipment will operate, only then we can derive the correct maintenance tasks for the equipment and assets.

10.5: Sources of Failure Modes

Before we discuss in detail how to prepare the PM tasks, we need to understand what failures are likely possible to occur on the equipment. Where do we source the failure modes? There are several ways to source the failure modes. Here are some of the sources of failure modes.

Figure 10.8: Sources of Failure Modes

OEM or PM Manual: When a piece of equipment was purchased from the OEM or vendor, the package includes an operating manual. Usually, the manual will contain a list of maintenance tasks that must be performed on the equipment. We can use this as a reference point but we must also consider the current operating context or how the equipment will be operated in the plant. We also need to assess the recommended interval proposed by the vendor as the dictated interval might be too conservative or too soon. **Downside:** The OEM can be a good source, but most OEM manuals are generic and conservative. The PM task indicated may differ from the current operating context or how the equipment will operate in the plant.

Previous Work Orders: This will also be a good source of failure modes as the failure had happened in the past. Typically, when a failure or breakdown occurs in the equipment, a Work Request will be generated by the operator. Once the equipment is repaired, the work order will be closed and the maintenance who performed the repair will go to the computer and update the system if CMMS or EAM is in place and the work had been completed. If these work orders and completion had been logged in the system, then they should be there on the computer. We can just type a piece of a particular equipment number and the lists of work orders generated and closed will be reflected in the system. **Downside:** This will be a good source of failure mode if previous work orders are written in detail especially if parts or items with codification or part numbers were used. The downside is when the report is incomplete or has not been closed completely in the system.

IIoT (Industrial Internet of Things): During my time, we have the Britannica Encyclopedia. Today we have Google. Almost all information is just a click of the finger on the World Wide Web. And they say that the internet is just the tip of the iceberg, as underneath lay much wider information on anything, called the Dark Web (just watching too many Netflix Movies). Anyway just disregard the latter. **Downside:** The disadvantage of using the internet to source the failure modes is that again, the operating context of the equipment might not perfectly match and the accuracy of the information may remain unverified. Almost all information is on the World Wide Web today, what is important is to check the accurateness of the report or if the source can be trusted and determine if the operating context is the same in your current organization.

Other Sister Companies: If you are a large organization with other business sites operating in other parts of the country or internationally, they might also have equipment and assets similar to what you are currently operating. Your PM team and planner can arrange an Online Meeting with them with today's technology such as Zoom or MS Teams, and share information regarding the PM task they perform on their equipment and assets. Of course, we also need to understand the operating context statement or how they are currently operating their equipment. **Downside:** This can be a good source of information if the current operating context is the same in your organization.

Operators of the Equipment: According to the book of John Moubrey on RCM II, the best source of failure modes are the operators of the equipment since these people will experience the failure first before the maintenance. This will be true if the operator knows their equipment intimately. If the operators are just mere switch flickers that start and operate their equipment without an intimate understanding of the equipment they're operating, then they may just provide little or no value whatsoever. This will apply if the operator knew their equipment intimately and they have a willingness to share what they know with maintenance regarding the failure. **Downside:** Operators can be a good source of information if they know their equipment intimately or a program of Autonomous Maintenance is in place, or they are part of the RCM Analysis team.

Plant's Rolling Stones: These are the maintenance people who have stayed in the plant for several decades and have witnessed the majority of the failures themselves. Usually, these people are silent, but seeing the wrinkles and the number of white hair they have, these people have a rich amount of knowledge and experience with them. What is important is for industries to have a system to capture their knowledge before they reach their retirement and leave the plant for good. Otherwise, they will just bring all they know to the grave. These people are good candidates to compose the RCM team since they have experienced the majority of the failures on the equipment being analyzed. **Downside:** By far, this is one of the greatest sources of information for failure modes that happened in the past. However, many industries have no program in place to capture what these people know.

Failure Mode and Effect Analysis: FMEA/FMECA reports already done on the equipment would be a great source of failure modes. Although there are several types of FMEA, such as Service FMEA, Process FMEA, Product FMEA, System FMEA, and Machine FMEA. We would be more interested in the Machine or Equipment FMEA in this case as the possible source failure modes and their causes are already indicated in the report. However, if a failure had been covered in the FMEA, or corrective action was already initiated and the failure is unlikely to occur, it does not mean this will not be included in the PM tasks lists as there might be routine inspections or lubrication activities that are still needed. **Downside:** The output of any RCM or FMEA depends on the maturity and experience of the team performing the analysis. If the team is less experienced, then they may not have written all the possible failure modes. Another point is that most corrective actions performed on FMEA can be redesigned or a change in procedure and what we are interested in is if there are any maintenance tasks to be carried out or if the failure has already been eliminated totally on the equipment.

Maintenance Forums: There are several maintenance forums on the internet today. This forum is like a coffee shop but the only customers are the maintenance themselves. You can ask any questions there and several people can respond if they know or have experienced your situation. There are also a lot of active experts in different fields such as Vibration Monitoring, and other PdM instruments. **Downside:** People in the forum may be willing to share information but again we need to check if the operating context is identical or not in your industry.

10.6: What the PM Task Should Include

Whether the PM task is derived through RCM, other streamlined versions, or other means, the PM task should be crystal clear enough for the crew who will execute the task. Our objective is to be ahead of failure. In deriving the PM task, the following details should be known.

Failure Modes: All maintenance tasks reflected on each specific task should address a particular failure mode. The reason for performing the task is to prevent a particular failure mode from occurring that will end up in a failure and downtime of the equipment. Try asking if these tasks will be performed on the equipment, will it prevent the failure from occurring? If the answer is yes, then the task should be included.

Inspection Items: Inspections can either be the use of human senses, gages, or smart sensors. Inspections using human senses will include activities on what to clean, lubricate or points to be inspected in the equipment to establish and sustain the basic equipment condition. As much as possible, inspections should be quantitative and not subjective. In case the inspection cannot be quantified, then a qualitative inspection should be written. If deviations are found during the inspection, there is an option on what to do in case anomalies and deviations are found. For example, upon inspecting the belt pulley, the belt's tension needs to be adjusted, and detailed instructions and instruments should be provided to the person conducting the inspection process.

Key Points and Method: Understand the reason for doing the task and the harm or impact it can generate when the task is neglected. Decide the simplest method of performing the items to be inspected, cleaned, lubricated, replaced, or overhauled. Also, indicate if there is any measuring instrument or device that will be needed during the inspection process, or the inspection will just be done using the human senses.

Tools Needed: Specify the tools, and materials, needed to perform the task. Tools must be prepared in advance with a minimum of 24 to 48 hours before the actual PM begins especially if they will be withdrawn from the storeroom.

Spare Parts and Consumables Needed: If parts replacement will be included in the PM tasks. All items, spares, and materials should be withdrawn from the storeroom a day or 2 before the actual PM execution. Storekeepers must be well informed of the schedule so that they can prepare for the parts and materials needed. Once these parts are staged, then the storekeepers can inform the PM crew that the parts are already available which they can withdraw from the kitting area.

Tasks Duration: The duration of conducting each task is usually estimated by the planner. This refers to the time needed to accomplish a particular task. The duration will be defined for each task which is identified in the Work Order. Although the duration of the task may vary from one person to another, as well as the experience of the person executing the tasks, hence it is recommended to have an average time for each task being accomplished. Once a particular task is done, the person can indicate the actual time performed so that the planner can make necessary adjustments during the next schedule.

Manpower Allocation: This will indicate the number of persons needed to accomplish a certain Preventive Maintenance task. There are also cases where a particular skill is needed and the person may or may not consume all the available given time but only the time required to accomplish a particular task. For example, if welding will be included and the duration of the task is indicated at 2 hours, then the welder will only perform the duration required and not the entire time to complete the PM. This must be indicated clearly in the Work Order plan.

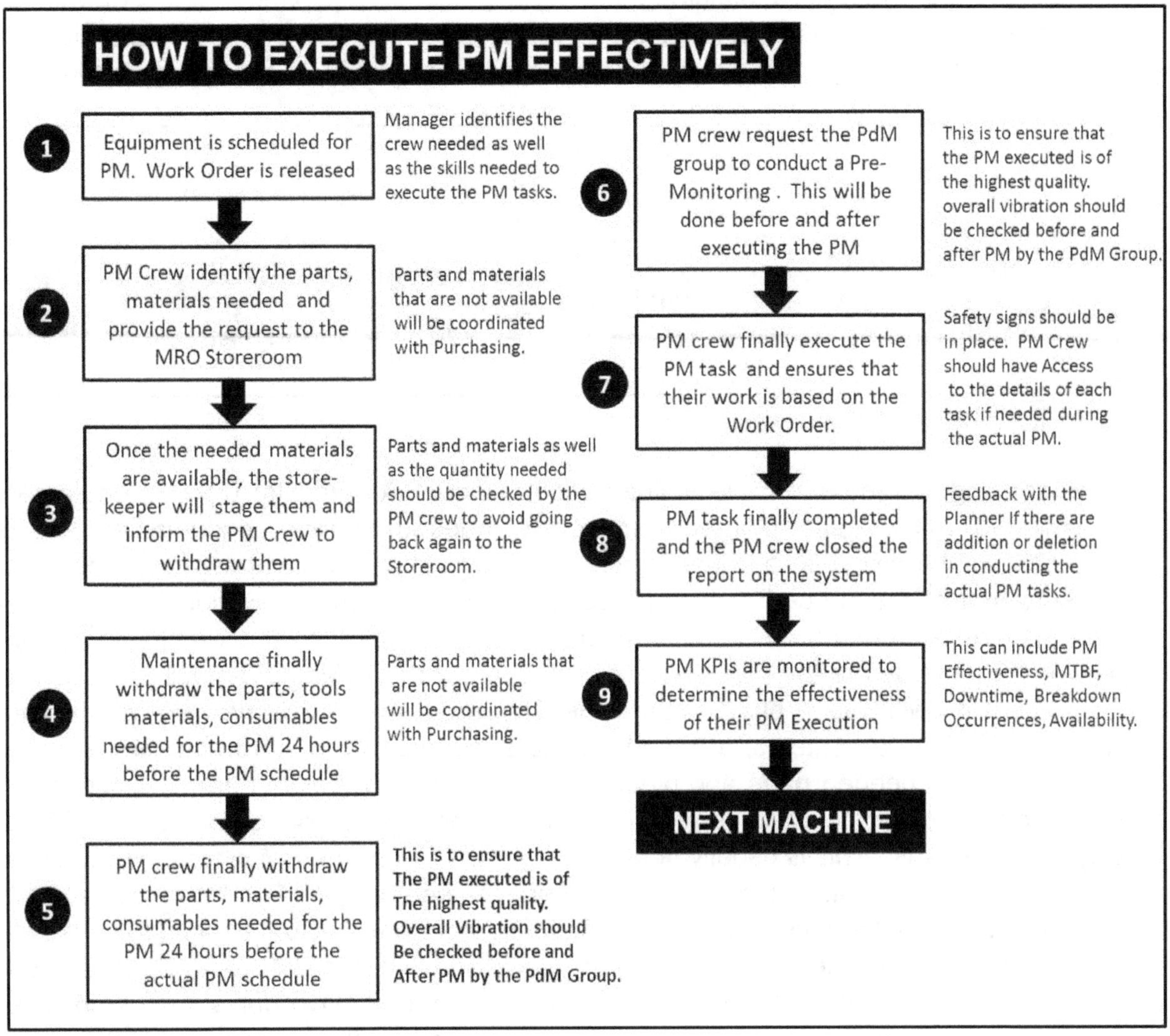

Figure 10.9: How to Execute PM Effectively

Interval: PM will require different intervals. It is important to specify the interval of each task. The unit for the interval will depend and vary from one industry to another. This may be in the form of calendar time, running hours, number of cycles, amount of volume produced, or any other means. It is also important to determine the occurrence of the failure and that the PM interval should always be shorter than the occurrence, otherwise the failure will be missed.

Detailed Procedure: The PM crew who will execute the task can bring their laptop so they can have access to the system or have a printout if they need to counter check their activity based on the procedure especially if the PM crew is new to ensure that they are in compliance on how the task should be accomplished or they can have their laptops available so that they can countercheck their actual activities performed based on the detailed procedures if this will be available in the system.

10.7: Collaboration Between the People Involved in PM

Planning for Preventive Maintenance for a piece of equipment or asset will require collaboration between different functions of the organization. Each of these people involved will be crucial, meaning if one function does not perform, then PM cannot be performed.

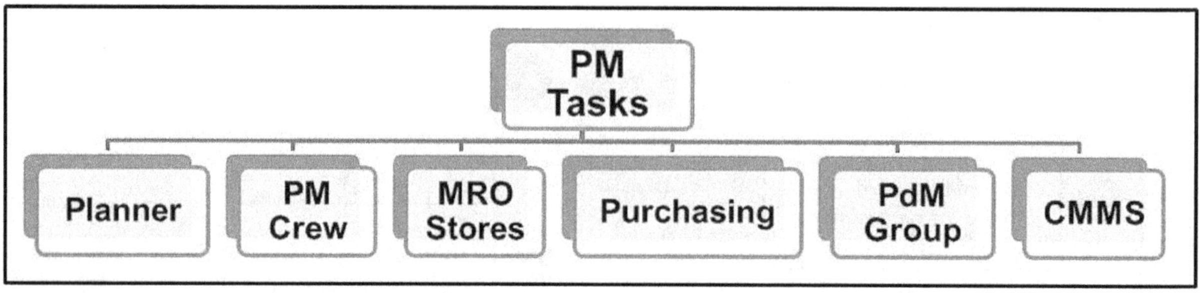

Figure 10.10: Different People Involved in PM

Everything will begin with the maintenance planner. A particular piece of equipment will be selected, and the scope of work, manpower needed, spares, consumables, and Bill of Materials will be prepared and the schedule will be set. The maintenance planner checks the system if all the needed items, spares, and consumables are available. If a part is not stored in the storeroom, the planner coordinates this matter to the Purchasing people to deliver the parts before the PM schedule is due to start. A Work Order will be generated by the system if CMMS is in place both to the Storeroom and PM Crew who will execute the work. Once the parts, items, and needed materials are in the storeroom, the storekeeper will prepare the needed parts and place them in a kitting or holding area. A cart is usually provided to hold the materials needed. The storekeeper will message or call the PM crew to withdraw the parts. This is usually done a day or two before the PM where the PM will stage all the needed materials in their PM bay or near the equipment. The PM crew can also call the Predictive Maintenance to do a Pre-Monitoring before the actual execution of work and a Post-Monitoring after the PM to ensure the effectiveness of the PM work. The work will now be executed by the PM crew based on the given scope of procedures, and guidelines are usually handy to ensure that they are performing based on the standards. At this point, the maintenance now closes the Work Order on the CMMS system indicating that the job had been completed. Special remarks may be added to the report and a feedback session is usually done if some addition, deletions, or changes must be added for future works. The planner makes the necessary adjustments and updates his plan based on the feedback of the PM areas for similar future works. This is usually how Preventive Maintenance is done.

Key Responsibilities of the Different Functions

- The planner will generate a Work Order to the Preventive Maintenance crew
- The planner or PM crew to check the MRO Storeroom on the spares, tools, items, materials, and consumables they need for the PM.
- For items or any materials that are not held in stock in the storeroom, either the PM crew or planner will coordinate with the purchasing to order these items needed.
- All parts, spares, materials needed should be withdrawn from the Storeroom at least 24 to 48 hours before the PM date

- The PM Master checklist will finally be checked before the actual PM execution.
- Pre and Post monitoring will be conducted by Predictive Maintenance to ensure the effectiveness of the PM work.
- The maintenance closes the report in the CMMS system after the completion of the PM tasks.

Chapter 11

Measuring PM Effectiveness and Performance

Measuring our maintenance performance indices and KPI's clarifies the need to focus on long term goals and strategies by comparing our actual performance against the goals we set forth. We can only be in control of the situation if we measure what is meaningful and important to us because measurements allow us to make the right decisions in our organizations.

11.1: Why We Need to Measure our Efforts on PM?

Almost all industries have their Preventive Maintenance activities, but many are not satisfied with the outcome. But regardless of how the industry performs its PM, it is important to measure the effectiveness of our activities on Preventive Maintenance. We simply cannot perform PM for the sake of just doing it. This means that if our efforts in Preventive Maintenance are both effective and efficient, then we can experience fewer failures and breakdowns in our equipment and assets. Remember that when we execute a Preventive Maintenance activity, we are using resources and the industry's money. Therefore to gauge whether our efforts on PM are effective or not, we need to have some form of measurement. These measurements and indicators will also tell us if we are performing the correct tasks at the correct interval. Performing Preventive Maintenance includes different activities at different intervals which can be performed by different groups. The PM tasks lists are the heart of the PM Program. The PM tasks lists indicate what the maintenance needs to do, and how to execute it. In its highest form, the PM tasks list represents the accumulated knowledge of the maintenance workforce or those who will execute these tasks on what to do to avoid a failure on the asset. These PM activities and tasks are assembled into lists and sorted by frequency of execution. They are directed on how the asset will fail. The rule is that these tasks should address the failure modes of the asset especially those with high consequences. But despite our efforts, there will still be failures and breakdowns even with the best PM structure. Our goal is to mitigate, reduce, control, and prolong the duration of the failure and convert the breakdowns that are left into a learning experience to improve the delivery of the maintenance function. This means that if the failure keeps on repeating itself despite the PM activities, maintenance can either perform an analytical problem-solving tool or derive the root cause of the failure.

Chapter 11: Measuring PM Effectiveness and Performance

As discussed in the previous chapter, Preventive Maintenance is composed of different activities from mere inspection to overhauling the entire equipment. It also includes the replacement of parts, cleaning, and lubrication. For scheduled replacement and overhauls, these PM activities should only be done if the part or item has a wear-out pattern and the schedule of performing these tasks must be based upon the useful life and not on the average life of the part or item. This means that there should be a wear-out mode. The important thing is that the interval should not be too soon or too late.

COMMON PM TASKS LISTS INCLUDES

TASKS	EXAMPLE
1) Inspection	- Looking for leaks in a hydraulic system
2) Cleaning	- Removing dirt/debris on the machine
3) Tightening	- Tightening of bolts w/ correct torque
4) Take readings	- Recording reading on gauges
5) Adjustment	- Adjust tension on drive belt
6) Lubrication	- Check level & health of oil
7) Parts Replacement	- Replace worn out parts
8) Overhaul	- Scheduled overhaul of pump
9) Functionality Check	- Check sensors if working
10) Predictive Maintenance	- Routine thermal scanning of motors

Figure 11.1: Common Tasks Performed on Preventive Maintenance

The amount of Preventive Maintenance needed at a facility varies greatly. It can range from walk-through inspection of facilities and inspecting equipment for deficiencies and deviation for later correction up to shutting down the equipment after a certain number of hours or after a certain number of units had been produced depending on the interval used by industries. These PM activities and tasks have two major objectives, first, to extend the life of the asset, and second to detect when the failure on the asset had begun its descent into a potential failure (not yet in a failed state) which is the start of wear-out mode. It is also the assumption of the PM design that when a problem is detected during inspection, the maintenance will respond with a corrective action to address the problem. Listed in this chapter are the different PM Indicators and KPIs we can use to track the effectiveness of how we execute Preventive Maintenance on our equipment and assets. What is important is that all the activities should be directed toward addressing a particular failure mode. The main reason for conducting Preventive Maintenance is to be ahead of the breakdown. It is always less expensive if maintenance is ahead of the failure instead of the failure always ahead of maintenance.

11.2: Machine Breakdown and Frequency

REMEMBERING ATP PLANNED MAINTENANCE 1998 - 2001

Sa Planned Maintenance, Isang Misyon, Isang Direksyon pa rin . . . . .

• ***Graph below shows actual results on the number of breakdowns after thorough implementation of Planned Maintenance 4 Phases To Zero Breakdown***

1st PILOT BDO TRENDING FOR PM PILOT MACHINES

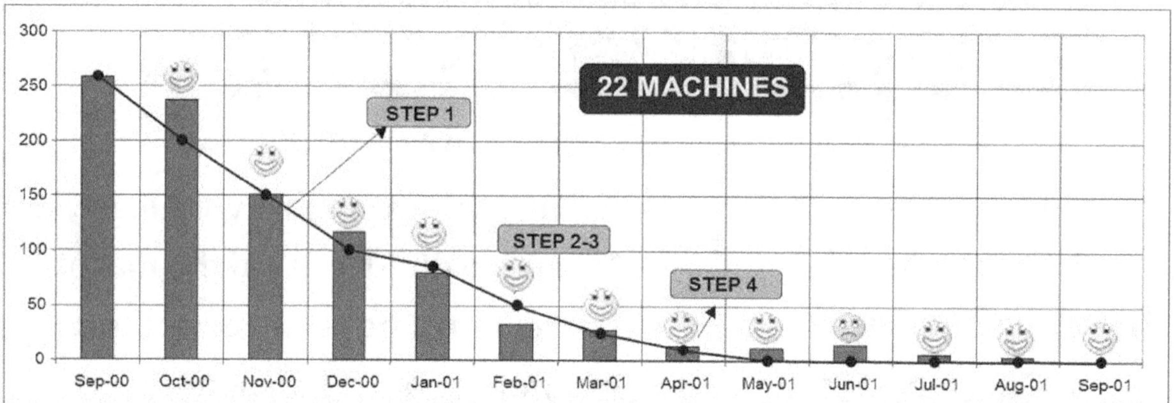

BDO		Sep-00	Oct-00	Nov-00	Dec-00	Jan-01	Feb-01	Mar-01	Apr-01	May-01	Jun-01	Jul-01	Aug-01	Sep-01
	PLAN	259	200	150	100	85	50	25	10	0	0	0	0	0
	ACTUAL	259	237	151	117	80	33	28	13	12	15	6	4	0

LEGEND : BDO - Number of Breakdown Occurrences

Figure 11.2: Actual Planned Maintenance BDO Tracking for Pilot

Downtime has many causes and can be considered Planned (Non-Machine-Related) and Unplanned Downtime (Machine Related downtime). Tracking this downtime should only be done for those caused by breakdowns and failures that will warrant a repair. This will be mostly expressed in hours as the best unit to use. Our goal definitely is to reduce both the number of breakdowns and downtime caused by these breakdowns in our equipment. It is also likewise important that all maintenance involved should have a unanimous and clear definition of what to include and not to include as a breakdown. Only function-loss breakdowns will be considered in tracking breakdowns. This means that breakdowns that caused the machine to stop since something failed will be considered. Exclude function-reduction breakdowns which are breakdowns and failures that can occur where the equipment is still capable of providing the primary function. For example, if the emergency stop is in a failed state yet the machine is still capable of delivering the products, this would not be listed in the tally for breakdown. Figure 11.2 was our actual results on the number of breakdowns as we journey the Planned Maintenance teams which piloted 22 Rank A machines composed of different types across all divisions and successfully reduced the breakdown recurrence through the 4 Phases of Planned Maintenance. BDO stands for Breakdown Occurrences.

When using this indicator in tracking the total number of breakdowns, first only function loss breakdown will be included which means that the machine stopped due to breakdown. The second is for the key people in engineering, maintenance, reliability, and other key people to have a sit-down and define clearly what will be included and not in the breakdown. This was one of our early mistakes. Failure to perform this separation on what to include and not as a breakdown will also affect other indices such as calculating reliability, MTBF, availability, and those indices that involved time.

MESSAGE: Do Not Commit the Same Mistakes We Did 2 Decades Ago [19]	
What We Included as Breakdown	**What We Excluded as Breakdown**
• Breakdowns caused by poor set-up and conversion where the maintenance stops the machine replaced parts which are damaged as a result of poor set-up and conversion • When the technician stops the machine due to breakdowns caused by defects and reworks such as worn-out punch, thereby producing product defects • Function loss breakdown due to no availability of spare parts • Unscheduled repair and overhauling of equipment • Unscheduled replacement of parts • Failure and breakage of tooling • Run to Fail items providing little or no consequences at all, like replacing the tower lamp • All cases of Function-Loss Breakdown **Failure to sit-down and discuss this will lead to each person having their own definition of breakdown** 	• Breakdowns caused by a material related problem • Actual conversion and set-up or the process of changing from one product to another • Machine stoppage caused by interruption, assists, and minor stoppages resulting from jamming of products due to dirt and so on. (Japanese call this chokotei) • Machine downtime caused by PM Schedule such as overhauling of parts • Machine Downtime caused by scheduled replacement of parts as reflected from the tool algo schedule • Repairs attributed by Predictive Maintenance findings • Machine stoppage caused by quality audits • Downtime caused by Facilities stoppages such as no air, no power, water, etc. • PM Schedule, the shutdown of equipment, all planned downtime • No inventory and all non-machine related downtime • Intermittent stops on the equipment such as assists and errors • Parts and items that have worn out and replaced during a Preventive Maintenance • Failures of secondary functions such as protective devices, leaks, broken gauges, sensors where the machine is still running and providing the primary function • All cases of Function-Reduction Breakdown

Figure 11.3: What Should be Included and Not as Breakdown

11.3: Preventive Maintenance Compliance

PM Compliance is one of the most common indices of Preventive Maintenance commonly used by industries which compares the number of activities completed to those that are scheduled to be done. This measurement is perhaps the most abused of all KPIs for maintenance. In an industry, a planner will provide the lists of maintenance tasks to be done on the equipment which will include the Bill of Materials, duration, manpower needed, MRO Spares for replacement, and in other instances will also include work done by third-party contractors. PM Compliance is a maintenance indicator that will measure how many PM tasks were actually completed and closed divided by the total number of tasks to be accomplished on or before the due date. The equipment will then be scheduled and another group of people will execute the Preventive Maintenance tasks. Hence, if we have 25 activities to execute in your Preventive Maintenance lists to be performed, let us say on a semi-annual schedule on a particular piece of equipment, and all 25 activities had been completed for a given period stated before the due date, then we can say that the PM Compliance is at 100%, which is perfect. Measuring PM Compliance or having a very high percentage of being compliant on every activity on Preventive Maintenance is useless if the equipment still encounters many infant mortality and random failures in which maintenance performs emergency work right after a Preventive Maintenance had been initiated on the equipment. If we are doing Preventive Maintenance activities, yet the equipment is always failing, then something is wrong with how we execute our Preventive Maintenance tasks, or perhaps, we need to overhaul not the equipment, but the lists of maintenance tasks that we perform on Preventive Maintenance. All activities on Preventive Maintenance should and must address a particular failure mode. The formula for PM Compliance is as follows:

$$\text{PM Compliance} = \frac{\text{Number of PM Tasks Completed}}{\text{Total Number of PM Tasks Listed}} \times 100\%$$

Performing Preventive Maintenance for the mere sake of complying will not improve the performance of the equipment. All PM activities must prevent a failure from occurring. If there are activities on the Preventive Maintenance lists that do not address any particular failure mode, then delete them in the PM lists since it is just a complete waste of time, money, and resources to execute such an activity. Remember the golden law on doing Preventive Maintenance. The law states that the cost of doing Preventive Maintenance must always have to be lower than the costs of consequences of failure it is meant to prevent. If the Bill of Materials for performing this annual Preventive Maintenance is, around 100,000 US dollars and the cost of failure it is meant to prevent, is 10,000 US dollars, then it is a good idea to disregard the PM. Having a high percentage of PM Compliance is next to useless if the cost of doing PM and replacing the equipment increases. PM compliance will be 100%, yet the cost of doing Preventive Maintenance almost tripled since we replaced the parts that are still functioning and we have not maximized the lifespan of these parts. PM Compliance can only be effective if we also measure and compare it with the emergency and repair hours on the equipment. If a PM activity does not need to be done, it is best to remove them from the list; otherwise, it might induce infant mortality failures into otherwise stable systems in the equipment.

11.4: Preventive Maintenance Effectiveness

As discussed, PM Compliance is the most used and abused metric. Measuring PM Compliance or having a very high percentage of being compliant on every activity on PM is useless if your equipment encounters a lot of infant mortality failures and emergency work right after a PM initiative. However, a better approach is to compare these indices with the number of hours spent on emergency work. We can use these indices to determine if the Preventive Maintenance tasks we perform are effective or not. This means that if the PM Compliance is high, yet the number of breakdowns or emergency work is also high, then something is wrong with how we perform Preventive Maintenance in the equipment since it should be the opposite. An effective PM strategy must reduce the amount of Reactive Maintenance in the plant by a margin of 30% or more. This means that if more time is spent on doing corrective and reactive maintenance compared to the Scheduled Planned Activities, then we need to evaluate the maintenance tasks in our PM lists and how it is being performed. Therefore, to measure PM Effectiveness is to plot the number of hours spend on doing PM and the number of failures or breakdowns experienced after performing PM:

- Number of PM hours
- Number of hours of Emergency Work

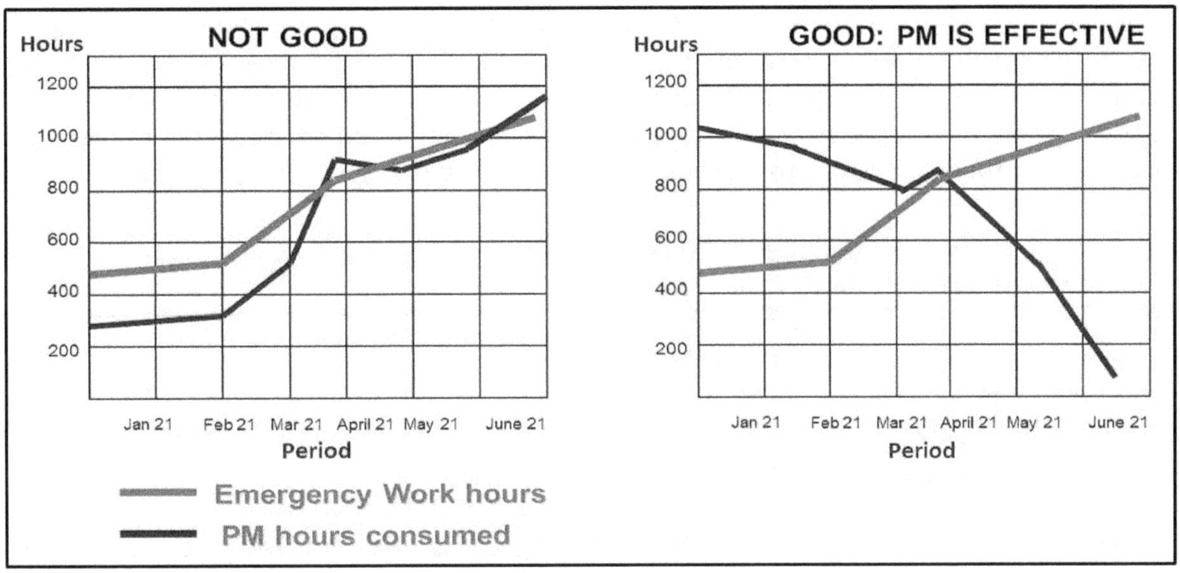

Figure 11.4: PM Effectiveness

If the Preventive Maintenance tasks were performed last January 15 to 18, 2022, the emergency worked hours should be measured starting from January 19 to February 18, 2022, and so on. Performing Preventive Maintenance for the mere sake of complying will not improve the performance of the equipment. All PM activities must prevent a failure mode from occurring, if not then delete those PM activities in your lists.

Another way of measuring PM effectiveness is to compare the number of Planned Work Orders versus Emergency Work Orders but I would prefer to measure this in hours instead of frequency. Although this will be the same for PM Effectiveness, the main difference is that this is based on the number of Planned Work Orders or Jobs generated compared to the number of breakdown or emergency repairs measured in frequency and not in hours. The word Planned in this case will include all activities on maintenance including Predictive Maintenance. The only problem I see in this PM indicator is that it is not measured in time but on its frequency. This means that if I have 10 PM Work Orders that have been fully completed, and 1 breakdown this might look good on record. However, if we convert this into time in which the 10 PM Work Orders took 10 hours to complete, and 1 emergency breakdown which took 48 hours to complete perhaps due to no available spare parts in the storeroom, then this is not actually a good performance from the PM Group. It would be better to use PM Effectiveness in hours to gauge the effectiveness of all the activities performed on Preventive Maintenance.

11.5: Maintenance Backlog

A maintenance backlog is a time indicator that consists of delays in performing the scheduled maintenance works, or those activities which have not yet been completed. This consists of pending scheduled planned activities allotted on maintenance. The maintenance backlog to be executed may include emergency work, Preventive, Corrective Maintenance, Predictive Maintenance, or improvement activities. To measure this index is to consider all the tasks and activities included in the maintenance planning. The maintenance backlog will be equal to the total sum of man-hours for the planned, pending, and completed work orders, divided by the total available man-hours. Maintenance backlog may be expressed by the number of hours. Although if we adopt this definition of maintenance backlog as those pending or delayed work where the due date had already been exceeded, then we might be missing something. There are various reasons for the delay in executing the work request such as;

- Delayed in the delivery of MRO Spare Parts
- Delayed in the delivery of consumables
- Operations waiving the Preventive Maintenance activities
- Maintenance performing other tasks
- Maintenance on leave
- Not enough manpower workforce on maintenance

One of the things that come to mind regarding the maintenance backlog is the maintenance tasks really need to be done, even if priority 1 is indicated in the work order, the question is, how do we define the priority in maintenance? The second question is, how about the interval? Are we sure enough to replace this part since the planner indicates to replace the part when it is still functioning? If we don't comply, then we have a backlog. My recommendation is that before measuring this index, maintenance must have a clear definition of what to consider and what not to consider as a backlog and not just base everything on the due date of the work order but rather on the criticality of the tasks. This means that if there are two pending work orders on different equipment where a failure on equipment 1 will totally halt the whole operations, while for equipment 2 which has a backup, then the priority should be given to equipment 1. Prioritization of executing these maintenance tasks should be based on the criticality of the equipment and the consequences of the failure if the tasks would have been neglected instead of basing the backlog on the due date of the tasks.

The criticality and consequences of the failure vary from one industry to another. A failure of a power plant may lead to a city blackout, where the plant will be on the nightly television news, and the company's lawyers are involved giving a lame excuse regarding the cause of the blackout. On the other hand, a failure from one carton box factory may mean a delay in the processing and your boss eventually asking for the root cause of the failure all the time. Here are some cases to indicate if the failure will be critical:

- If the equipment is in an unmanned location
- If the failure can result in environmental consequences
- If the failure can result in safety consequences or can kill or injure someone
- If the failure can shut down the whole operations temporarily
- If the failure can permanently shut down the whole operations (loss of job)
- If the failure can result in industrial accidents
- If the failure can breach any known environmental regulations and laws
- If there is an absolute certainty that the failure cannot be detected

- If there is no redundancy or standby for the equipment
- If the cost of the failure will be extremely high and can result in a penalty
- If the company lawyers and insurance people will be involved in the failure
- The worst case will be if the media and those crooked old politicians will be involved

If any of the activities specified in the tasks will lead to the above situations, then we can define a Priority 1 basis and without a doubt must be included in the maintenance backlog if the tasks had not been done on the due date provided. On the other end, for lesser critical failures, we can reschedule the date to a more convenient time and remove them from the backlog. Every failure has a specific set of consequences associated with it, and if the consequences are very serious, all efforts should be done to prevent the failure from occurring. Proactive maintenance has much more to do about avoiding or reducing the consequences of failure rather than it has to do with preventing or eliminating the failure itself. Therefore, a proactive task is only worth doing if it deals successfully with the consequence of failure which it is meant to prevent.

The maintenance backlog provides a list of maintenance tasks that must be accomplished over a given period. Although a small amount of maintenance backlog is generally acceptable. These maintenance backlogs must likewise be reviewed and prioritized if they really need to be performed or not. If a certain maintenance backlog is irrelevant, then it should be deleted from the lists. Prioritizing these backlogs and analyzing the causes for the delays will definitely improve the future scheduled maintenance plans of the plant. People involved in generating the activities such as planners must continuously be in constant communication with those who will execute the task itself. The idea is not to eliminate the backlog but to manage it effectively. The formula for Maintenance backlog is as follows:

$$\text{Backlog} = \frac{\text{Sum of All Planned, Pending, and Completed tasks}}{\text{Total Available Man-hours}}$$

Generally, these maintenance backlogs will always be present since performing maintenance is a never-ending process needed to sustain the equipment. They can also be tricky as any maintenance tasks that are written by the planner with a due date and have not been accomplished will always result in a backlog. Maintenance people must review all the tasks and intervals performed on their equipment and assets. First on the list will be the frequency or interval of conducting the tasks. Are we doing these tasks too much or not? For example, in building A, the window-type aircon needs to be cleaned monthly as indicated in the plan, however, if the room temperature doesn't fluctuate, then we can ask the planner to make this activity every quarter or even semi-annually or just based the cleaning if there will be an increase or decrease of temperature from + /- 5 ° C.

11.6: Should We Measure Wrench Time on Maintenance?

According to studies, most industries will have a wrench time of around 30 to 35%, yet this can be increased from 50 or more. Wrench time started its roots in 1910 when the IE (Industrial Engineer) conducted a time and motion study for the daily production operators in a plant. Although this practice is still being carried out today as industries hire IE (Industrial Engineers) to conduct a time and motion study. A clever operator upon seeing a person with a stopwatch approaching them will opt to go to a slow-motion mode because whatever the time the IE recorded for operators in producing a product will now become the standard and management are keen to question operators on why the target has not been hit.

Similarly, wrench time is an indicator that is used to measure how much time the maintenance people spend on doing actual maintenance work on the equipment. Others refer to this as the tool time or when maintenance is actually holding a tool which of course is not only the wrench. Although according to its definition, the wrench time will not include the travel time, work order assessment, tools organization, encoding report on the CMMS, locating spares in the system, attending meetings, task communication, working on an improvement, or taking a break time.

In one of my training on World Class Maintenance – The 12 Disciplines, a couple of guys working in an automotive plant approached me during a coffee break and asked me if what they are doing on maintenance is right or wrong. I asked them precisely what they are doing so I can provide them my thoughts. One guy said that at the start of their shift, they need to get a form and write down all the activities they did on an hourly basis and at the end of the shift, they need to submit the form to their supervisors. I told them that supposed that I am your supervisor and at the end of the shift, you wrote 10 pages report stating what you have repaired every single hour, another person submitted a blank report with only his name on it with just a 1-hour activity on performing greasing on 20 rotating equipment. My question is, who performed maintenance more effectively and efficiently? Does it mean that if you repair an average of 5 machines every hour making your report very long indicates that your wrench time is better than the person who just performs a 1-hour greasing during his shift?

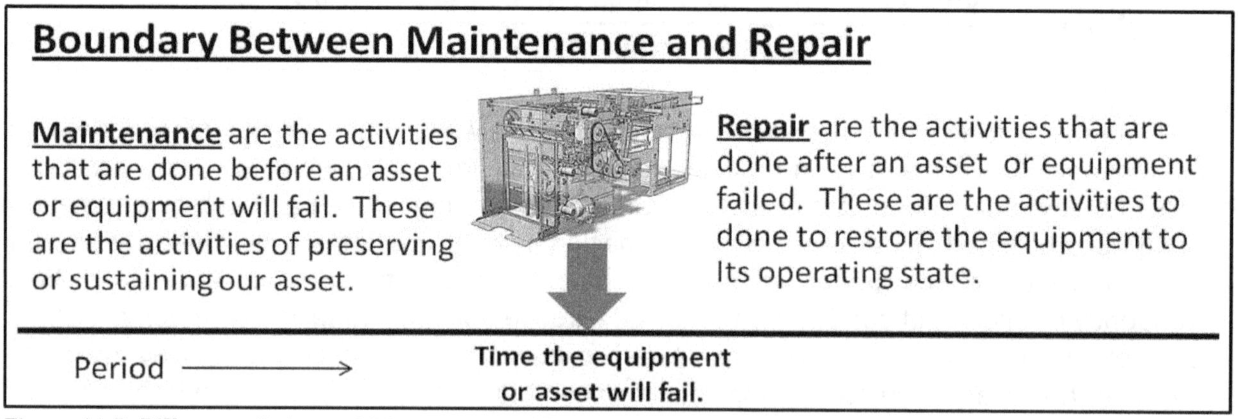

Figure 11.5: Difference Between Maintenance and Repair

My thoughts may differ from other consultants but when we look up the word maintain in the dictionary, it means to sustain or to preserve something. Therefore, the term maintenance are the activities that are done before an asset or equipment will fail. These are the activities of preserving or sustaining our equipment and assets. Repair are the activities that are done after an asset or equipment failed. These are the activities done to restore the equipment back to its operating state. Maintenance and repair are the opposite of both worlds. While others think that maintenance and repair are the same. They are not, since there is a boundary and a thin line that separates them. We need to create this mentality with our people. Do we learn from the failures? Industries will never learn from their problem by just repairing the equipment as it will just cause a repeat of the failure. The problem with wrench time is that it will measure both Reactive and the Planned Work together. Does it mean that if what we did for the week is to repair failures all the time, our wrench time is high? The answer of course is yes, this means that if there are a lot of repairs done, the wrench time is high, which also contributes to a higher downtime and maintenance cost.

Another point I would discuss in wrench time is that doing improvements, and performing root causes will not be included in the wrench time. The job of maintenance is not only to preserve but also to improve the reliability of their equipment and assets by identifying parts with inherent design weaknesses and modifying them. Although other experts say that maintenance can only preserve and sustain their equipment since no amount of maintenance can really improve the reliability of the equipment. My take on this is that the word maintenance can mean 2 things. We have the task and we have the people. If we consider maintenance as a task, then I agree that maintenance tasks cannot improve reliability. The task can only sustain it. But if we consider maintenance as human beings who are the maintenance people themselves, then they can definitely improve the reliability of their assets by understanding the inherent design weaknesses of their equipment and adopting continuous improvement strategies and modifications. Although some industries would want to measure these indices, caution must be taken in measuring them, and I would strongly recommend including them in the Planned Maintenance Work.

11.7: Percentage of Maintenance Cost to RAV

RAV or Replacement Asset Value is also called the Estimated Asset Value (EAV). It is the cost of maintaining the asset which is measured against the value of the asset. It can be said that as the percentage of the cost to replace the asset. As we continue to operate our equipment and assets, the value of our equipment depreciates. The main cause of depreciation is that our assets deteriorate over time. There will always be the subject of wear and tear for mechanical parts. Deterioration is given and will happen in our equipment and assets for as long as the asset is loaded. The best that maintenance can do is to prolong its process, however, when the time comes, then the lifespan of the equipment had been reached, and the equipment will be decommissioned which finally ends its lifespan. With this, the cost value of our equipment will be reduced based on the amount of time we use our equipment and assets. The book value of equipment or asset refers to the book from the Accounting of Finance Department and is reflected on the industry's financial statement. This will be equal to the Total Cost of the Asset minus its liabilities. On the other hand, the Market Value will be the value or cost of the asset according to the stock market. This will be the cost of the asset when sold to the market.

If we speak about the Percentage of Maintenance cost to RAV which is expressed as a percentage, the lower its value the more effective we are in maintaining and preserving our equipment and asset. This can be used by maintenance managers and decision-makers to decide whether to still continue operating the equipment, modify it, or just retire the equipment for good and purchase a piece of new equipment. This is like owning a car for 5 years. This will be one of the deciding factors on whether we still continue using the car or just sell it at the current Market Value and purchase a new one.

Imagine you plan to purchase a car, hence, you went to a car dealer and plan to purchase a brand new Toyota Vios 1.3 at PHP 681,000.00 or $ 13,266.00. You also heard from a colleague of yours that her friend is selling her Toyota Vios rush as she will be immigrating to another country. You and your wife went to see the lady and the car as well. The cost of the car was PHP 395,000.00. You learned that the woman just bought the car exactly 1 year ago and checked the speedometer which is at 12,050 kilometers. You interviewed the owner and asked her about the expenses she incurred throughout and the owner said that she changed the oil, filter, car wash, and the consumable which is the fuel. There was no scratch on the car and it still looked brand new. You also checked the engine and seems everything is all good. So the question, in this case, would you purchase the brand new car or go on to purchase the 1-year-old car from the lady driver?

• Brand New Car	= PHP 681,000.00
• Lady Selling Car Rush	= PHP 385,000.00
• Difference	= PHP 286,000.00

• Lady Selling Rush Car Expenses	
• Change Oil (just once)	= PHP 3,000.00
• Car Wash (Monthly)	= PHP 2,400.00
• Fuel (Consumable)	= PHP 11,685.00
• Maintenance Cost	= PHP 17,085.00

For industries, we need to define clearly what will be included in the Overall or Total Maintenance Cost since this can vary from one industry to another. Second, we need to be consistent in only including the cost that is spent on maintaining the asset during its period. This indicator, Percentage of Maintenance Cost to Replacement Asset Value can provide us information and decision on whether we need to maintain or replace the asset with a brand new one. The lower the value of this percentage, the better. This means that if your assets' Percentage of Maintenance Cost to RAV is at 20% or more, then your maintenance is too expensive. Mostly the target value will be at 3% but lowering this value means that we have lowered the cost of maintaining the asset and it would be more economical to keep the asset operating instead of selling it at a depreciated value and purchasing a new asset or equipment. If we compute the Percentage of Maintenance Cost to RAV for the second-hand car that we decide to purchase, then this will be equal to:

- Percentage Maintenance Cost to RAV (%) = (Annual Maintenance Cost ($) x 100) / RAV ($)
- Percentage Maintenance Cost to RAV (%) = (PHP 17,085.00 x 100) / PHP 385.000.00
- Percentage Maintenance Cost to RAV (%) = 4.43% (for the second-hand car)

The Percentage of Maintenance Cost to RAV can also provide us information on which assets we need to replace, modify, and those that we need to maintain. If these indices will be included in the maintenance function, this should be discussed with the IT to integrate this into the system if the plant currently has an EAM or CMMS software as this will be very tedious to do individually for all equipment and assets in the plant. Although we can have a decision on whether to continue operating or sell the equipment whose percentage will be 20% and beyond, what is important is we also need to consider the non-tangible or qualitative factors such as the following:

- Will the new equipment be easier to operate?
- Will there be less movement on the operator?
- Will the new equipment be more ergonomically convenient for the operators?
- Will the training for the new equipment be easier than the old one?

Case Study: An OEM is proposing to management to purchase new equipment. The cost of the new equipment is $ 50,000.00, and the maintenance cost is projected at $ 8,000.00 annually. The cost of the current equipment is $ 65,000.00 and the average maintenance cost is $ 12,000.00. The Replacement Asset Value today of this machine is $ 35,000.00. A TPM Planned Maintenance Team had concluded its prototype for evaluation on the current equipment and is proposing a modification cost of $ 5,000.00 which can increase productivity by 30% and is projected to reduce their maintenance cost by 60%. It is forecasted that the demand in the next 3 months would be increased by 20%. Would it be feasible to buy the new equipment as recommended by the vendor or proceed with the modification by the Planned Maintenance?

Given:

- Cost of New Machine = $ 50,000.00
- Projected Maintenance Cost Annually for New Machine = 12,000.00
- Cost of Old Machine = $ 65,000.00
- Projected Maintenance Cost Annually for Old Machine = 12,000.00
- Projected Maintenance Cost Annually for Old Machine with Modification = $ 12,000 - (12,000.00 x 0.6)
- Projected Maintenance Cost Annually for Old Machine with Modification = $ 4,800.00
- RAV for the Old Machine = $ 35,000.00
- Modification Cost = $ 5,000.00

For the New Machine without Modification

- Percentage Maintenance Cost to RAV for New Machine = (Annual Maintenance Cost ($) x 100) / RAV
- Percentage Maintenance Cost to RAV for New Machine = $ (12,000.00 x 100) / $ 50.000.00
- Percentage Maintenance Cost to RAV for New Machine = 24 %

For the Old Machine without Modification

- Percentage Maintenance Cost to RAV for Old Machine = (Annual Maintenance Cost x 100) / RAV ($)
- Percentage Maintenance Cost to RAV for Old Machine = $ (12,000.00 x 100) / $ 35.000.00
- Percentage Maintenance Cost to RAV for Old Machine = 34 %

For the Old Machine with Modification

- Total Maintenance Cost = [12,000 – (12,000 x 0.60)] = $ 4,800.00
- Percentage Maintenance Cost to RAV (w/ Modification) = (Annual Maintenance Cost ($) x 100) / RAV

• Percentage Maintenance Cost to RAV for Old Machine = $ (4,800.00 x 100) / $ (35.000.00 – 5000)
• Percentage Maintenance Cost to RAV for Old Machine = 16 %

The conclusion, in this case, is to proceed with the Planned Maintenance modification as this indicates the lowest value regarding the Percentage of Maintenance to RAV and disregard the proposal of the vendor to purchase the new equipment. This indicator can be a good indicator of whether we continue to sustain the equipment or retire it for good and purchase a new one.

11.8: Mean Time between Failures (MTBF) Explained

By definition, MTBF is an average measure of reliability that the equipment will run without failing. The most common units used will be in hours. Its origin can be traced to the US Military Standards (MIL-STD-217), and since then, it has been widely used in other applications in various types of industries. The original Reliability Prediction Handbook was MIL-HDBK-217, the Military Handbook for the Reliability Prediction of Electronic Equipment which was published by the Department of Defense, based on work done by the Reliability Analysis Center. The MIL-HDBK-217 handbook contains failure rate models for various electronic systems and components such as integrated circuits, transistors, diodes, resistors, capacitors, relays, switches, connectors, and so on. These failure rate models are based on the field data that can be obtained for a wide variety of parts and systems. MTBF can also be defined as the average time between two successive failures, and it is a measure of the time the equipment is operating until the time it encounters a failure. By empirical testing or allowing a part to fail which is a form of destructive testing, the length of performance or the functional life of a population of items can be divided by the total number of failures to achieve the MTBF of a part or component. By the word mean, we speak about an average amount of operating time between the occurrences of breakdowns or failures that require repair.

MTBF will be easy to compute if the equipment is suffering only from breakdown losses, however, if a machine is experiencing other losses, these must be excluded in calculating the MTBF value of the equipment. Several equipment losses will also constitute a downtime such as minor stoppages, set-up, or conversion, defect losses, start-up losses, and quality defects. What is important in calculating the MTBF value is to consider only the downtime caused by breakdown only.

MTBF is also defined as the average time between two successive failures and therefore is a measure of the trouble-free time. It is an average measure of reliability that a device will run without failing. It is a reliability engineering term that means the average amount of operating time between the occurrences of breakdowns that requires repair. It means the average amount of time between two successive failures and is based on historical data or estimated by the vendor, which is used as a benchmark for reliability. MTBF trend will be the higher the value, the more reliable the machine or asset is performing. Depending on how you intend to measure MTBF since this can be measured on a system-based level, sub-system based, component-based, equipment-based, several groups of machines, sub-assembly based, or even to the spare part level itself. The formula for MTBF will be as follows:

$$MTBF = \frac{\text{Operating Time}}{\text{Number of Failures}}$$

Where: Loading Time = Available Time – Non-Machine Related Downtime
Operating Time = Loading Time – Machine Related Downtime

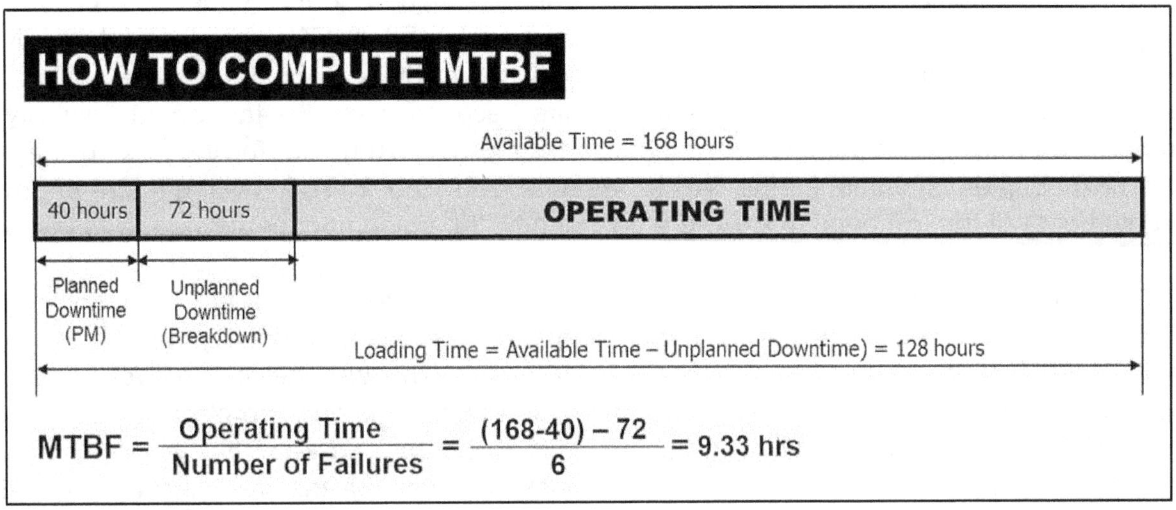

Figure 11.6: How to Compute MTBF

This means that if we are performing the correct Preventive Maintenance on our equipment, then we expect fewer breakdowns which means achieving higher MTBF on our equipment and assets.

11.9: Reliability Explained

Reliability is the probability that the equipment will perform its specified function under specified operational and environmental conditions under a given and specified time. Reliability can be said as the probability that no failure will occur throughout a prescribed operating period. According to Bazovsky, the modern concept of reliability in popular language is the capability of equipment not to fail or break down in operations. When equipment works well and performs to do its job for which it was designed to perform, such equipment is said to be reliable. Reliability is the probability that an item will operate without failure throughout a specified interval. For example, if we schedule the next week's production, the equipment's reliability or probability will be our main concern. Since reliability is a probability, we need to have some historical data on how the equipment performed, perhaps within the past couple of months, to better understand how the equipment will run in the future. This means that to know the future of the equipment, we need to know the past of how the equipment performed. One problem with Production Planning is that it assumes that the equipment will run without a single failure or downtime, which, in reality, is hard to achieve. Second, breakdowns or failures are not the only types of equipment losses that can be experienced on the equipment. What is important is determining the type of losses the equipment is suffering and assigning the correct indicator to measure its performance. My point of view, in this case, is that failures and breakdowns must not only be the indicator to determine if equipment or asset is reliable or not since failures and breakdown is just one part of the losses. This means that if a piece of equipment experienced no breakdown, the MTBF is high as well as its reliability, but if the equipment's speed is reduced by 15% would we declare it to be reliable? The problem with the definition of reliability is that it assumes that failures and breakdowns are the only losses on the equipment which is not actually the case.

If I explain reliability in the simplest way I can, assuming that the reader is in his youth or high-school days today, and you have your current girlfriend, but if we think about your life 20 years from now your current sweetheart may or may not be your future wife. What I am saying is that reliability is about your future wife because when we talk about reliability it is about the future. How will my equipment run for the next 10 hours, for the next 10 days, for the next couple of months, for the next six months, and so on? To know the future of how the equipment will run, we need to know the past of how the equipment performed.

In figure 11.7, let us say that we want to determine the reliability of this equipment for the next 10 days, and reliability is at 90%; this means that there is a probability that the equipment will run 90% of the time with a 10% chances of failure. This will answer how your equipment will function in a given set of times, but again this calculation is only a probability to give us an idea of how the equipment will run for the next 10 hours. To know the future of how our equipment will operate, we need to have a background about the past and asks questions such as:

RELIABILITY FORMULA

$$R(t) = e^{-\lambda(t)} \times 100\%$$

Where : R = Reliability
t = time
λ = lambda or 1 / MTBF

- If t = 10 hours and MTBF = 100 hours, then lambda = 1 / MTBF = 0.01

$$R(10) = e^{-(0.01)(10)} \times 100\% = 0.9 \text{ or } 90\%$$

- If t = 20 hours and MTBF = 100 hours, then lambda = 1 / MTBF = 0.01

$$R(20) = e^{-(0.01)(20)} \times 100\% = 0.818 \text{ or } 81.8\%$$

- If t = 100 hours and MTBF = 100 hours, then lambda = 1 / MTBF = 0.01

$$R(100) = e^{-(0.01)(100)} \times 100\% = 0.3678 \text{ or } 36.78\%$$

Figure 11.7: Formula for Reliability

- What is the MTBF of this equipment for the past period?
- What is the Failure Rate of this equipment for the past period?
- How well did the equipment perform over the past couple of years?
- What failures were mostly encountered?
- Were we able to eliminate the failure?
- What is the probability of failure this time?
- Is the failure likely to occur in the equipment?
- Are we prepared for the failure this time?

This indicator can determine which equipment has high and low reliability at a given time, we can use this indicator to focus our attention on which equipment to prioritize in terms of Preventive Maintenance.

Chapter 12

Automating Preventive Maintenance through CMMS

In planning for automation for maintenance, we need to identify what are those that are important that needs to be automated. It should be understood that the CMMS software should adjust to the requirements needed by maintenance and it is not the maintenance who should be adjusting to the software.

12.1: Maximizing the Use of CMMS

What I wrote in my first book on World Class Maintenance, automation and CMMS should be the last maintenance discipline to be done as we need to prepare and understand first what are those important to us that need to be automated. The main objective of CMMS, EAM, or whatever software or system is used is to automate and streamline the maintenance process so retrieval of needed information should be fast, accurate, and consistent. While many industries already have their own form of system or software for maintenance in for form of CMMS (Computerized Maintenance Management Software) or EAM (Enterprise Asset Management) in place, many of the features of this software are underutilized. What is important is that it should serve its purpose and maintenance is benefitting from its use. The key is that maintenance should implement the different disciplines before they can start automating.

Although during my employment days more than a couple of decades back working in a large semi-conductor industry, our Facilities was using CMMS software to automate their work orders and Preventive Maintenance. As time travels, so as technology and automation itself. Today's CMMS has further been improved, enhanced, and upgraded due to stiff competition in the market as well as to satisfy the needs and wants of the maintenance function. This means that today's CMMS is not only about automating the Work Orders or the Preventive Tasks but other important and key information as well, but CMMS software also serves just more than its primary function. Throughout the years, further enhancements and upgrades were done which led to the birth of another kind of software which we now called the EAM or Enterprise Asset Management Software.

If we discuss the history of CMMS, it started during the early 60s with very limited capabilities with the IBM mainframe computers. Everything was big and slow during that time. In fact, the first inception of the hard disk drive originated from IBM. IBM called it the IBM Model 350 Disk File which was a huge device. This drive contains fifty sets of 50 24-inch disks contained inside a cabinet that was as large as a bookshelf and can store around 5 MB of data.

This storage can store 5 MB storage

Figure 12.1: The Inception of the Hard Drive

The difference between these two is that Enterprise Asset Management (EAM) software provides a holistic view of an organization's physical assets and infrastructure throughout their entire lifecycle, starting from the design, procurement, installation, commission, operations, and finally its disposal. While a CMMS will focus more on the maintenance needs. EAM software will be much bigger and more comprehensive as it will include all the multiple business functions in an entire organization. Finance, Purchasing, Warehouse, Supply Chain, Human Resources, and other functions will also be included in the EAM software. Most of them have not maximized their use. Although preparing work orders for scheduled and repair works is just one of them. Whether the industry is using a CMMS or EAM, the software is just as good as what we populate in them. EAM software is designed for managing an organization's physical assets across the whole enterprise. They were developed after the CMMS and provide industries to link computer systems across multiple business sites for large business organizations. EAM software not only includes the maintenance management capabilities but likewise will also consider the Total Cost of Ownership (TCO) for a company's physical assets. They provide a wider range of features to monitor, manage, analyze, and evaluate its asset performance as well as the costs incurred throughout the whole asset's life cycle, from its acquisition to its final decommissioning and everything between them. EAM is much bigger in scope compared to CMMS, as well as its price. A CMMS will exclusively focus on maintenance, while an EAM system or software includes all the other functions of the organization that is entirely linked to the asset. A CMMS starts tracking after an asset has been purchased and installed, while an EAM system can track the whole asset's life cycle, starting with the design, installation, and its entire life span until its time of decommissioning.

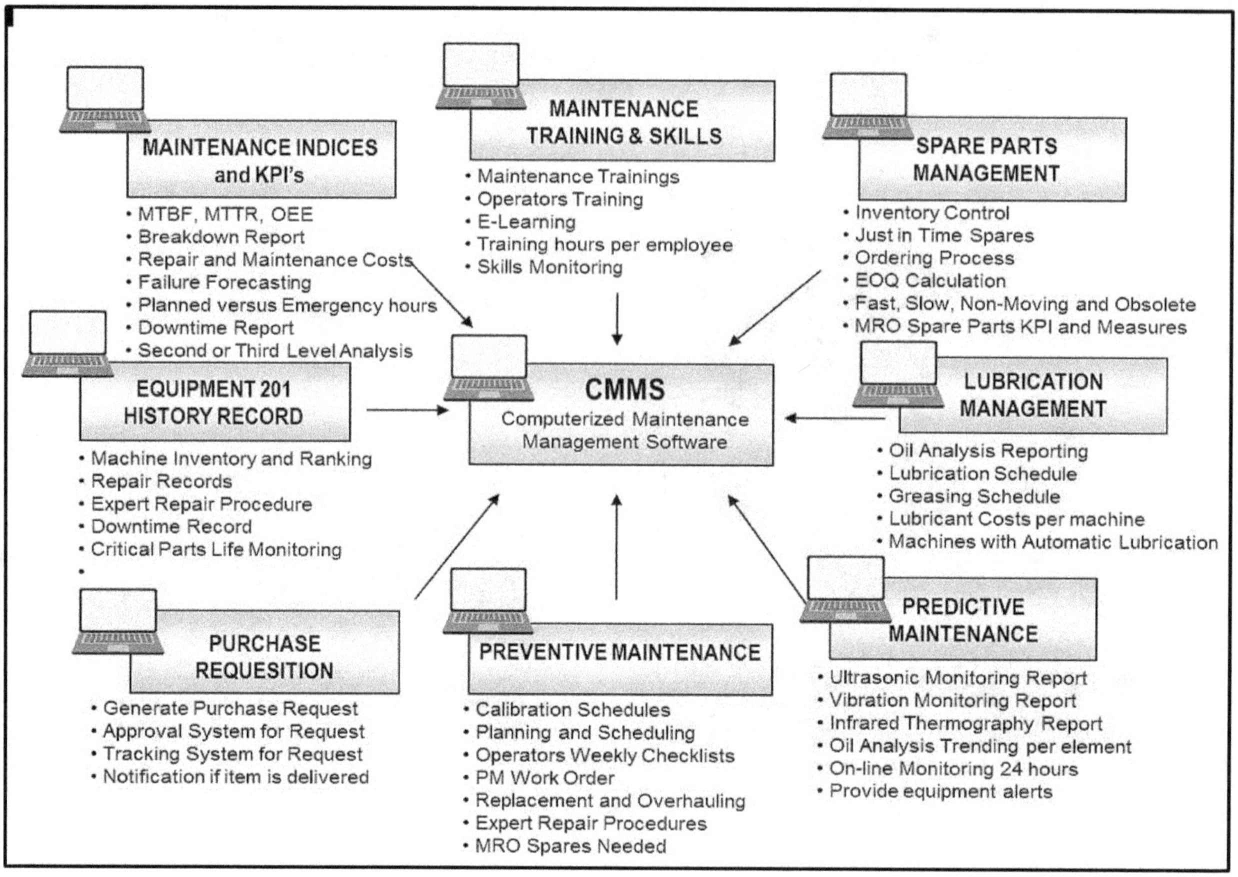

Figure 12.2: Understand What the CMMS Can Do

Although even with today's CMMS and EAM software with all its functions, features, capabilities, and apps, there is a common problem I see with industries. It seems that despite the capabilities of today's CMMS and EAM, most users in industries underutilized their capabilities. Perhaps from a scale of 1 to 10, around 2 to 3 percent of its capabilities and features are often used in industries while the rest of its features remain untapped. That is why in my original conception of the 12 Disciplines on World Class Maintenance, CMMS should be the last strategy for maintenance as maintenance needs to identify what are the most important facets of each discipline that needs to be automated so that it will fully serve its purpose. In most industries today, what happens is that the Decision Makers purchased the software, which can either be a CMMS or EAM, then tell everyone to use it. The result of this will just end up with the software being underutilized and the person using the software will just be the IT or the CMMS administrator. The maintenance once again will be doing everything manually not knowing that the software can do it more efficiently, and their thinking is that this thing can cost me my job. If one of the industry's objectives is for the CMMS to generate the KPIs or indices for maintenance, then the first thing we need is to educate all people on how to decode each of these losses. A code should be provided for every breakdown, minor stoppage, or if the downtime is considered Planned or Unplanned downtime. Operators or whoever will populate the system should ensure that they enter the correct code otherwise these indices being measured will have the wrong value.

- What will be included as Planned and Unplanned downtime?
- Do operators and maintenance have a universal definition of what is a breakdown and not?
- Are all breakdowns properly coded?
- Are minor stoppages properly coded and separated from breakdown loss?
- When the machine stops, and downtime occurs will it reflect on the system in real-time?
- Do operators know what codes to enter when downtime occurs?
- How do operators populate the data?
- Can a second or third level be generated? This means if OEE is being tracked, can the software generate the three components, Availability, Performance Rate, and Quality Rate? If Availability is low, can it generate the third level on what losses does the equipment suffer?
- How about the other machine downtime such as conversion, start-up cutting-blade loss, and defects?

While CMMS or EAM vendors claim the extensive features and benefits their software can provide which is true, this can only benefit the maintenance function if they are used and the data we populate is important in our day-to-day work. CMMS software or system will not reduce the downtime or number of breakdowns in our assets, but if used correctly, it can provide maintenance the necessary information and data on what seems to cause the downtime. Users together with their IT and CMMS vendors should review all the main features, and capabilities of their software, and ask questions if the software is capable of handling these kinds of things, or if not, can we do something about it. For example, if we discuss the capabilities of the CMMS software for the MRO Spares, ask the IT or vendor regarding the following just to name a few;

- Is the approval system for Purchase Requisition automated?
- Can the software be capable of providing a photo or explosive view of the part or spare?
- Is the CMMS capable of providing information if a substitute or alternative part exists just in case the current inventory level is zero?
- Is the software capable of providing the parts catalog?
- Can bar-coding be integrated into the software?
- Is the software capable of producing bar code stickers for every spare or item in the storeroom?
- Is the software capable of providing MRO Analysis such as Fast, Slow, Non-Moving, or Obsolete?
- Can the software derive the minimum or maximum quantity for fast and slow-moving items?
- Is the software capable of providing an EOQ (Economic Order Quantity) for fast and slow-moving?
- Can the CMMS automatically provide the lead time for all items stored or an SDE Analysis?
- Can the software identify all parts that need a buffer or safety stock?
- Is the software capable of providing KPIs or indices for the storeroom and maintenance?
- Can EDI (Electronic Data Interchange) be integrated into the current system?
- Can the storekeeper have access to the PM Schedule to identify the parts needed?
- Is the CMMS capable of automatically generating Purchase Order for fast and slow-moving items?
- Can the software automatically generate a BOM (Bill of Materials) for PM?
- Can the parts, items, tools, and consumables needed in a PM together with the part number be sorted by equipment?
- Can the software include substitute or alternative parts if the original part has zero stock?
- Can the software flag down the items or parts that need to be replenished?

These are just some of the questions that need to be raised when planning to move to automation. What are those that are important to us that need to be automated and is the software or system capable of providing that information?

12.2: What CMMS Should Provide for PM

When a piece of equipment will be scheduled for Preventive Maintenance, a Work Order will usually be generated by the system. A PM Work Order is a document containing all the information needed to perform and accomplish the maintenance task and provides the process for completing that task. PM Work Orders include the scope of the work to be done, who will perform the task, and other relevant details. The PM Work Order will include the name, model of the equipment to undergo PM, the scope of work to be done, and Bill of Materials (BOM).

SHOULD BE AVAILABLE ON CMMS	WHO CAN ACCESS
PM Schedule for the Week or Month	PM Planner, PM Crew, Operations, CMMS, PdM Maintenance Manager, Storeroom, Purchasing
Scope of Work and Bill of Materials (BOM)	PM Planner, PM Crew, Operations, CMMS Maintenance Manager, Storeroom, Purchasing
Items, Spares, Materials, tools, Consumables Needed	PM Planner, PM Crew, CMMS, Maintenance Manager, Storeroom, Purchasing
Detailed Instruction of Each PM Tasks	PM Planner, PM Crew, CMMS, Maintenance Manager, Engineering
Summary of Overall Cost on Doing PM	Accounting, PM Planner, PM Crew, CMMS, Maintenance Manager, Engineering, Reliability
List of Alternative Parts if Original Part is Unavailable	PM Planner, PM Crew, CMMS, Maintenance Manager, Storeroom, Purchasing, Engineering
Equipment 201 File, Past PM, Breakdowns	PM Planner, PM Crew, CMMS, Maintenance Manager, Engineering, Reliability
Vendor or OEM Profile	Accounting, Finance, PM Planner, PM Crew, CMMS, Maintenance Manager, Engineering
PM Contracted Work	Accounting, Finance, PM Planner, PM Crew, CMMS, Maintenance Manager, Engineering
Predictive Maintenance Monitoring	CBM Group, PM Planner, PM Crew, CMMS, Maintenance Manager, Engineering, Reliability

Figure 12.3: What the CMMS Should Include for PM

The maintenance manager assigns someone responsible for the task. The person assigned coordinates with the MRO Storeroom regarding the needed items and materials before the schedule is due. The storekeeper should be able to access the PM Schedule and prepares the kit needed and inform the maintenance that everything is waiting at the gate or holding area of the storeroom. Items that are not stocked will be coordinated with the Purchasing and the purchasing will order the parts from the vendor. The PM Crew assigned will withdraw the parts from the storeroom and staged them near the equipment before the actual PM schedule. PM is finally executed and the maintenance closed the report on the system that the equipment completed its PM based on the schedule. If there are adjustments to be done or special instructions, which will be encoded in the system, it should be reflected in the system and the

PM Report is finally closed. For the system to function efficiently and effectively, the following in figure 12.3 should be available in the system. Here are some of the things I can think of on what to automate on Preventive Maintenance. With today's software capabilities, and as new features are added and the software continuously upgraded, there is no limit as to what we really want to automate. But one thing of importance is we automate what we think is important to us. As I have mentioned in my first book on World Class Maintenance Management – The 12 Disciplines, the CMMS software should adjust to the maintenance people's requirements on what they want to automate and not the maintenance people adjusting to the CMMS software.

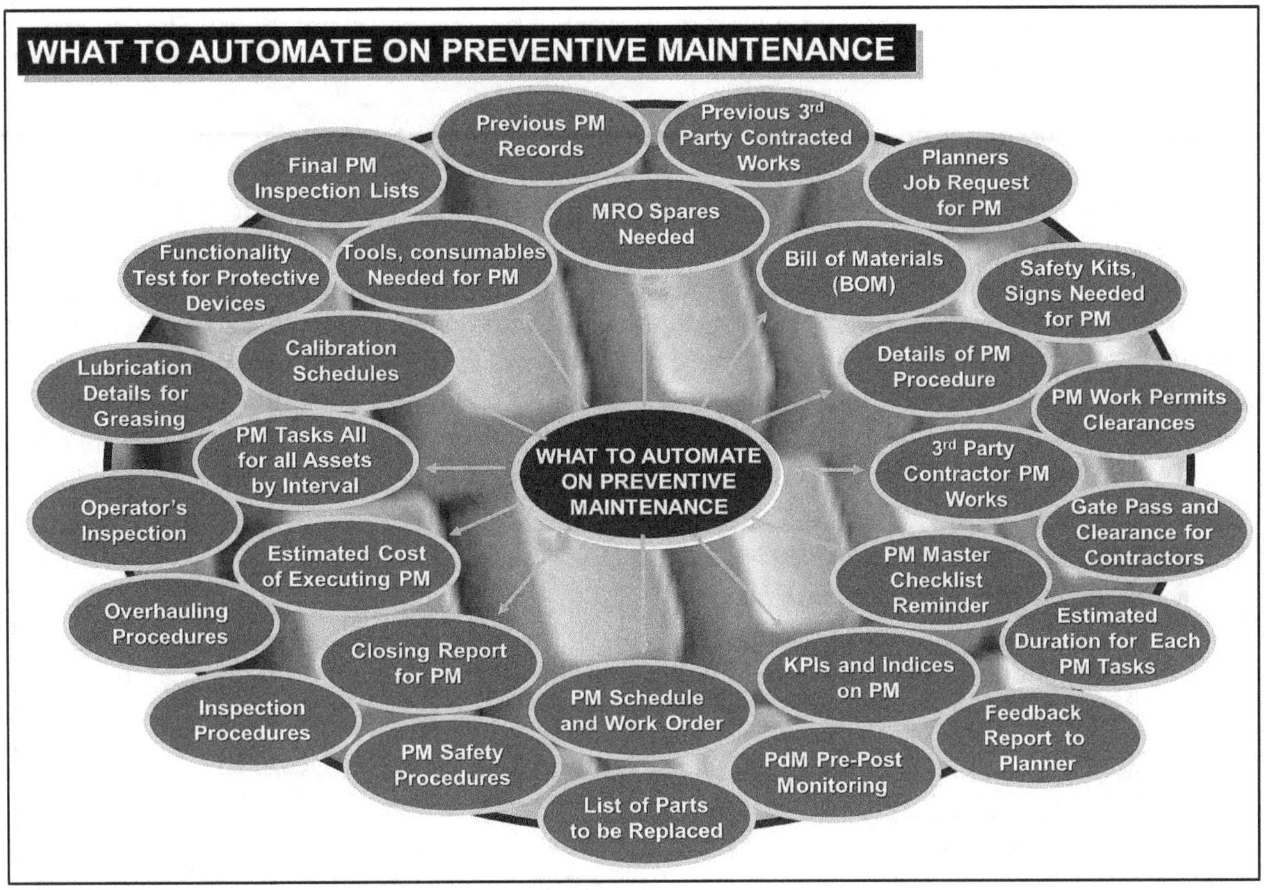

Figure 12.4: What to Automate on Preventive Maintenance

CMMS software can improve many aspects of our daily maintenance activities if correctly applied. It can provide industries with the tools to understand and analyze their maintenance needs and requirements. They can eliminate manual data entry, incorporate alerts, triggers, and escalation procedures, and shift their focus from manual administrative tasks to performing automated maintenance activities. They can also assist in planning and predicting future needs, prolonging the life expectancy of equipment while managing the different processes. To get CMMS to provide useful information, it must be well-configured, important data must be populated, and maintenance must be motivated in using the CMMS software for their day-to-day needs. The amount of work that goes into the CMMS implementation is the population of data, data migration, system configuration, and testing. The data populated into the CMMS software should be as complete and accurate as possible. CMMS is one tool that can genuinely enable an organization to meet its goals and targets. Like any other tool, the CMMS will be plain useless if it will not be used and just sit idle on the table, or perhaps if maintenance is not prepared to use it as intended, then it can provide us more harm than good to the organization.

12.3: CMMS Access for the Storekeepers

Although the majority of CMMS and EAM software covers the storeroom. Both maintenance and storekeepers must have access to the availability of the spares and items needed for a PM schedule. MRO Spare parts that are needed for Preventive Maintenance which have a long lead time and are not stocked in the storeroom should be well-coordinated with Purchasing so they can arrive before the actual PM scheduled date.

Both the storekeeper and PM Crew should have access to the schedule, especially on what parts, tools, items, or consumables will be withdrawn or borrowed from the storeroom so that the storekeeper can prepare them in advance and the PM Crew can withdraw them before the actual scheduled PM for a particular machine. The PM Crew should have access to the part's lead time, especially to those that are not stocked in the storeroom. The maintenance can coordinate with the IT department to include the lead time for every single part that is stocked and not stocked inside the storeroom. There are two ways to calculate the lead time, which can either be by using the average or exponential. I would recommend using the exponential since the lead time is actually based on the last couple of deliveries. Every time a new delivery has been made, the lead time should be adjusted automatically in the system.

LEAD TIME TO ORDER

- **There are 2 ways to determine the lead time to order, first is by getting the average dates and second doing it exponentially.** Getting it exponentially is more accurate as it focused more to the most recent lead time

ACTION	DATE	LEADTIME
Ordered	Mar. 5, 2013	37 days
Received	April. 5, 2013	
Ordered	May. 17, 2013	40 days
Received	Jun 26, 2013	
Ordered	Oct 11, 2013	37 days
Received	Nov. 17, 2013	
Ordered	Nov. 20, 2013	22 days
Received	Dec 12, 2013	

Getting the Average

$$\text{Average Lead Time} = \frac{37+40+37+22}{4} = 34 \text{ days}$$

Computing Exponentially

(37 + 40) / 2 = 38.5 days

(38.5 + 40) / 2 = 39.25 days

(39.25 + 37) / 2 = 38.125 days

(38.125 + 22) / 2 = 30.06 days

Order should be done on a monthly basis

Figure 12.5: Average and Exponential Lead time Calculation

It is also important for the system to flag down these parts that need to be ordered before the PM date is due. These are those parts that are not stocked in the storeroom. The maintenance can seek the help of IT or the CMMS vendor to include the lead time for every single part that is stocked and not stocked inside the storeroom. If today is March 22, 2022, and a compressor will be having a major Preventive Maintenance on December 15, 2022, the system should allow us to see and identify what parts are not currently stocked in the storeroom as in figure 13.5. They will be flagged down and ordered 2 or 3 months before December 15, 2022, to ensure that they will be available during the PM schedule. CMMS system should be capable of providing us this important information for all equipment that will undergo major Preventive Maintenance in the future.

ITEMS NOT CURRENTLY STOCKED IN THE MRO STOREROOM						
Machine Number	Parts Desciption	Part Number	Quantity Needed	Vendor	Leadtime Exponential	PM Schedule
Compressor MC-101	Cylinder Head,	103485.00	1	Taiwu	60 days	15-Dec-22
	Suction Valve,	103657.00	1	Taiwu	90 days	15-Dec-22
	Intake Valve	104879.00	1	Taiwu	62 days	15-Dec-22
	Delivery valve,	105968.00	1	GNS	85 days	15-Dec-22
	Cooling water jacket	105843.00	1	GNS	60 days	15-Dec-22
	Connecting Rod	106789.00	1	GNS	60 days	15-Dec-22
	Main Bearing	109294.00	1	Taiwu	30 days	15-Dec-22
	Crank shaft	101932.00	1	Taiwu	90 days	15-Dec-22

Figure 12.6: Items Not Currently Stocked in the Storeroom Needed for PM

Chapter 13

FAQS and Tips on Preventive Maintenance

> ***PM is one of the basic disciplines performed on the equipment and facilities. The main goal of performing this task on a scheduled basis is to extend the equipment's life and to assure its capacity in support of the plant's goals and targets. In performing Preventive Maintenance, the basic law to consider is that the cost of doing PM must always have to be lower than the overall cost of the consequences of failure that it is meant to prevent.***

13.1: Frequently Asked Questions on PM

Here is a list of Frequently Asked Questions (FAQs) and answers on Preventive Maintenance. Some of which have been raised in my previous training classes.

Figure 13.1: Preventive Maintenance FAQs, and Tips

1. **Why does Infant Mortality exist after executing PM?** Although there are many causes of Infant Mortality Failures, the primary cause will be human errors. Usually, this occurs when

we come in direct contact with the equipment, especially during overhauls. According to the study of James Reasons and Alan Hobbs in their book Managing Maintenance Errors, overhauling contain two activities, dismantling and reassembling the equipment. Human errors occur when maintenance starts to reassemble the equipment. This is when human errors are most likely to happen.

2. **Is it possible to eliminate Infant Mortality Failures after PM?** We can reduce or mitigate the chances of human error by integrating both Precision Maintenance as well as Predictive Maintenance on PM. For Predictive Maintenance, we can perform Pre-PdM monitoring before and Post-PdM monitoring after conducting a major Preventive Maintenance especially when both overhauling and replacement are involved in the equipment. Integrating Precision Maintenance simply means being consistent and precise in how we execute the PM tasks. This means that whoever is performing the PM task should produce the same outcome whether the person is the most or least experienced in the craft.

3. **If a part or spare is still in working condition, should it be replaced if this is indicated in the maintenance plan?** Although there will be an exemption to this question, the main point to consider is if a part is still working then it should remain in service. We also need to consider if this part can be under the watch of the Predictive Maintenance side of the house, and they will be the best people to recommend the best time to replace it. However, on the other side of the story, if the equipment is regulated and should comply with government laws and regulations, then we have no option but to comply. As my good friend, R. Keith Mobley once said to me that it takes a considerable amount of time to pass a law in Congress and there is no existing Predictive Maintenance instrument on the planet that can provide the potential failure of one or two years ahead. This means that if the equipment is undergoing a major PM shut down right now and the next PM will be 2 years from now, there is no Predictive Maintenance instrument that can predict the failure 2 years in advance.

4. **In Question number 3, if a part or spare is still in working condition, and it was decided that it will not be replaced since it is still working, should it be included in the Maintenance Backlog indicator?** Officially and technically the answer is yes since it is originally been stated in the plan. However, it can be removed from the maintenance backlog once the planner agrees and revised the original plan to exclude that particular task.

5. **What is the primary objective of doing Preventive Maintenance?** The primary objective of doing Preventive Maintenance is to be one step ahead of failures. Although not all failures can be prevented, there will be failures that will exhibit a wear-out pattern. These parts must be identified and a corresponding task should be performed. It will always be less costly if maintenance is ahead of the failure rather than the failure always being ahead of maintenance.

6. **Is Predictive Maintenance part of the Overall Preventive Maintenance strategy?** If we talk about the approach from the West, both Preventive and Predictive Maintenance are separate strategies. However, if we discuss the concept from the east, especially TPM or Total Productive Maintenance, both Predictive and Preventive are part of a bigger strategy which is called Planned Maintenance. Planned Maintenance is one of the major pillars of

TPM which supports Autonomous Maintenance activities and improves the way maintenance is performed on the equipment. The activities of Planned Maintenance will include Preparatory Phase, Restoration activities, Improvements and Redesign, and a Stabilization process so that the equipment will not revert back to its deteriorated stage.

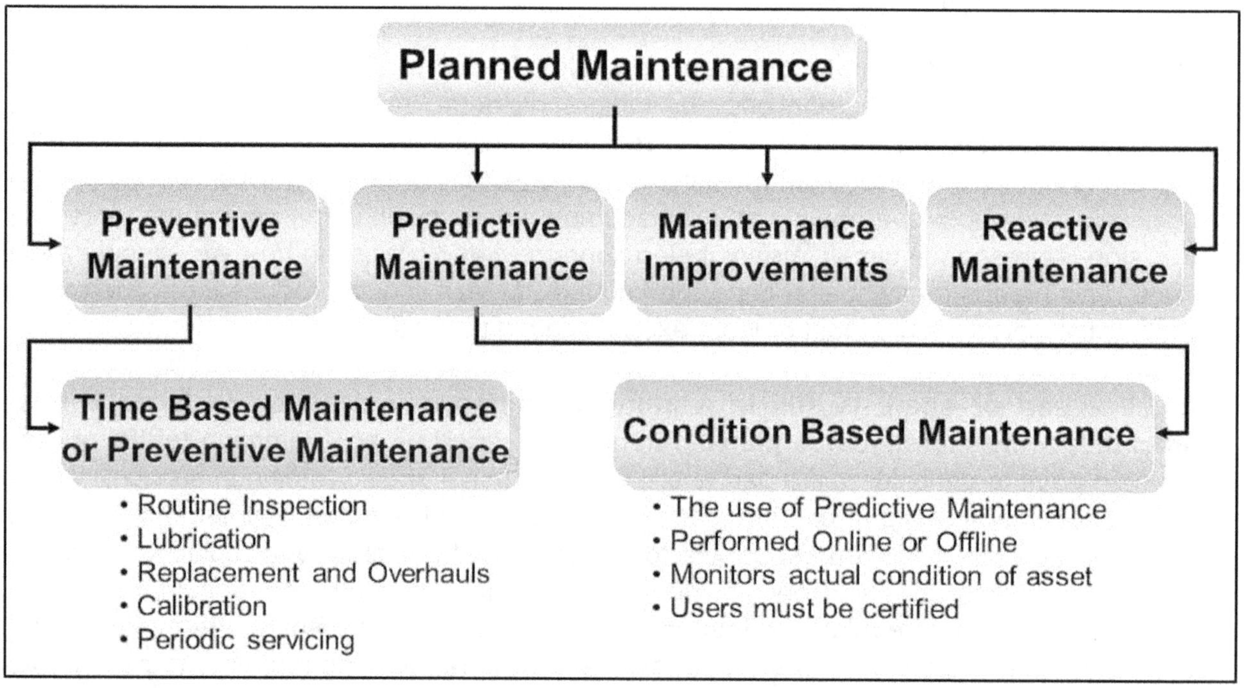

Figure 13.2: What Planned Maintenance Includes

7. **Why is it that Reliability-Centered Maintenance (RCM) only considers Scheduled Restoration and Scheduled Discard as Preventive Maintenance tasks?** Scheduled Restoration refers to overhauling activities, while Scheduled Discard refers to parts replacement. Although human Inspections are part of the overall Preventive Maintenance tasks, in RCM they are considered as On-Condition Tasks which means any activity which entails checking the actual condition of the equipment using Predictive Maintenance instruments, gages or human inspection will default to this task. Another form of inspection for RCM is the Failure Finding Task or Functionality Inspection which refers to inspecting protective and hidden devices. I think this is a matter of nomenclature or what term we want to call them.

8. **How do we eliminate human error in PM?** Human error is part of being human. We cannot eliminate human error but we can reduce or mitigate them especially when the consequence of failure is high. If we look at maintenance errors, it is not random, indeed. Most maintenance errors happen during installation, reassembly, overhauling, and repair activities or when maintenance is trying to experiment with how to repair the failure, especially if this is the first time they have experienced this kind of breakdown. We can only benefit from failure if we can learn from it. Similar to question number 2, integrating Precision Maintenance on PM will lessen the chances of human error committed during PM.

9. **What can we do if Operations would not allow the equipment for PM and always waive PM on the equipment?** Although this is a common issue, especially in manufacturing industries, what we can do is to sit down with these people and explain to them that the equipment is not a plug-and-play device just like our TV or Mobile phone since there are movements inside. When there are movements of mechanical parts, then after an (n) amount of movements, it will stress out and completed its cycle. Perhaps one of the reasons why operations waive the equipment is that they experience a lot of problems after PM which are called infant mortality failures. Operations people should be guaranteed that the equipment should perform well after PM and those who execute PM should avoid further delays and waste in executing their task. Maintenance must also explain to operations that failure to perform Preventive Maintenance may result in much higher costs if equipment breaks down.

10. **How would we know that the Preventive Maintenance task we prepared will be effective enough when executed?** Although there are PM indicators that will tell us if our PM task is effective, this is after executing Preventive Maintenance. Hence, if the PM task were listed and finalized, how can we know whether the tasks we have written are effective or not before executing them? My response to this is to try to look at everything in reverse. Review each of the PM tasks listed and ask what particular failure mode does this task addressed? The more failure modes addressed by these PM tasks, logically and pragmatically, the more effective the PM will be when finally executed.

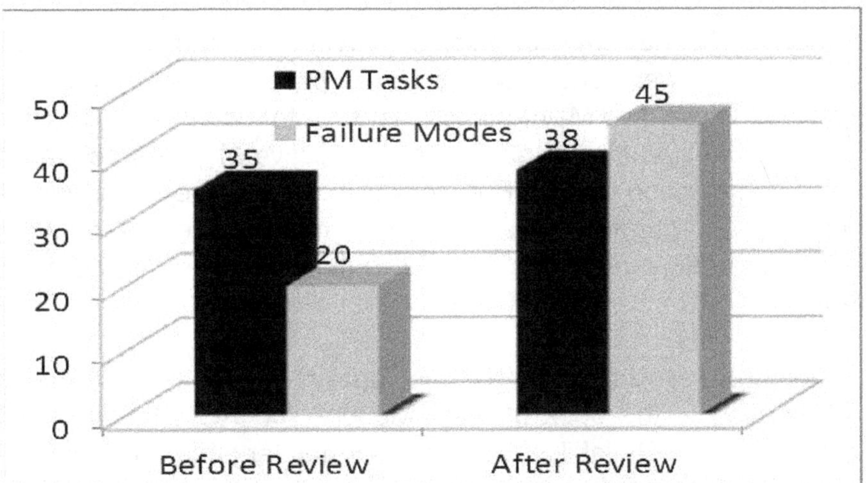

Annual PM	Before Review	After Review
PM Tasks	35	38
Failure Modes	20	45

Figure 13.3: Compare and Review PM Tasks and Failure Modes

11. **If a plant is implementing RCM and the analysis has been completed, should the current Preventive Maintenance task still be done on the equipment?** Again, we can use the graph in 13.3 to determine which task is more comprehensive if finally done on the equipment. If RCM will address more failure modes than the existing Preventive Maintenance tasks, then the current PM task should be superseded by the RCM task. This is also to avoid

redundancy in performing the task. Although RCM is more detailed and comprehensive since it will utilize all the available maintenance tasks which include run to fail, Preventive, Predictive, Failure Finding tasks, and modification.

12. **Should Autonomous Maintenance tasks be included in the maintenance planning?** In Autonomous Maintenance, the operator's tasks will include three standards, cleaning standards, lubrication standards, and inspection standards. They will be identified and consolidated with the PM tasks to avoid redundancies and duplication of tasks when Autonomous Maintenance reaches Step 5. These standards will be finalized by the operator, therefore, the maintenance tasks for operators should no longer be part of the maintenance planner's responsibility.

13. **When executing PM, what if there are other activities found that were not originally listed in the plan, should this be done and included in executing the PM task?** The maintenance planner may not always be perfect in the plan, this will depend on several factors. First, the availability of the parts or if the parts can be delivered immediately which will be on an emergency basis, second, the additional time to be spent, and third, its criticality, or if it will still last until the next scheduled outage. If this should be addressed immediately, then operations people should be informed regarding the delay due to additional work done on the equipment. This should also be revised and added to the next scheduled PM of the equipment.

14. **Can Predictive Maintenance eliminate Preventive Maintenance tasks?** No, some failures will be subject to wear out since the failure pattern is age-related, hence this will be under the care of the Preventive Maintenance tasks. Although more failures can be predicted, than prevented, industries must strengthen their Predictive Maintenance tasks. Several tasks can be done in a combination of both Preventive and Predictive Maintenance. For example, the tasks to prevent and predict a bearing seizure may be the use of Vibration Monitoring, human inspection for audible noise, performing routine greasing, and temperature monitoring using a thermal gun

15. **Why is Redesign / Modification the last option on the RCM Algorithm or Decision Diagram?** [17]According to the book of John Moubray on RCM II, quote, which comes first redesign or maintenance? Reliability, design, and maintenance are inextricably linked. This can lead to a temptation to start reviewing the design of existing equipment before considering its maintenance requirements. In fact, the RCM process considers maintenance first for two reasons. First, most redesign and modification takes from six months to three years from conception to commissioning depending on the cost and complexity of the new design. Hence maintenance has to maintain the equipment as it exists today and not what should be there in the future. Second, most organizations are faced with many more apparently desirable design improvement opportunities than are physically or economically feasible.

16. **When do we consider contracting PM?** Contracting Preventive Maintenance to 3rd parties should be considered if the cost of performing PM in-house by regular employees will

[17] Moubray, John, ***Reliability-Centred Maintenance II,***(Butterworth Heinemann., 1997), Page 189-190

be less expensive if done by a third-party provider. Second, we can consider contracting maintenance if the current level of skills to perform a particular work does not exist. For example, perhaps a Corrosion Expert who by profession is a qualified person capable to engage in the practice of corrosion control on buried or submerged metal piping systems and metal tanks. Such a person must be accredited or certified as being qualified by the National Association of Corrosion Engineers or be a registered professional engineer who has certification or licensing that includes education and experience in corrosion control. The industry may be needing his services since a lot of corrosion exist in the plant and recommend works to prolong the process of corrosion.

17. **What machines should we prioritize on PM?** The reason for conducting a Machine Ranking for all equipment and assets is to determine which machines should be prioritized for Preventive Maintenance as this equipment once identified will have an impact on operations when it failed. We can also include those machines which are used frequently and decide the interval based on running hours or calendar days. The maintenance planner should have the lists according to the rank of each piece of equipment so that the planner can start planning for all Rank A equipment.

18. **What is the best advice you can provide on PM?** Preventive Maintenance is a good strategy that can help benefit industries when implemented correctly. This means adapting PM correctly no more, no less with just the right amount of resources needed. We can only realize the benefits of doing PM if we stop abusing, misusing, or overusing this strategy. This means that both maintenance and top management must understand and accept what Preventive Maintenance can and cannot do.

19. **When is the best time to replace the oil in our equipment?** While many OEMs and vendor will recommend their clients to change their oil based on running hours, a better approach is through the use of Oil Analysis. Several Oil Analysis instruments can provide this specific information such as RULER or also called the Remaining Useful Life Evaluation Routine. Another way is through what is called the TAN (Total Acid Number) and TBN (Total Base Number) crossover. This means that when the value of both TAN and TBN is the same, then it means that the oil should be changed. Other industries may be using the ISO 4406 Cleanliness Codes as a means of determining if their oil needs to be changed or not. Contamination is also the reason for oil degradation, hence, the need to change the oil on a time-dominated (running hrs.) frequency is essential. More contamination means more failures and frequent oil changes, therefore it is important to analyze oil based on the number of contaminants as well as what elements are present and not by the frequency of changing oil itself based on running hours. By knowing this information, maintenance can strategize measures to improve fluid cleanliness and lengthen its drain interval. This simply means that if oil can be maintained clean, then there is no reason to change it.

20. **How can we perform PM in a toxic and reactive culture if management cut cost on every corner and operations always waive the equipment for PM?** First, World Class Industries were not born that way, they were also reactive. The key to their success is that they initially challenge themselves to become better starting with the same amount of resources and time. For a start, work on a plan and execute the PM based on the plan. Just

start with one machine. If the machine improves as a result of the PM performed, then do it again on other machines. If the same results were obtained, just do it without hesitation until everyone is on board. The key is to gain their trust and confidence.

21. **Can OEE (Overall Equipment Effectiveness) be used to track the effectiveness of executing PM on the equipment?** No, a good PM program will reduce the number of breakdowns in the equipment. Reducing the number of breakdowns and failures will increase the availability of the equipment, but availability is just one of the three components of OEE. If breakdowns are the only losses in the equipment then it can improve OEE, but we also have to consider the other losses on the equipment as well.

13.2: Tips on Implementing Preventive Maintenance

Here are some tips that the industry can avail of so we can get the maximum value of our efforts in executing Preventive Maintenance to our assets and equipment. These tips if implemented will provide the industry a better outcome in executing their Preventive Maintenance on their equipment and assets.

1. **Integrate Precision Maintenance into Preventive Maintenance:** The main reason for integrating Precision Maintenance in PM is to minimize the chances of human errors, especially during overhauls so that we can reduce the chances of infant mortality failures from occurring during start-up activities. What is important is that whoever is performing the PM task whether the person is the most or the least experienced in the craft should yield the same outcome and result.

2. **Perform a Pre and Post Predictive Maintenance Monitoring:** A day or two before the actual execution of PM activities on the equipment, the Predictive Maintenance group should conduct a Pre-monitoring on the equipment. Once the Preventive Maintenance task has been completed a Post-monitoring will again be performed by the Predictive Maintenance group before endorsing the equipment back to operations. If for example the reading from Vibration Monitoring had been reduced before and after executing the PM task, then we can be confident that the chances of infant mortality occurring after PM will be minimal.

3. **Free the Maintenance Planner from Reactive Works:** Maintenance planners should be focused on future works and not on doing job orders for reactive work. In most industries, when the equipment fails, the operator will write a Work or Job Request and give it to maintenance. The maintenance inspects the equipment if a spare or item will be needed. If a spare is needed, then the maintenance will check the system. If the part is available, then the maintenance will withdraw the spares from the storeroom. Once the repair job is complete, the maintenance should close down the report on the system. Complete details such as if there are any tools, spares, or consumables used and what activities have been done to repair the equipment. If there are important mentions that need to be added, then they should be written in the report. What is important at this point is for the maintenance to close the report on the system completely. The planner will have the time to prepare the plan for every critical piece of equipment in the plant that will be subject to Preventive Maintenance.

4. **Be Consistent in Greasing:** When performing routine greasing, be consistent with the type of grease and thickener used. In short, never ever mix two types of grease with different thickeners as we will be having problems with incompatibility. This means that if the current grease used is Lithium Grease, and the grease available is Calcium Grease, do not mix them on the same grease gun. Even if other Grease Compatibility Charts indicate that Lithium and Calcium grease can be mixed, the additives and base may still be different. Always be consistent on the grease used to avoid incompatibility issues. Advise purchasing on just using one vendor for your grease. Do not trust the compatibility charts from the grease vendors since those charts are only for their products and only for the thickeners. To have grease compatibility is to be compatible not only with the thickener but also with the base and additives.

5. **Conduct a Machine Ranking to Prioritize PM:** There should be a complete inventory of all the equipment used in the plant including the Facilities / Utilities. Once the inventory has been completed create a category and determine all worse and problematic equipment and conduct a Machine Ranking for all equipment and assets as discussed in Chapter 3, Section 1 of this book. The reason for conducting a machine ranking is to determine all Rank A as this will be the focus of future Preventive Maintenance works. Priority for conducting Preventive Maintenance should be given to all Rank A or worse equipment in the plant.

6. **Storekeepers Should Have Access to the PM Schedule:** Storekeepers must have access to what equipment is scheduled so that they can prepare the needed kits in advance a minimum of 24 to 48 hours before the actual PM schedule. Once all the items, tools, spares, and consumables are available, the storekeeper can call the PM crew for them to withdraw the parts in the storeroom and place them near the equipment. Once the last product had been run through and the machine is finally stopped, then everything needed is already in place saving valuable time for the PM crew.

7. **Do Not Assume That Similar Machines will have Similar PM:** There may be identical or similar equipment in the plant with the same model, specs, and OEM. If this is the case, then do not assume that their PM task would be identical in every way. We also need to consider the equipment's operating context or how they are operated. Try asking the following questions. Do these equipment run at the same time which means that their running hours are identical? Does the equipment serve as a backup or redundancy? Do the operators have the same skills? Are they twins or do they have the same DNA?

8. **Prepare Everything Before the Actual PM:** Generate a PM Master Checklist on all spares, items, tools, consumables, safety signs, and lubricants that will be needed from the storeroom for the actual Preventive Maintenance work one or a couple of days before the actual PM. Withdraw them from the storeroom before the actual PM and store these items as much as possible near the equipment. Storeroom people should have access to the PM Schedule so that they can prepare the needed items for the PM. Refer to figure 4.14 in Chapter 4 for an example of a PM Master Checklist.

9. **Every PM Task Should Address a Failure Mode:** Have a check and balance and review the task that will be undertaken on Preventive Maintenance. Each of the tasks is listed

for a specific reason and that is to prevent an unexpected failure in the future. This means that each Preventive Maintenance task listed should address a particular failure mode. If a task is listed that does not address a particular failure mode, then this should be deleted in the PM task since it will just add up to wasted manpower and resources.

10. **Involve Operators in Maintenance:** There are maintenance tasks that should be done by operators. For an industry implementing TPM, as the operators move from one step to another, three standards will be generated which include cleaning, lubrication, and inspection standards. Visual Controls will also be prepared by operators so that anomalies and deviations can easily be detected. The most important reason why operators should be involved in maintenance is that these people will be the first line of defense against equipment-related failures since they will be the ones who will experience and witness the failure first before the maintenance. With the correct knowledge and coaching from maintenance, they can sense the failure earlier for those failures with symptoms. This means that they can inform maintenance if they sense problems and actions can be taken before it ends up in a catastrophic failure. Failures can be addressed when it is small which makes them less costly. Remember that small problems matter most and this is the main essence of establishing Autonomous Maintenance.

11. **Implementing Autonomous Maintenance is Always a Top-Down Approach:** It is very clear in the TPM Developmental steps that this is a Top-Down Approach. This means that implementation of this pillar should come from the CEO or President of the industry. In the book on TPM World Congress in 1990, 3000 plus industries implemented TPM and only 10% of them were successful. This means 90% of them either fail, discontinue, or the program just fades away for good. When it was analyzed why 90% of them failed, the primary reason was the lack of management support and commitment. In implementing TPM, what we need from management is both commitment and ownership.

12. **Success of Autonomous Maintenance Implementation Depends on Planned Maintenance:** While Autonomous Maintenance is the most talked about and popular pillar of TPM, Planned Maintenance should be the strongest pillar in any TPM implementation. The reason behind this is that if Planned Maintenance is weak, then Autonomous Maintenance will collapse. Autonomous Maintenance will only move if the maintenance provides them with the information and knowledge they need on the equipment. Maintenance will serve as their official coach and mentor for operators.

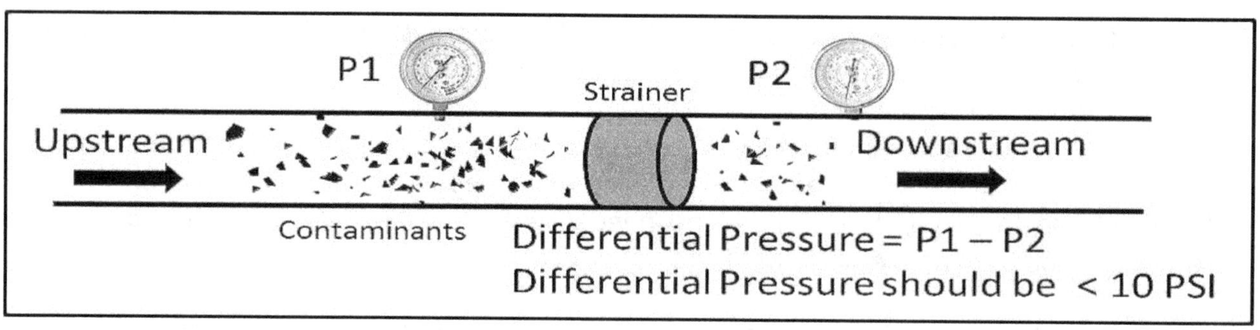

Figure 13.4: Inspections Should be Quantitative

13. **Never Use the Word Check on Inspection:** The word check is subjective. When conducting an inspection, as much as possible, quantify it and determine the threshold or limit. If an operator is inspecting the difference between the upstream and the downstream pressure gages, determine the limit of the pressure difference. If the limit set is set at 10 psi, then this will be the signal for the operator to write a work request for the strainer to be cleaned. If the inspection cannot be quantified, then a qualitative measurement must be identified and should be clear and precise to the person conducting the inspection process.

14. **Planning is a Continuous Improvement:** The main purpose of planning is to organize our human resources to perform scheduled work. The plan may not always be perfect, however, the good thing is we can make it better through constructive feedback from the people who execute the PM work. What is important is that there should be continuous cooperation, communication, coordination, and collaboration between the planner and the people involved in executing the PM work.

15. **Never Use MTBF to Determine the Interval for Overhauling and Replacement:** Collecting failure data to calculate MTBF to determine the maintenance tasks for replacement and overhauling will not apply since MTBF is only about the average life. Determining the interval to perform replacement and overhaul should be based on the useful life and not on the average life. This means that PM overhauls and replacement tasks will only apply to age-related failures. Remember that it is not MTBF but rather the remaining useful life that is significant when attempting to determine the best replacement tasks to avoid failures. PM should be based on the useful life and not on the average life which is MTBF.

16. **There are Two Ways to Determine the Interval for Replacement and Overhauling:** The first is through the use of the Age-Exploration Method which means that if there are no traces of wear-out pattern on the part that will be replaced or overhauled, then we can increase the interval by 10% on the next PM. Do this again until a wear-out pattern is visible on the part or item to be replaced. Second, conducting an Oil Analysis test on Wear Metal Debris Analysis will provide maintenance on what metal elements are increasing over time. This indicates that the wear-out process is starting to happen. What is important for maintenance to interpret the report is for them to know what particular element their spares and mechanical parts are made of. Revert back to the vendor of the spares and asked them for this information since they are the ones who fabricated the parts.

17. **Include and Involved Operation Managers in PM Scheduling:** When scheduling equipment for Preventive Maintenance, involved operations managers in the discussion as these are the people responsible for the load of the equipment and they know the equipment and assets that are planned to operate continuously on the production floor. Although we may not agree 100% with the equipment that will be released by operations for PM, we need to provide relevant information to them such as the criticality of conducting PM, and monitoring records from Predictive Maintenance especially if a potential failure is spotted on the equipment. Let them understand that if a failure is imminent to happen then it will end up with more problems for both operations and maintenance.

Chapter 14

Improving Existing Preventive Maintenance Program

Maintenance must understand that PM overhauls and replacements can introduce Infant Mortality failures. Preventive Maintenance overhauls and replacement should be done only by skilled craftspeople with the right knowledge, tools, and skills. If you have the slightest doubt about reassembling the equipment back again in one piece, then think not twice but thrice before dismantling it. Remember that reassembly is when the majority of human errors can occur.

14.1: Improving Existing Preventive Maintenance in our Industry

Although the majority of industries currently have their existing Preventive Maintenance that they perform routinely on their equipment and assets, this chapter can help in enhancing your current Preventive Maintenance activities performed in the plant. Performing Preventive Maintenance on our equipment and assets should have an effect by having lesser failures and downtimes experienced on the equipment. If there are still many breakdowns experienced right after Preventive Maintenance then we need to review and evaluate the current activities that we perform on the equipment as well as the interval on when these tasks are being executed. This chapter provides the reader with basic guidelines and information on how to improve their existing Preventive Maintenance.

14.2: Improving Routine Preventive Maintenance Inspection

1. **Visual Inspections Should be Quantitative:** As much as possible, inspections should be quantitative. This means that there should be a limit or threshold which should be understood by the person conducting the inspection whether this will be done by the operator, maintenance, or a 3rd party contractor. If a pressure gauge is being inspected for a drop or difference, specify the maximum difference in pressure for example limit is at +10 psi. If this is not possible then a qualitative standard should be set. For example, the tank should have no leak of any form. Avoid using the word check and be specific about what you want the inspector to do. If

operators are performing the inspections, then they should be taught about the impact or consequences of the failure.

2. **Specify Precisely What to Do if Limit is Reached:** Provide specific instructions on what to do if the limit or threshold has been reached. Should the machine be stopped? If the operators are performing the inspections, are they capable of correcting the problem, or do they need to call maintenance? Indicate if there are tools or measuring devices that will be needed for the inspection. If a bolt needs to be retightened then indicate the correct amount of torque and that a torque wrench will be used. Also, indicate what happens when excessive torque will be applied to the bolt. Be specific once the limit or threshold has been reached.

3. **Review the Interval or Frequency of Inspection:** Can the inspection be done daily or every 2 days instead of doing it every shift? Consult with the maintenance and engineering if there is any harm in prolonging the duration of the inspection. If we have been inspecting this every shift for a couple of years, would it be possible to lengthen the interval? The same goes for those done weekly, and monthly inspections.

4. **Review Operator, Maintenance, and 3rd Party Contractor's Checklist for Redundancy:** Review the inspection checklist done by the operator, maintenance as well as 3rd party contractor activities for any duplication or redundancy. Note that a duplication means that the same task is performed by different people or groups. Decide on who is the best and most qualified to perform the task. This is different from performing a combination of tasks where the different task is done by different people to address a common failure mode at different intervals. For example, if the failure mode is a bearing failure, operators may be performing lubrication, while the Predictive Maintenance group is performing Vibration Monitoring, while the operators are carrying out a daily reading of temperature using a thermal gun. If Autonomous Maintenance is in place the consolidation of the operators and maintenance tasks will be carried out in Step 5.

5. **Use of Visual Control or Visual Management during Inspection:** The purpose of using visual control is to make the problem, deviation, abnormalities, or anomaly from standards much more visible to everyone. Visual Control allows us to see the problems more easily. They are also meant to provide instructions and convey information. For example, when a bolt is properly tightened to the correct torque, a reflectorized sticker is placed on the head of the bolt and the nut. This means as the machine vibrates and the bolt starts to loosen, the marker (sticker) will no longer be aligned to the nut which can be easily be seen by the operator and corrected immediately. In one of the plants I visited, the operator's break time was staggered which means that their 15 minutes break times in the morning and afternoon for operators are not the same and there is an overlap. This means that some operators will break from 10:00 am to 10:15 am, the next will be from 10:15 to 10:30 am and another group will be from 10:30 to 10:45 am. During break time, they can have some coffee and watch the fish in the aquarium. This 200-gallon aquarium tank is not only for display but it serves 2 purposes. The bubbles in the tank are supplied by the compressor located in the basement which is unmanned. If no bubbles are coming out from the aquarium, it will be a signal for the operators to call the facilities of their plant. The water supplied in the tank is wastewater being treated. If toxic chemicals are present in the water, the fish will not live long.

Figure 14.1.: Aquarium Serving as a Visual Control

6. **Ensure Operators Understand the Reason for Conducting Inspection:** When assigning inspections to operators, ensure that the operator knows the reason for conducting the inspection and what harm it can provide if the inspection will be neglected. Again if possible, set a limit or threshold and indicate precisely what to do once the limit is breached.

7. **Inspection Should be Precise and Complete:** When conducting Visual or Human Inspection, it should be clear, precise, easy to understand, and complete. Inspection should include the following information:

- Are there any measuring devices, tools, or instruments to be used?
- Determine the Interval for conducting the inspection.
- Estimated duration for each inspection process.
- Map on which inspection should come first, then the next until the last
- Responsible person to conduct the inspection
- Key points and methods on how the inspection will be conducted
- Limits or thresholds should be known by the inspector
- What to do if the limit or threshold has been reached

14.3: Improving Major Preventive Maintenance Shutdown

Major Preventive Maintenance is done over a longer period. Usually, the task involved in this case will include overhauls and replacements. Scheduled replacements are also called scheduled discard, while scheduled overhauls are also termed scheduled restoration. Scheduled discard tasks entail discarding an item or component on or before a specified age limit regardless of its condition at the time. The frequency of scheduled discard or replacement tasks is governed by the age at which the item or component shows a rapid increase in the conditional probability of failure. Wear-out characteristics most often occur when equipment comes into direct contact with the product. Age-related failures also tend to be associated with fatigue, oxidation, corrosion, and evaporation. Scheduled restoration or overhaul tasks entail re-manufacturing a single component or overhauling an entire assembly on or before a specified age limit, regardless of its condition at the time. The frequency of a scheduled restoration task is governed by the age at which the item or component shows a rapid increase in the conditional probability of failure. Scheduled restoration tasks are technically feasible if there is an identifiable age at which the item shows a rapid increase in the conditional probability of failure. This means that there is a dictated life and most of the items will survive to that given age and they should restore the original resistance to failure of the item. The frequency of conducting scheduled restoration tasks can be determined satisfactorily on basis of historical failure data or through the age-exploration method. Listed below are ways to improve our existing PM shutdown.

1. **Include a PM Master Checklist on Things Needed before the PM Schedule** as in figure 4.14 when Scheduling PM. List all items that need to be prepared before the actual PM is due. Make this a habit whenever there is a major Preventive Maintenance schedule to avoid delays that can prolong the duration of executing PM on the equipment.

2. **Integrate Precision Maintenance in Executing PM:** The reason for integrating Precision Maintenance is to minimize human errors that can lead to infant mortality failures when operators start the machine after an extensive PM has been performed. As discussed in this book most human errors are not random, they occur when we start to reassemble the equipment back again to its original state. It is best to have a sequence on how to reassemble the equipment and whoever is performing the overhaul should comply with the sequence.

3. **Perform Pre and Post Predictive Maintenance Monitoring:** Collaborate with the Predictive Maintenance group such as Vibration, and Thermography to conduct a pre-monitoring before and after executing Preventive Maintenance. The readings gained from monitoring should be improved after executing Preventive Maintenance. This will also ensure that the chances of incurring infant mortality failures will be reduced. Integrating both Precision and Predictive Maintenance will reduce the chances of infant mortality failures from happening in the equipment.

4. **Ensure All Maintenance Tasks Should Address a Failure Mode:** For existing Preventive Maintenance tasks, check each task and answer the question on why this task is performed and what failure mode are we addressing? If a maintenance task does not address any failure mode, delete them from the list. When writing a specific task, ask yourself if by doing this maintenance task, can we be one step ahead of failure? Are these maintenance tasks sufficient? Do we need to add another maintenance task to ensure the failure will be captured?

5. **Review the Current Interval:** For overhauls and replacements, use the concept of age exploration recommended by the late Anthony Smith. [18]There is a proven technique that we can

[18] Anthony Smith, ***Reliability-Centered Maintenance, Gateway to World Class,***(Mc Graw Hill Inc., 1993), Page 126-127

employ to refine that guesstimate over time and to determine more accurately the correct task interval. It is called the age exploration method. Let us say our initial overhaul interval for a fan motor is 3 years, when we do the first overhaul, we meticulously inspect and record the condition of the motor and all of its parts and assemblies where aging is thought to be possible. If the inspection reveals no such wear-out or aging signs, when the fan motor comes due for overhaul, we automatically increase the interval by 10% and repeat the process, until we see incipient signs of wear-out or aging. This means that if no wear out is found on the motor, then the next overhaul will be as follows:

- Age Exploration = (3 years x 365 days / year) + (1095 days x 0.1)
- Age Exploration = (1095 + 109.5) days
- Age Exploration = 1204.5 days

14.4: Improving Scheduled Greasing Practices

Whether one is using just a manual grease gun or the most advanced and modern Ultrasonic grease gun today, there are several correct practices that every maintenance or operator must understand before applying grease to their equipment and machines. Excessive grease can lead to different problems, such as increased temperature or damage to electric motors' seals. Insufficient grease applied can also lead to high operating temperatures, and collapsed seals, and in the case of greased electric motors, it can result in energy loss and breakdowns. Over greasing the motor's bearing will only agitate the grease while pushing it away, spreading the grease to other portions and parts inside the motor. Here are correct practices on greasing.

1. **Educate your People on Lubricating Oil and Grease:** Training is always the backbone and foundation of any cultural change before starting anything in the industry. As I have always mentioned in my previous books, starting any reliability or maintenance strategy in the plant will start with training and education. The maintenance people and lubrication users need to be educated regarding the correct knowledge they need to build their skills to do their jobs right the first time around. There is simply no shortcut to this. Learning anything from experience is costly as equipment, and different machinery types need to fail before we can even learn something.

2. **Know the Operating Temperature of the Equipment:** Every grease on the market have a maximum allowable temperature at which it can be used. The dropping point is the temperature at which the grease will drip. As a rule of thumb, subtract 50 ° C from the dropping point and this will be the maximum allowable temperature that the grease can be used. The temperature limits for use of grease are therefore determined by the dropping point, oxidation, and stiffening at low temperatures. Figure 14.2 is a Grease Application Guide that can be used to determine the correct grease that can be applied to our equipment and components.

3. **Map out Your Greasing Sequence:** Provide a Grease Mapping Sequence as in figure 14.3. This will provide the person responsible for carrying out the greasing on which machines are scheduled to be greased and not. This also provides the sequence on which machines or components to start with until the last component so that the time spent to grease all scheduled machines will be maximized with minimum manpower movement loss.

Grease Application Guide

Referrence NLGI Lubricating Grease Guide 4th Edition

Properties	Aluminum	Sodium	Calcium Conventional	Calcium Anhydrous	Lithium	Aluminum Complex	Calcium Complex	Lithium Complex	Polyurea	Organo-Clay
Dropping Point (Deg F)	230	325-350	205-220	275-290	350-400	500+	500+	500+	470	500+
Dropping Point (Deg C)	110	163-177	96-104	135-143	177-204	260+	260+	260+	243	260+
Max. Usable Temperature F	175	350	200	230	275	350	350	350	351	352
Max. Usable Temperature C	79	93	110	135	177	177	177	177	177	177
Water Resistant	Good to Excellent	Poor to Fair	Good to Excellent	Excellent	Good	Good to Excellent	Fair to Excellent	Good to Excellent	Good to Excellent	Fair to Excellent
Work Stability	Poor	Fair	Fair to Good	Good to Excellent	Good to Excellent	Good to Excellent	Fair to Good	Good to Excellent	Poor to Good	Fair to Good
Oxidation Stability	Excellent	Poor to Good	Poor to Excellent	Fair to Excellent	Fair to Excellent	Fair to Excellent	Poor to Good	Fair to Excellent	Good to Excellent	Good
Rust Protection	Good to Excellent	Good to Excellent	Poor to Excellent	Poor to Excellent	Poor to Excellent	Good to Excellent	Fair to Excellent	Fair to Excellent	Fair to Excellent	Poor to Excellent
Principal Use	Thread Lubricants	Rolling Contact	General Uses	Military	Automotive Industrial	Multi Service Industrial	Automotive Industrial	Automotive Industrial	Automotive Industrial	Hi Temp Freq Relube

Figure 14.2: Grease Application Guide for Selection

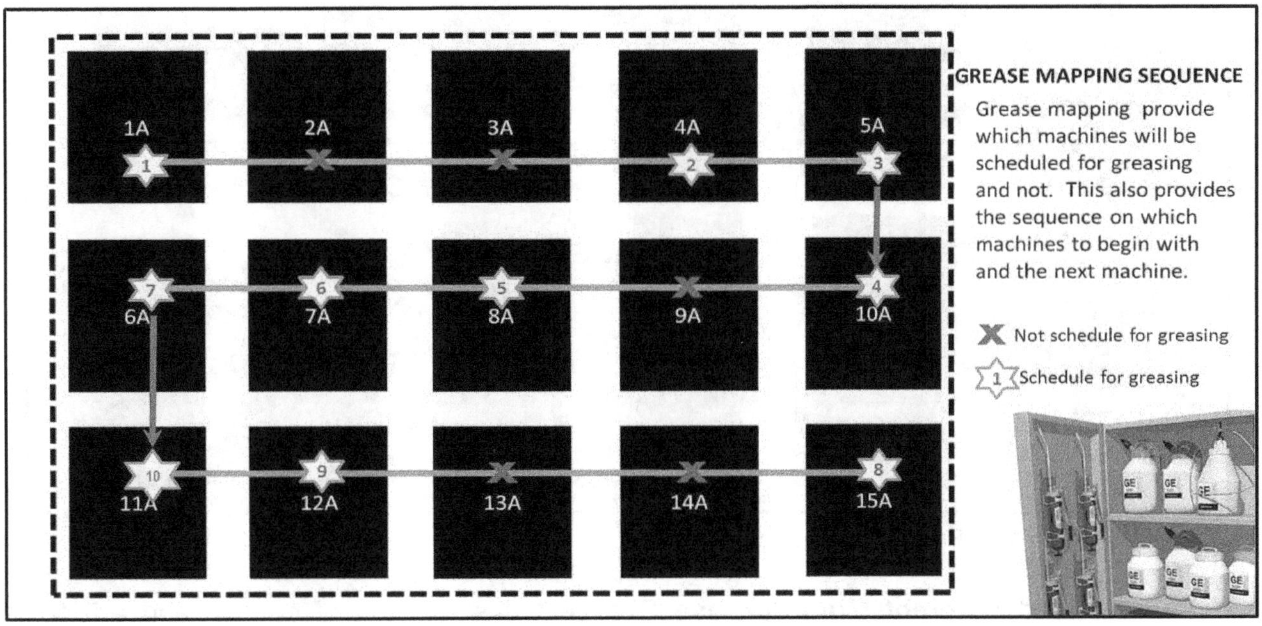

Figure 14.3: Generate a Grease Mapping Sequence

4. **Selecting the right grease:** Use a thermal gun or Infrared Thermography camera to determine the maximum operating temperature of the system that is subject to be grease. Note that the grease's dropping point is not the same as the grease's useful operating temperature. The temperature at which the grease can be used will always have to be lower than the dropping point temperature of the grease. Determine the type of grease to be used based on the Grease Application Guide Table provided in figure 14.2. Determine all the number of systems, machines, equipment, assets, and components to be greased and note their actual operating temperature. Knowing the operating temperature, deciding the type of thickener of

grease to be used, and keeping a record of all greasing activities. Dedicate the grease gun per type of grease to be used to avoid incompatibility issues. For example, use the same grease gun for lithium grease and another separate grease gun for calcium grease, and so on.

5. **Consolidate All Grease Guns Used in the Plant:** Organize a team and consolidate all greasing activities. Make a survey on how much grease and grease guns are used in the plant. Use the same type of grease gun and know the output per stroke of each of the grease guns used in the plant. What is important in greasing is to pump the correct amount of volume of grease to the subject. In case you do not know the output per stroke or dispensing capacity of your grease gun, try getting a weighing scale and pump one full stroke of your grease gun to the weight scale. Confirm the amount of grease expelled on the grease gun per shot. Pump the grease 10 times and measure the total weight of the grease dispensed on the weighing scale. After doing this, divide the weight by 10 to determine the grease gun's actual pumping capacity per stroke, either in grams or in ounces.

Figure 14.4: Different Multi-Purpose Grease

6. **Never Mixed Different Multi-Purpose Grease:** Just a word of caution and advice to maintenance and lubricating users in industries. One common mistake I see maintenance do is mixing different grease in their grease guns since it says the same thing, "**<u>Multi-Purpose Grease</u>**." Even if two grease containers indicate a multi-purpose, their thickener can be different (for example, calcium, lithium, sodium, and so on). The base can be different, which can be synthetic or mineral. Their NLGI number may also be different (which is the hardness of the grease) 0. 1, 2, 3, and so on. The additives can also be different since they are of different brands. Their properties may also be different such as operating temperatures, dropping points may also be different, so it is not recommended to mix different grease in your grease guns. Maintenance people must also be responsible for informing this matter to the Purchasing and to

the Storeroom people to be consistent on the grease they purchase by sticking to the same type and brand at all costs.

7. **Consolidate All Types of Grease Used:** Summarize all the grease types used for each application and refer to the table in figure 14.2 to determine the maximum range of operating temperature the grease can be used. Select the lubricant for every application, together with the correct amount of quantity. For every type of grease used, provide a dedicated grease gun for each of them. It is important to consolidate every type of grease used and specify all NLGI numbers for every application and greasing interval or frequency. It is good to have one or two vendors rather than having so many vendors on grease. This will also be the same as in the case of lubricating oil.

8. **Apply Color Coding for Different Grease:** When using several grease types, it is a good practice to provide color-coding for every type of grease, grease gun, and grease nipple used to avoid the chances of human error on mixing different types and brands of grease. As explained previously in this chapter, their thickeners, additives, or bases may or may not be compatible with each other. Grease incompatibility creates many issues, including oil base separation, thermal breakdown, thinning of the grease, formation of hard deposits, and depletion of additives. All this can accelerate the early failure of the part or subject it lubricates. The use of color-coding is no rocket science. Our main goal is simply to reduce any chances of human error and failure occurring.

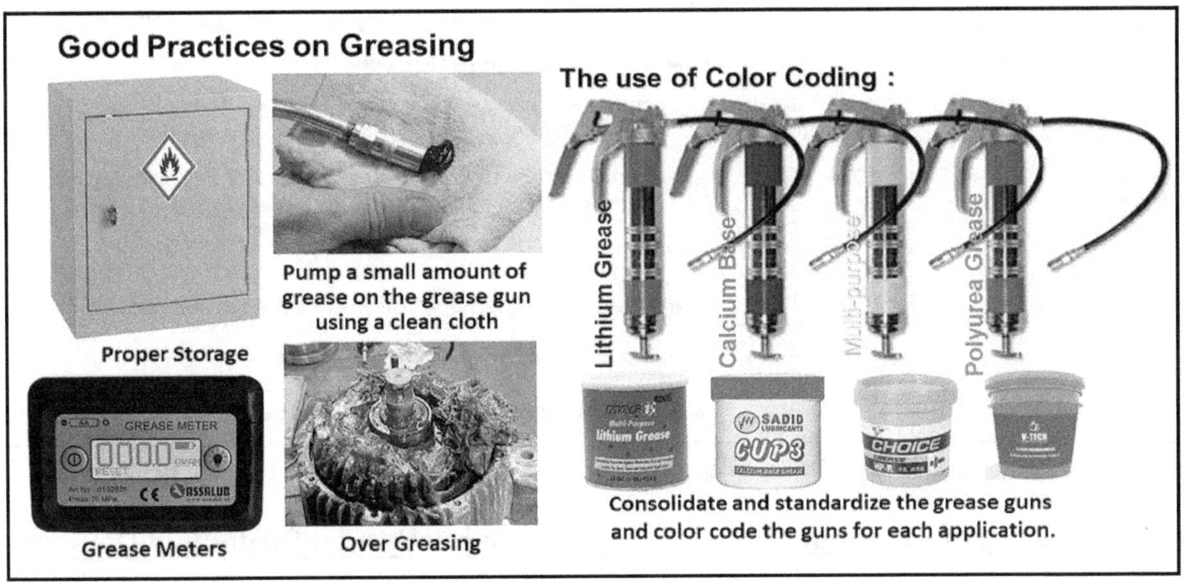

Figure 14.5: Correct Practices on Greasing

9. **Clean the Nozzle before Greasing:** Always make sure that the dispensing nozzle of the grease gun is clean before pumping the grease. Try to pump a small amount of grease out of the dispensing nozzle into a clean cloth, then wipe off with a clean lint-free cloth before attaching the grease gun to the machine's grease fittings. The purpose of doing this is to remove the contaminants attached to the grease. Do the same on the fitting or nozzle, which you will be greasing. Remember that when contaminants are mixed with the grease, the grease's contaminants cannot be removed and will be there permanently, unlike lubricating oil.

10. **Determine the Correct Interval for Greasing:** The interval for greasing depends on several factors such as temperature, amount of contamination in the atmosphere, humidity, vibration of the equipment, position of the component if vertical or horizontal, and the type of bearing used. Use the table and formula in Chapter 6 section 5 on determining the interval for greasing.

11. **Provide a Detailed Greasing Inspection:** Provide a detailed precision greasing as in figure 14.6 for all components, machines, and assets that will be scheduled for greasing. Discuss with the CMMS administrator or IT if this can be populated on the system so that whenever there is a greasing, we can have a printout regarding the details of lubrication on each component.

SAMPLE OF PRECISION MAINTENANCE ON GREASING

LUBRICATION DETAILS

1	Interval	42 days	Can be rounded off to 40 days or monthly.
2	Operating Room Temperature	41 C	Ambient temperature is the air temperature of the environment where equipment is placed.
3	Equipment Operating Temperature	150 C	Refers to the temperature of the equipment Measured with a thermal gun.
4	K Factor	1	Greasing Interval Factor derived from Handbook of Lubrication and Tribology Volume 1 (Less Harsh)
5	Lubricant Type	Lithium Grease	Type of grease to be used
6	Lubricant Brand	Shell EP3 Lithium Grease High Temp	Brand of grease to be used. For a different type of grease, refer to Grease Compatibility Table
7	Grease Dropping Point	177 - 204 C	Temperature in which the grease will become liquid. Note that this is not the usable temperature of grease
8	Gease Max Usable Temperature	177 C	Refers to the maximum temperature this grease can be use
9	Amount of Grease to be applied	150 grams	Quantity of grease to be pumped from the grease gun
10	Type of Grease gun to be used	Trico Brand with Meter Reading	Grease Gun to be used with meter
11	Responsible	Mr. Rolly Stones	Person responsible to perform the greasing

Figure 14.6: Precision Maintenance for Greasing

14.5: Procedure for Conducting Functionality Inspection on Protective Devices

This procedure will refer to determining the Failure Finding Interval or when Functionality inspection will be conducted on Hidden Protective Devices and redundant assets. The objective of conducting a Failure Finding Interval or Functionality inspection is to avoid any chances of multiple failures where the protected function fails since it was not detected because its protective device also failed. Note that Failure Finding Interval will only apply to Hidden Failures and to consider a hidden failure, a secondary failure should take place. Failure Finding Interval does not apply for Evident Failures. These maintenance tasks will check if the protective devices present in the system are still functional or not.

Identify All Protective Devices: Protective devices act as the equipment defense for multiple failures. Usually, the failure of protective devices is hidden. For hidden protective devices, a functionality inspection or failure finding task should be conducted to determine if the protective device being inspected is still functional or not. These are placed on the equipment as defenses against failure and to protect the protected function. The more critical the protective device, the more frequent will be the inspection. Use the table in Chapter 6, section 7 of this book to determine the interval of conducting functionality inspection for protective devices.

Major Step	Task Description and Instruction	Responsible
1 Identify all the Protective Devices and Redundancy for all equipment and assets.	a. Determine the boundary of the system/subsystem which is analyzed on the RCM process and identify all protective devices.	PM Team
2. Determine the average life or the MTBF of the protective device.	a. Determine the average life or MTBF of the protective device. Refer to OEM or base the average life on the experiences of maintenance.	PM Team
3. Determine the Availability required for the protective devices.	a. Determine the availability required for the protective device using the table on Failure Finding Tasks in chapter 6. b. Note that if the protective device is critical aim for higher availability and if it is less critical aim for a lower availability.	PM Team
4. Determine the Failure Finding Interval based on the table provided	a. Determine the Failure Finding Interval based on the availability of the protective device	PM Team
5. Calculate the Failure Finding Interval of the protective device.	a. Calculate the Failure Finding Interval to determine the frequency of performing a functionality inspection.	PM Team
6. Round of the interval or frequency and include in the CMMS.	a. Round of the interval is to be done daily, weekly, monthly, quarterly, yearly, and encode in the CMMS. b. Ensure that the PM Crew or person responsible complies with the dictated interval of performing the functionality inspection.	PM Team

Figure 14.7: Procedure for Conducting Functionality Inspection

Chapter 15

The Conclusion

Maintenance is not about shifting from Preventive to Predictive Maintenance nor transitioning from Reactive to Proactive Maintenance. In maintenance, there is no transition; it is about having an understanding of when to use the different maintenance tasks concurrently depending on the consequences of failure. The key is knowing when to use the different maintenance tasks, and this is possible with the aid of an algorithm or decision diagram.

15.1: Treat Maintenance as a Business

My thoughts before is that maintenance should be a profit center, but as time passed by I slowly realized that whether we like it or not, there will always be cost in doing maintenance. It is given. If industries apply the right maintenance strategies, then their equipment will run better producing more revenue and profit for the industry. Maintenance will always contain a cost, but if done correctly then whatever the cost of doing maintenance should always have a return far greater than the cost. This is where maintenance now becomes a profit center. Just like in any other business, there is always an investment, but just like in any business we need to ask that if we invest in something, would there be a return on investment or payback? If there is, then we move ahead with the investment.

For the industry to benefit from maintenance, ask what will be the investment needed? If training is needed, then ask what happens after we train our people? What is expected from them? If Predictive Maintenance instruments are needed how would it benefit the industry when we invest in these instruments and in the certification of the users? What is important is to treat maintenance as a business investment for industries. Maintenance should no longer be treated as a cost or profit center but it should be treated as a business. As the late Konosuke Matsuchita said profit is a reward for the appreciation given by a customer, it should not be the ultimate goal of the company, because if profit is the goal, then any means to get the profit can be made. This means that we can profit from maintenance by doing all the wrong things or by cutting costs which is merely temporary and may also have repercussions.

Chapter 15: The Conclusion

Any business needs an investment in time, money, and resources. If you are in the business of providing products, then the customers must be satisfied with your products. If you are in the business of providing services, then the customers must be satisfied and happy with the services being provided. Doing this will ensure profit for the business. What I am saying is that profit is just the result of a happy customer. It should not be the ultimate goal of any business because if this will be the goal, then any means to get the profit can be obtained which is not the right thing to do. Cost-cutting schemes just to please the profits will have their day of reckoning, especially for maintenance.

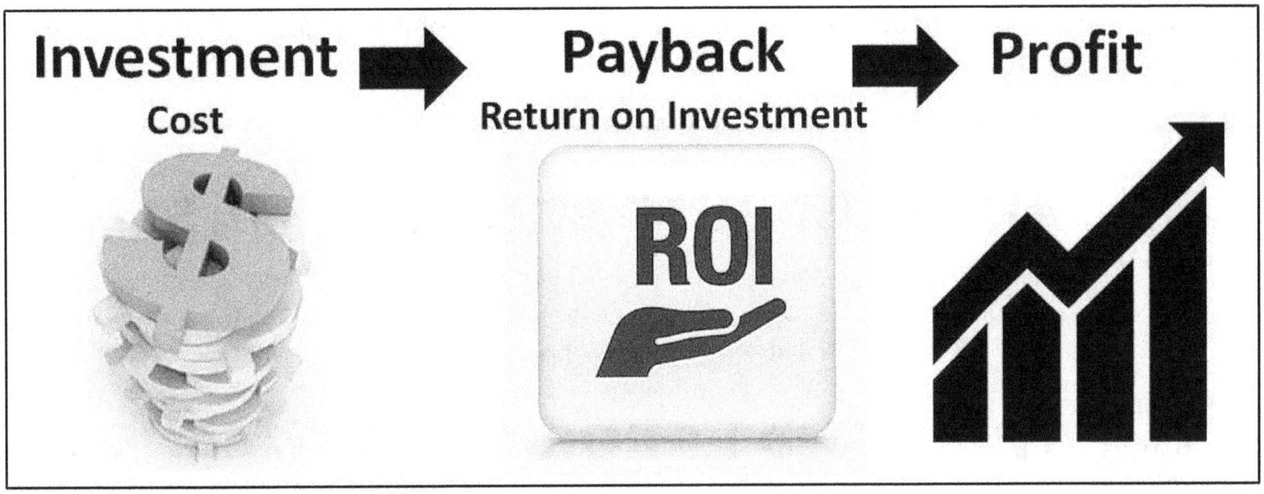

Figure 15.1: Investment – Payback - Profit

All functions of the organization depend on its equipment to produce revenue. Each of these functions of the organization will have a direct or indirect role in the equipment. All employees within the organization depend on their equipment to deliver the goods or services they provide as this will be the main source of their salary, benefits, and the reason for the industry to continue. If the equipment cannot deliver the services or products it provides, then it's just a matter of time before the company can declare bankruptcy. But the thing is the equipment will only operate as expected if we invest in our people. What I meant is that we need people who can operate and maintain the equipment correctly. Before we can do anything, we need to be certain that our people are capable. If we need to improve the equipment, then we need to improve the people first and not the equipment because it is the people that will improve the equipment and not the other way around. This means that this cannot be reversed.

Top Management must understand that people involved in reliability and maintenance need investment in knowledge. The main role of the training department is to identify the level of knowledge their employees need to perform their jobs correctly. Knowledge will be the basic need. Once the knowledge is provided to the people, then this will be used to develop the skills of every employee in the entire organization. Skills will be the ability to do their jobs correctly the first time around. Once the skills are developed and the people are capable of teaching others what they know, then we have achieved the level of mastery. But what happens is that the needs may not be identified or perhaps they are identified but the time and budget to be spent on their people is cut by management. In this case, what happens is that these people will do all sorts of trial and error learning how to cope with their day to day problems then learn everything by experience, which later on will be integrated into their company's culture as the correct way of doing and perceiving things around here in the plant. In short, this is how we do things in the industry since the very beginning of time.

Everyone involved in industries depends on their equipment and assets to earn revenue to derive profits. This can only happen if we apply not the best but the correct practices on maintenance. Remember that the best can always be made better. The bottom line is that the equipment will only run smoothly if the correct maintenance is applied and the different functions of the organization know their role and will work together in achieving a common goal and objective. This is what industries must realize and do. Each of these functions has its share of responsibility for the equipment.

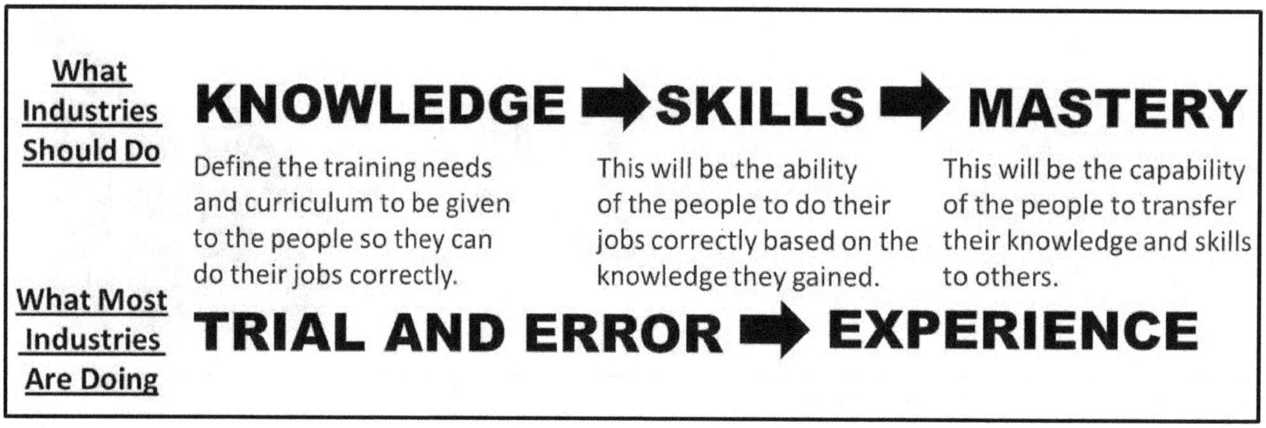

Figure 15.2: Knowledge, Skill, and Mastery

If we have 8760 hours in a year and a downtime or 3942 hours where 950 hours include all the Planned and Preventive Maintenance, while 2992 hours refers to machine-related downtime. Then the cost of revenue lost in 1 year will be $ 1,645,600.00

Case 1: Trial and Error - Experience

- Total hours in 1 year: 8760 hours
- Total Downtime: 3942 hours
- Planned Downtime: 950 hours
- Unplanned Downtime: 2992 hours
- Cost of Product: $ 550.00
- Cost of Product Lost in 1 year: $ (550.00 x 2992) = $ 1,645,600.00

Case 2: Knowledge – Skills - Mastery

- Total hours in 1 year: 8760 hours
- Total Downtime: 950 hours
- Planned Downtime: 550 hours
- Unplanned Downtime: 400 hours
- Cost of Product: $ 550.00
- Cost of Product Lost in 1 year: $ (550.00 x 400) = $ 220,000.00
- Difference: $ (1,645,600.00 – 220,000.00) = $ 1,425,600.00

Assuming that the correct practices were applied to maintenance, and the Planned and Preventive Maintenance activities were reduced from 950 to 550 hours by removing all intrusive activities and reviewing the intervals. The Unplanned Downtime was reduced from 2992 to 400 hours in a year. If the cost of the product is $ 550.00, then an additional $ 1,425,600.00 products were produced. And in this case, we are just speaking about 1 piece of equipment. If the plant has 1000 pieces of equipment then this will be $ 1,425,600,000.00 in one year. This case still does not mention the cost of performing maintenance, man-hours, and other resources. Just imagine what it would be like if all employees especially operators and maintenance have the correct knowledge to perform their jobs correctly. There are so many ways to save cost on maintenance but everything will start with its very basing foundation and that is knowledge. If Top Management people and Decision Makers want to save money big time, then they should be the first to understand what maintenance and reliability are all about and what they can do if done correctly. The bottom line is that industries should invest in acquiring the right knowledge for their people so that they can perform their jobs correctly. Learning everything from trial and error will be costly in the long run.

15.2: Solutions to the Top 10 Problems on PM

In Chapter 1 of this book, we have listed the Top 10 Problems on PM that most industries experience. Let us discuss some solutions to these problems

1) Add on PM Checklists Syndrome: Maintenance must review the tasks they perform on Preventive Maintenance and remove all those intrusive tasks which are not feasible for Preventive Maintenance, especially for overhauls and replacements. If the part or item is not age-related then other tasks such as Predictive Maintenance can address this problem. Reviewing the maintenance tasks and asking why this particular task needs to be done will provide us some insights into whether the task needs to be done or not on the equipment. Determine what failure mode this maintenance task is meant to prevent or predict and use the RCM Algorithm or RCM Decision Diagram as a guide. We also need to check if any duplicated tasks are already done by third-party contractors and decide who is the best person to execute this particular task.

2) Introduction of Infant Mortality Failures: Infant Mortality Failures are usually introduced during Preventive Maintenance when overhauling or any form of contact is made with the equipment. The number one cause of infant mortality failure is human errors and these usually occur during the reassembly process of overhauling. During overhauls, two activities are involved which include disassembling and reassembly. When we disassemble the equipment, human error is minimal, but the chances of human error increase once we start to reassemble the equipment, especially when different people will be executing the tasks. Integrating Precision Maintenance in executing Preventive Maintenance will reduce the chances of human errors committed. Likewise, having Pre and Post Predictive Maintenance monitoring can ensure that the Preventive Maintenance execution was correctly done.

3) Replacement of Good Parts Just to conform to the PM Specs: Parts and defective spares should be replaced during Preventive Maintenance. The current plan for PM should be reviewed together with the interval. If there will be parts that will be replaced, the PM can seek the recommendation of Predictive Maintenance if this is under their watch. If the part is still in perfectly good condition, then it should not be replaced. Reviewing the history of the part can provide maintenance some information on deciding whether to replace the part or not. The only problem, in this case, is if the equipment is under regulatory laws from the government that a scheduled outage is required such as in power plants for compliance. If a scheduled outage will be done every 2 years, then there are no Predictive Maintenance instruments that can detect a potential failure 2 years before it will fail. In this case, my hands are tied and I cannot recommend more but to comply with government laws and regulations even if the law itself either needs to be replaced or overhauled.

4) The Case of Random Failures: One of the measures of determining if the PM was successful or not is to measure them. Several measurements and indices will indicate if our PM was successful or not such as PM Effectiveness, downtime, availability, MTBF, number of breakdowns, and others. One of the main reasons why random failure exists is that the basic equipment condition has not been well established on the equipment. These very basic tasks such as keeping the equipment clean, completing bolts with the correct torque, applying the correct lubrication, and equipment free from any form of leaks will reduce random failure dramatically. TPM books authored by Japanese consultants seldom use the word random failures, but rather they use the term accelerated deterioration. It means that if the basics are taken care of, then the item will eventually wear out naturally.

5) Ageing Workforce and Nearing Retirement: Industries must have a program for these people since they have stayed in the plant for quite some time. They can be used as SMEs or Subject Matter Experts on teaching the younger maintenance generation about the topics they are familiar with. What industries do not want to happen is for these people to just leave or retire for good, where their knowledge just goes with them to the grave. The company hires new graduates who just recently graduated from college with little or no experience and find their way into the plant through trial and error and end up again in the same mode. What we want is to permanently break this cycle. Whatever these Rolling Stones know should be extracted and documented.

6) Lack of Training in the Maintenance Function: Training is the source of knowledge and essential to any industry are competent people who understand their equipment intimately. This is the missing link ingredient in any change or continuous improvement effort. Training is a venue for acquiring knowledge so both operators and maintenance can perform their jobs correctly. Knowledge is the starting point of building the skills of our people. This is needed to perform the required work correctly. The training department must define the gap between the current status and what is expected from our people. Our ultimate goal is to achieve mastery in which a person is capable of transferring correctly the skills and knowledge to others. This is what we want to achieve with our human workforce. Management people must also talk with the people before sending them to the training and demand what is expected from them after completing the training. Management must not only support but be committed to implementing whatever was learned from the training if this will be beneficial to the plant.

8) Frequent Reorganization in the Plant: Although this is a Top Management prerogative, there will be cases where this will be beneficial to the plant as employees learn new things, but having a frequent reorganization may also be detrimental to the organization. If the plant is running smoothly, then this should not be the case. Reorganization or restructuring of employees is no guarantee of success. This is usually done if an industry has problems financially. During my employment days, I have experienced this where I see projects and improvements abandoned since the person is transferred to another responsibility. We also have cases of operators in which some have already reached Step 7 of Autonomous Maintenance and the members were disbanded and transferred to other areas of the plant demoralizing the entire team. Although management might think that this is for the good of the organization, the problem is that if the person has some hesitation, then the plant might lose some of its valuable people as they start looking for other jobs. If a plant is operating well, then there is no need for any reorganization.

9) Lack of Poor Documentation on PM: Documentation is very important for the maintenance function. This can serve as a good reference for the planner in planning future works to be done. What is important is if a system such as CMMS or EAM is in place in the plant, then we can have paperless documentation and maintenance can just encode the correct details such as closing a Work Order. Documentation for maintenance must be simple. Maintenance must also understand the consequences of either not doing or faking it in some cases. Documents referred to in this case will include maintenance transactions with the storeroom, closing Work Orders, populating data, recording downtime, and generating reports which can either be done manually or electronically.

10) PM is waived by Operations: What is important is for both operations and maintenance people to have a sit down just like the La Cosa Nostra of Mafia and discuss this matter thoroughly. Communication and having a mutual understanding of the equipment and assets they are operating will be beneficial for the industry. We all know that the equipment needs to operate to generate revenue and profits for the plant. On the other end, operations people must also understand that their equipment is not a plug-and-play machine that when you turn it on will be absolutely maintenance-free. For as long as parts move inside the equipment, these parts can be subject to wear, and in most cases, premature failure occurs if the basics are not well established. Maintenance needs to gain the trust and confidence of the operations people that the chances of having an infant mortality failure would be minimum if PM will be performed. This can be done by integrating Precision Maintenance in executing PM. Maintenance must also understand that we just cannot stop the equipment for a scheduled PM if a particular lot needs to be shipped within the day. What I believe is if the equipment runs well after Preventive Maintenance is executed in the equipment with no infant mortality failures, then the operation's trust can be gained.

15.3: Every Function has Their Responsibility on PM

Preventive Maintenance is discipline number 5 in my book on World Class Maintenance I have written in 2009. This is part of the basic discipline. This discipline is not only for those who will execute or plan the task. It is much more than that. To make PM effective, other functions of the entire organization should also realize and understand their involvement with PM. Preventive Maintenance is a collaboration between the different functions from the Storeroom, IT (in charge of CMMS or EAM), Finance, Operations, Safety, Quality, and the rest of the people in the entire organization. These functions must collaborate, cooperate, coordinate and ensure that they know their role and responsibility. What I am saying is that Preventive Maintenance can only be implemented successfully if the different functions understand the technical side of performing PM. Our equipment and assets can only run consistently if they are sustained and well maintained. There are mechanical parts movements inside. Each movement represents a cycle and these parts are not designed for an infinite cycle. In layman's terms, we cannot run the machine to fail, it needs to be stopped for some time so that maintenance can perform the tasks.

When silos exist, then there is always the doubt that equipment will perform its intended function and that is to produce either a product or service for their clients which in return provides us the revenue and profit. Silos create a separation and a loss of chance to collaborate with one another to become better. If silos exist in an organization, then expect operators just to operate, equipment to fail, maintenance to repair, and if the part is not around, cannibalize parts somewhere since an item is not stocked in the storeroom, and a lot of other problems to occur. Industries depend on their equipment and assets to function so that they can earn revenue and that can only happen if their equipment is operating. While others talk about managing assets, the truth is that assets, machines, and equipment can only be managed if they are well maintained. It does not take a rocket scientist to figure this out. All we need is just a modicum of common sense.

Figure 15.3: PM is Not Only for the PM Crew and Planner

On the other side, the main protagonist which includes the planner and the PM Crew can only function and perform their jobs correctly if they are equipped with the correct knowledge and skills to perform their job right the first time around. If Top Management does not invest in their people's needs, then nothing will ever change. All I can say is that mistakes and human errors in maintenance are costly. In fact, human error is the leading cause of Industrial Disasters and accidents around the world. These accidents and disasters are not picky as they don't choose industries which means that they can happen in any business type.

World Class Industries were not born that way, in fact, they started reactive themselves. What they realize is that maintenance is the activities needed to preserve, conserve and sustain their equipment and assets. This can only be made possible if the different functions of the organization will not work in isolation but in constant communication, cooperation and collaboration. Once maintenance can successfully sustain and preserve their equipment and assets, their job is not yet complete. It is also the role and responsibility of maintenance to observe their equipment for any inherent design weaknesses and challenge themselves to improve them so that the life-cycle of the equipment can be extended. And I think that is all I have to say about that.

Appendix A: Answers to Preventive Maintenance Quiz

Take Quiz on Preventive Maintenance Part 1

1. b
2. c
3. Bonus
4. b
5. d
6. c
7. a
8. a
9. a
10. d
11. a
12. b
13. c
14. c
15. Bonus
16. d
17. a
18. b
19. b
20. c
21. a
22. b
23. d
24. c
25. d
26. d
27. e
28. b
29. d
30. c

Take Quiz on Preventive Maintenance Part 2 (True of False)

1. b
2. b
3. b
4. b
5. b
6. a
7. b
8. b
9. b
10. a
11. b
12. a
13. b
14. a
15. a
16. a
17. a
18. b
19. b
20. b

Note: Bonus: means any letter is correct

Appendix B: RSA Maintenance Courses

RSA Reliability and Maintenance Consultancy Firm have been around for 17 years, and through these years, we have upgraded and developed more courses suited for our reliability and maintenance people in industries. Here is a complete list of maintenance courses and services that we offer in-house, public or online training in your plant which your industry can avail and benefit from.

RSA Courses on Total Productive Maintenance

1. Total Productive Maintenance (2 or 3 days)
2. Planned Maintenance 4 Phases to Zero Unplanned Breakdown (2 or 3 days)
3. Understanding Autonomous Maintenance, Operators 7 Steps to Empowerment (2 or 3 days)
4. Understanding Focused Improvement-Kobetsu Kaizen (1 day)
5. Relationship between OEE and Equipment Losses (1 day)
6. Advance Maintenance Strategies on Planned and Autonomous Maintenance (2 days)

RSA Courses on Reliability and Maintenance Strategies

7. Lubrication Strategy-Understanding Tribology, the Importance of Oil Contamination Control (2 days)
8. Reliability-Centered Maintenance for Industries (2 or 3 days)
9. Condition-Based Maintenance, Total Approach to Failure Prediction and Analysis (2 days)
10. Root Cause Failure Analysis-Understanding Equipment Failure (2 or 3 days)
11. Optimizing Equipment Reliability-Streamline RCM Approach (2 days)
12. World Class Maintenance Management - The 12 Disciplines (2 or 3 days)
13. Understanding MRO Spare Parts and Storeroom Management (2 or 3 days)
14. Failure Mode and Effects Analysis (FMEA/FMECA) (1 day)
15. Practical Best Maintenance Practices (2 or 3 days)
16. Advance Maintenance Leadership in TPM, RCM, LUB, and RCFA (5 days)
17. Advance Maintenance Strategies on RCM and RCFA (2 days)
18. Advance Maintenance Strategies on Lubrication and CBM (2 days)
19. Cutting-Edge Maintenance Management Strategies (3 days)
20. Implementing Preventive Maintenance for Industries the Right Way (2 days) New
21. Understanding the Concept of Life-Cycle Management (2 days) New

RSA Courses on Reliability and Maintenance Concepts

22. Meaningful Measures of Equipment Performance - Understanding MTBF, MTTF, MTBA, MTTR, MTTS, Failure Rate, OEE, and Weibull Overview (2 days)
23. Basic Maintenance Concept, Understanding Reactive, Preventive, Predictive, and Proactive Maintenance (1 day)
24. Understanding Proactive Maintenance (1 day)
25. Maintenance Best Practices on LUB and CBM (2 days)
26. Preventive and Predictive Maintenance Strategies (1 day)
27. Proactive and Precision Maintenance Strategies (1 day)

RSA Facilitation, Guidance, and Consultation Services Includes

- Facilitation and consultation on Total Productive Maintenance implementation
- Facilitation and consultation on Root Cause Failure Analysis Investigation
- Provide guidance on starting up a Predictive Maintenance strategy in the plant
- Facilitation and consultation on Reliability-Centered Maintenance
- Conduct initial assessment on maintenance Technical Training Needs Analysis
- Facilitation of Strategic Planning for Maintenance and Reliability professionals
- Provides assessment on World Class Maintenance-The 12 Disciplines
- Consultation on setting up an Oil Contamination Control in your plant
- Implementation of TPM Planned Maintenance Pillar
- Implementation of TPM Autonomous Maintenance Pillar
- Implementation of World Class Maintenance Management-The 12 Disciplines
- Maintenance Assessment to determine where your industry currently stands

RSA Reliability and Maintenance Consultancy Firm accept in-house training services for industries in local and international overseas countries. Special arrangements can be offered for overseas countries on any of our maintenance courses selected. These courses are what industries need to improve how we maintain and sustain our equipment and assets in the plant.

The following are lists of training I offered to plants and industries concerning reliability and maintenance courses. My mission is to uplift our maintenance human resources' technical competence in industries searching for ways to achieve maintenance excellence by capturing the industry's reliability and maintenance correct practices. These courses provide in-depth details and a wealth of information regarding maintenance correct practices from the most basic to the most advanced maintenance and reliability strategies. These powerful courses have been proven by industries that the best way to reduce their maintenance cost is to sustain their equipment's reliability. With consistent focus on output and productivity and secondary to maintenance and reliability, the latter results in frequent failures, costly unscheduled repairs and unexpected downtime, and an inevitable high cost of doing maintenance that calls for the adoption of a more rigorous and more effective maintenance strategy that is truly world-class.

Finally, it is also undisputed that every maintenance manager's challenge is maximizing the equipment's reliability through a traditional and often self-designed centered Preventive Maintenance system. This practice is why the approach to maintenance management seems to remain reactive rather than proactive. Truly, these courses are designed for all maintenance managers, engineers, and professionals whose mandate is to optimize their equipment capacity and reliability at the lowest possible cost. In the preceding light, we designed these courses that can be tailored fit and made available to be conducted in-house in your plant for your people's wider participation. Should you be interested in any of these courses, you may reach me at https://www.rsareliability.com/ or email me at rollyangeles@rsareliability.com.

Appendix C: PMS Training Course Details (2 Days)

Figure App. C1: PMS Training Module

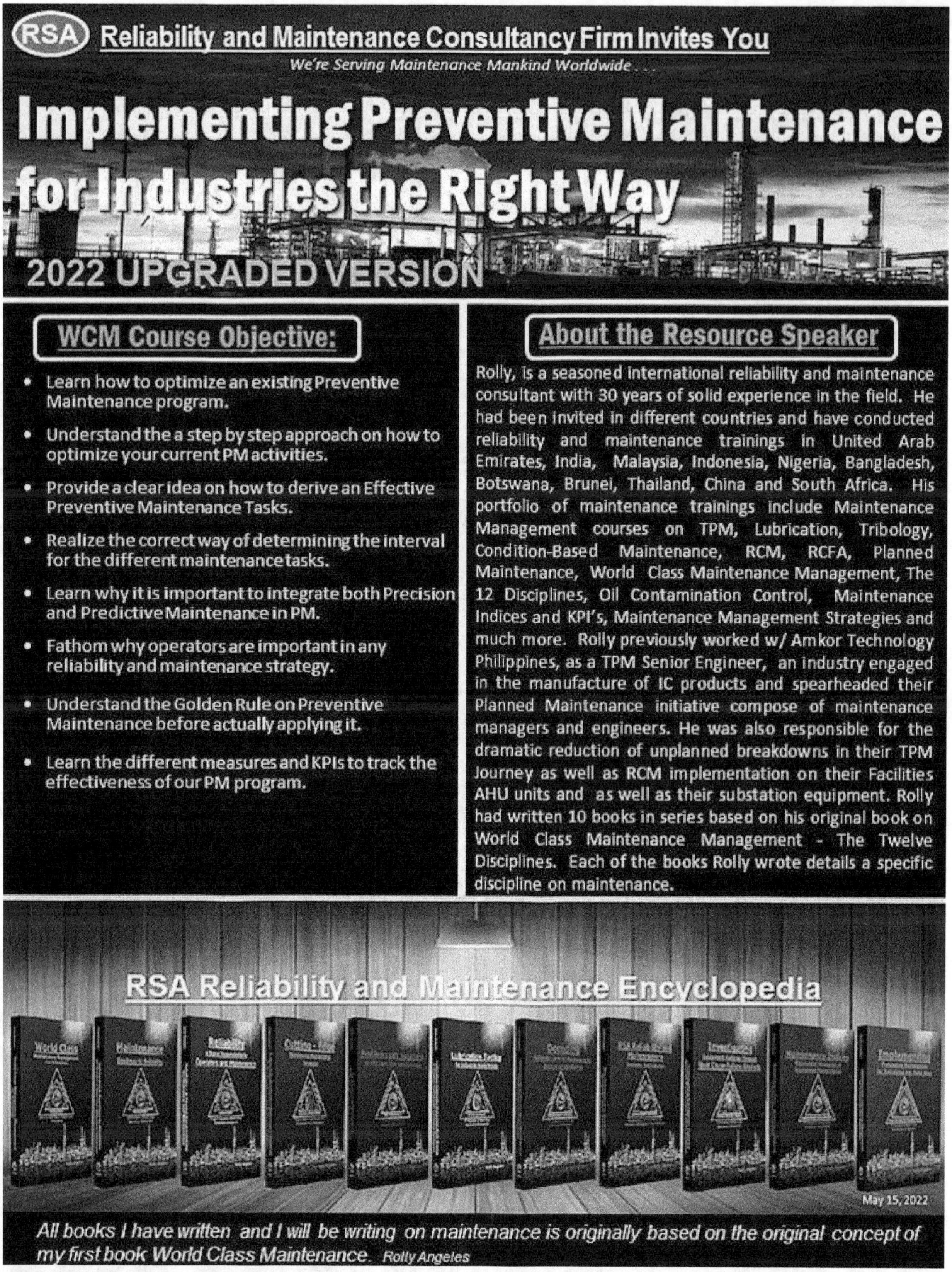

Figure App. C2: PMS Training Module

Appendix C: PMS Training Course Details

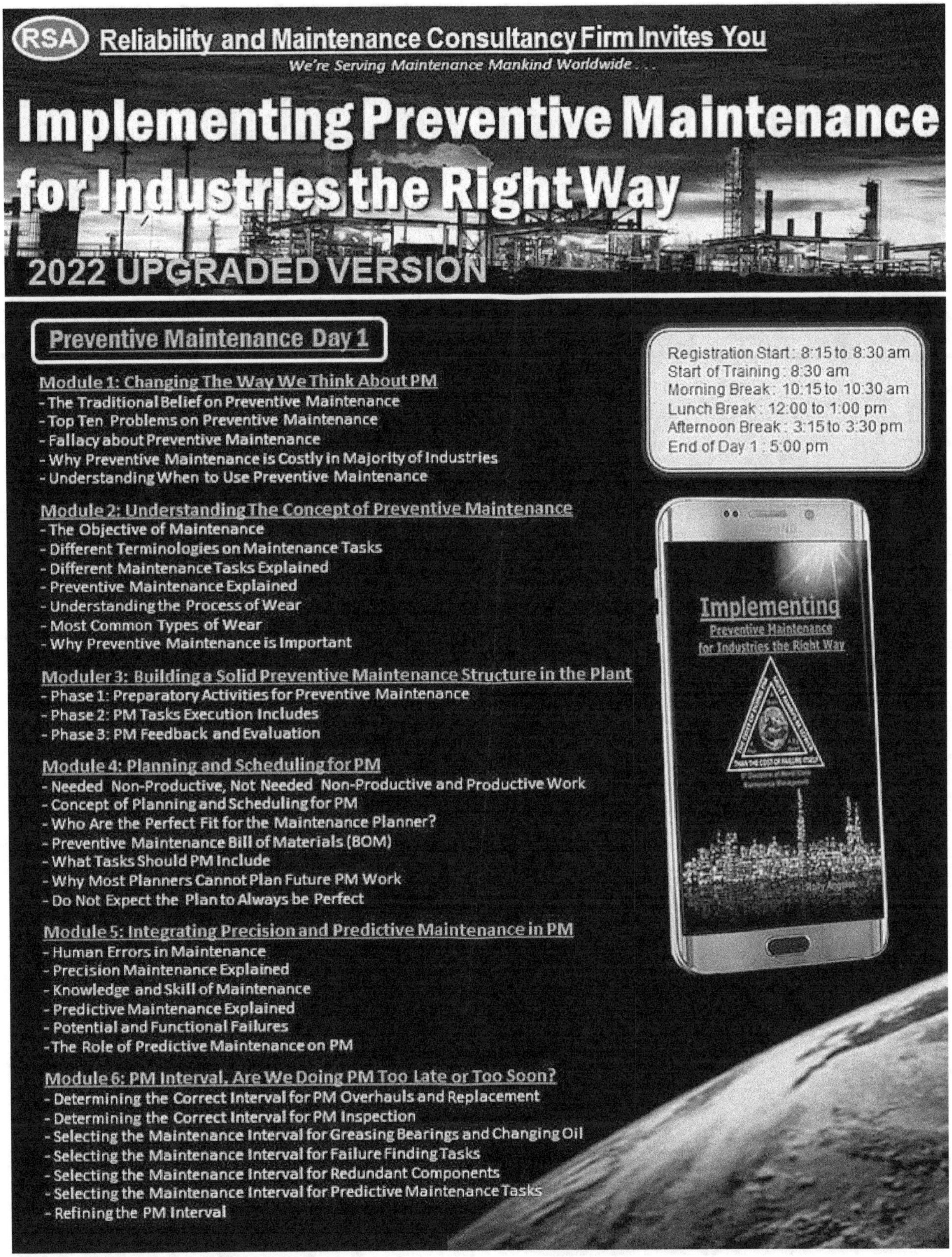

Figure App. C3: PMS Training Module

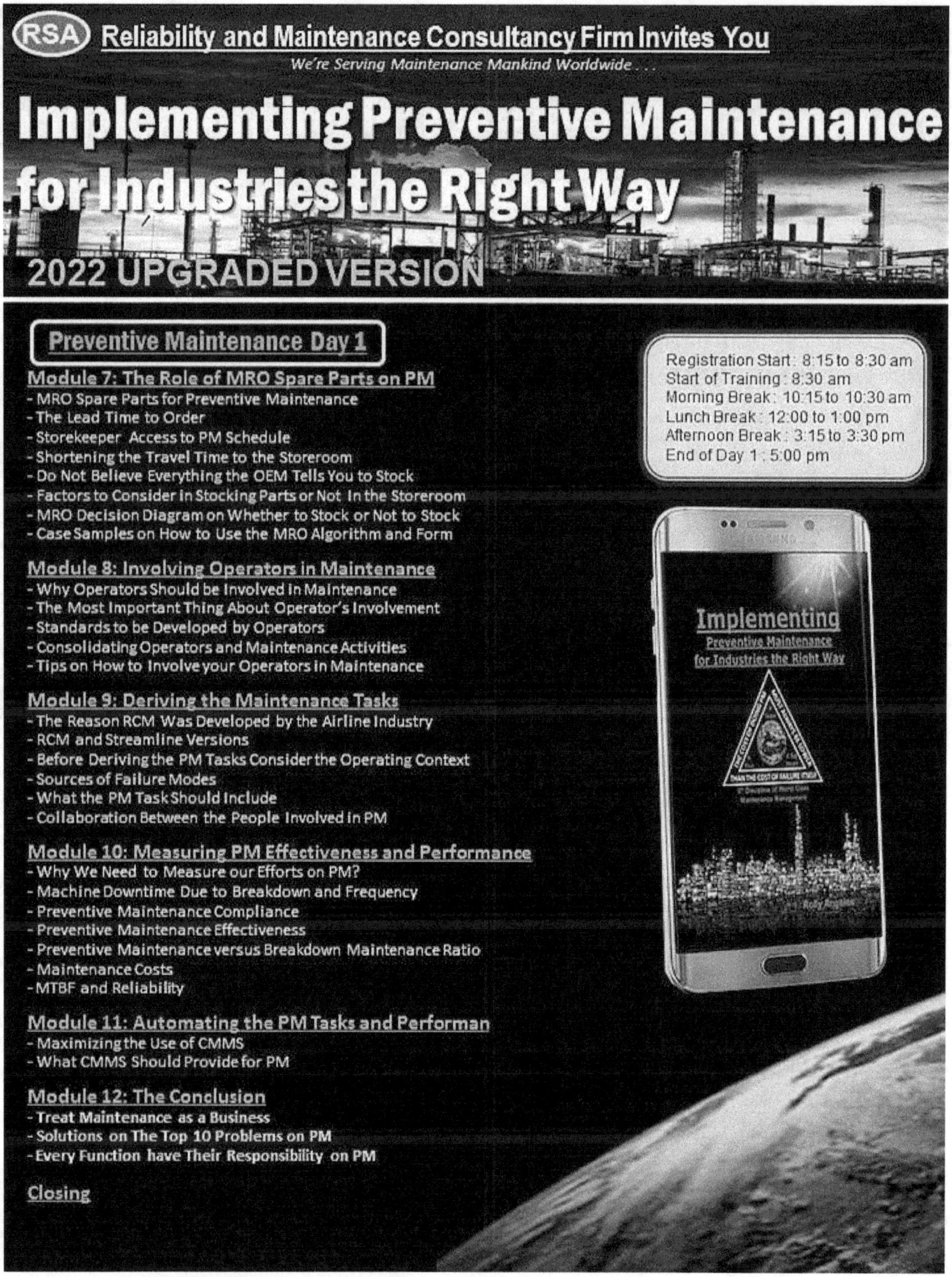

Figure App. C4: PMS Training Module

Serving Maintenance Mankind Worldwide

This book is dedicated to all "Maintenance" out there in industries. My mission is to reach out to industries searching for ways to improve their maintenance human resources. I believe that the key to improving reliability is to provide our maintenance people the knowledge and education they need to maintain their equipment and assets. At my small firm, we serve maintenance mankind worldwide.

World Class Maintenance Management - The12 Disciplines Training

WCM in Thailand
October 2013

WCM in Indonesia
June 2013

WCM in South Africa
June 2012

WCM in Malaysia
August 2010

WCM New Delhi, India
December 2010

WCM at Meralco, Philippines
August 2016

WCM in Bangladesh
November 2009

WCM in India, Chennai
July 2008

WCM at Bharat Petroleum, India
September 2008

WCM Kolkata, India
February 2018

WCM in Mumbai, India
December 2016

WCM in Malaysia
July 2013

We're Serving Maintenance Mankind Worldwide

Bibliography

- Angeles, Rolly, **RSA Reliability and Maintenance Newsletter Vault Collection, Subscriber's Edition,** Central books Printing, Philippines 2021
- Angeles, Rolly, **Investigating Equipment Failures through Root Cause Failure Analysis,** Central books Printing, Philippines 2021
- Angeles, Rolly, **Maintenance Indices, Meaningful Measures of Equipment Performance,** Central books Printing, Philippines 2022

Angeles, Rolly, **Cutting Edge Maintenance Management Strategies,** Central books Printing, Philippines 2020

- Angeles, Rolly, **Decoding Reliability Centered Maintenance Process for Manufacturing Industries,** Central books Printing, Philippines 2021
- Angeles, Rolly, **Lubrication Tactics for Industries Made Simple,** Central books Printing, Philippines 2020
- Angeles, Rolly, **Maintenance Roadmap to Reliability**, Central books Printing, Philippines, 2016.
- Angeles, Rolly, **Problems, and Solutions on MRO Spare Parts and Storeroom,** Central books, Printing, Philippines 2020
- Angeles, Rolly, **Reliability, A Shared Responsibility for Operators and Maintenance,** Central books Printing, Philippines, 2018.
- Angeles, Rolly, **World Class Maintenance Management – The 12 Disciplines**, Central books Printing, Philippines, 2009
- Bazovky, Igor, **Reliability Theory and Practice,** Dover Publications Incorporated, 2004
- Campbell, John Dixon, **Uptime, Strategies for Excellence in Maintenance Management,** Productivity Press, 1995
- Gross, John, **Fundamentals of Preventive Maintenance,** Amacom Books, 2002
- Levitt Joel, **Complete Guide to Preventive and Predictive Maintenance**, Industrial Press Inc. 2003
- Masaji Tajiri and Fumio Gotoh, **Autonomous Maintenance in Seven Steps – Implementing TPM on the Shop Floor**, Productivity Press, 1999
- Moubray, John, **Reliability Centred Maintenance II, Second Edition**, Great Britain: Butterworth, Heinemann, 1997.
- Mobley, R. Keith, **An Introduction to Predictive Maintenance**, Butterworth-Heineman Elsevier, 2002, Heinemann, 2002.
- Nakajima, Seiichi, **Introduction to TPM, Total Productive Maintenance**, Portland Oregon, Productivity Press, 1984
- Nowlan, Stanley and Heap, Howard. **Reliability-Centered Maintenance**. Springfield, US Department of Commerce National Technical Information Service, 1978.
- Pall Industries Hydraulics Co. **Contamination Control and Fundamentals**: New York, 1994.
- Palmer, Richard, **Maintenance Planning and Scheduling Handbook Second Edition,** The McGraw-Hill Companies, Inc., 2006
- Reason, James and Hobbs Alan, **Managing Maintenance Error, A Practical Guide**, Ashgate Publishing Limited, 2003
- Robinson, Charles, and Ginder, Andrew, **Implementing TPM, The North American Experience**, Portland Oregon: Productivity Press, 1995.
- Smith, Anthony, **Reliability-Centered Maintenance, Gateway to World Class**, Mc Graw Hill Inc, 1993
- Wireman, Terry, **Preventive Maintenance**, Industrial Press Inc. 2008

Glossary on Maintenance

This glossary is a collection and compilation of all my books on reliability and maintenance which includes the following;

- Volume 1: World Class Maintenance Management – The 12 Disciplines
- Volume 2: Maintenance – Roadmap to Reliability
- Volume 3: Reliability – A Shared Responsibility for Both Operators and Maintenance
- Volume 4: Cutting–Edge Maintenance Management Strategies
- Volume 5: Problems and Solutions on MRO Spare Parts and Storeroom
- Volume 6: Lubrication Tactics for Industries Made Simple
- Volume 7: Decoding Reliability-Centered Maintenance Process for Manufacturing Industries
- Volume 8: RSA Reliability and Maintenance Newsletter Vault Collection, Subscriber's Edition
- Volume 9: Investigating Equipment Failures through Root Cause Failure Analysis
- Volume 10: Maintenance Indices – Meaningful Measures of Equipment Performance
- Volume 11: Implementing Preventive Maintenance for Industries the Right Way

8-Disciplines is an analytical problem-solving tool designed to determine the probable or most likely cause of the problem. This method can be applied to defects and equipment-related failures. This method establishes a permanent corrective action based on data and provides the probable cause of the problem.

ABC Analysis also called inventory stratification is a technique used to classify and optimize inventory levels. Inventory is based on the value and usage of an item. Class C refers to standard parts, consumables, or commodity items that can be delivered by the vendor on a regular schedule or can be easily made available from a few hours to a couple of days where the cost is less than $100/unit. Class B are standard parts and can be made available in a few days to weeks. Parts are mid to high costing around $100 or more. Class A: are one-of-a-kind parts with long lead time, high cost, and low demand. Can range from $500 to $100,000 or more. Items classified in this category are critical and expensive.

Abnormality can be defined as any deficiency, disorder, slight irregularity, defect, bug, flaw, or any unwanted condition that could lead to other equipment problems. This is addressed during Autonomous Maintenance Step 1 activities.

Abrasive Wear is a type of wear that can be categorized by a single keyword, cutting. Abrasive wear occurs when hard particles are suspended in a fluid or projections from one surface roll or slide under pressure against another surface, thereby cutting the other surface. Abrasive wear can be either two-body or three-body abrasion.

Absolute Filtration refers to the smallest size of particle that will be removed during filtration. This means that if the filter's size is 5 microns and the rating is absolute, almost all the contaminants in the range of five microns and above should be removed by the oil filter. The filtration efficiency for absolute rating usually is high at 99% and above, which is the capability of the filter to remove contaminants.

Absolute Viscosity measures the resistance to flow when an external force is applied like a spindle driven by a motor. For example, when we buy water-based paint and open it, it takes some force to stir the paint. However, if we add water and mix it with the paint, it is much easier to stir. This means that the viscosity decreases.

Accelerated deterioration means that the part, item, or spare had failed prematurely and has not reached its lifespan. We can state that we have a case of premature wear.

Additive is any blended chemical added to a lubricating oil or grease to improve its performance and properties to protect the base. It is also mixed with the oil to counter any negative effects of the oil from heat and contamination. Additives usually represent around 1 to 30 % of the lubricant with engine oil having the most amount of additives.

Adhesive Wear occurs when a peak or asperity from one surface comes in contact with the other surface's peak or asperity. There may be instantaneous micro-welding caused by the friction or heat involved due to the lubricant's loss of film.

Advance Discipline refers to the last 2 disciplines on World Class Maintenance that includes the state-of-the-art non-destructive tools and diagnostics instruments for Predictive Maintenance and software such as CMMS or EAM for the maintenance requirements.

Age is the measure of a unit, item, or spare's total exposure to stress which can be expressed as the number of operating hours or other stress units since the asset started operating until its retirement or decommissioned.

Aging is the process where certain parts of the equipment deteriorate or wear out because of stress-induced over a given period. It is also termed as wear and tear or deterioration. TPM refers to this as natural deterioration.

Age Exploration Method is a method for determining the interval for age-related items by increasing the interval for Preventive Maintenance overhauling and replacement by 10 % if aging or wear-out is not yet evident when it will be overhauled or replaced.

Analytical Ferrography (ASTM D7690) is an oil analysis test where solid debris suspended in a lubricant is separated and systematically deposited onto a glass slide called a ferrogram. The slide is examined under a microscope to distinguish particle size, concentration, composition, morphology, and surface condition of the ferrous and non-ferrous wear particles

Anthropometric Errors are a type of human error that occurs because the person simply cannot fit into the space provided, cannot reach something, or is not strong enough to move or lift something. Industries aiming for multiskilling should also consider anthropometric factors to avoid human errors.

Antifoaming or Defoamer is the process of removing entrapped air or bubbles in the lubricating oil. Almost all lubricating oil systems contain some form of air. Air is found in four phases: free air, dissolved air, entrained air, and foaming. A defoamer or an anti-foaming agent is a chemical additive in oil that reduces and hinders foam formation in industrial process liquids.

Anti-Wear additives (AW) provide the protective layer on moving parts that minimize metal contact effects. Anti-wear additives are used in many lubricating oils to reduce friction, wear, and scuffing under boundary lubrication conditions, where full-film lubrication cannot be fully maintained. Anti-wear additives are also called boundary lubrication additives.

Asperity, in tribology, on a micro-scale are topographical irregularities that are similar to peaks and valleys of a solid surface. This can only be seen if the surface is enhanced through a Scanning Electron Microscope (SEM). Once the two surfaces' asperities come into contact, lubrication is displaced, and there is metal-to-metal contact and fracture later occurs.

Asset Hierarchy is an index of all your maintenance equipment, machines, and components, and how they work together. This explains the parent-child relationship of assets in the plant.

Asset Management is defined as executing that set of processes to realize value as the organization defines it from those assets.

Attrition Rate also referred to as a churn rate, is the rate at which people leave the industry. This refers to the number of people who have left the company, divided by the average number of employees over a given period.

Autonomous Maintenance is one of the main pillars of TPM which refers to the activities in which operators perform their daily inspection, lubrication, parts replacement, minor repair and troubleshooting, accuracy checks, and so forth on their own equipment, aiming to keep the equipment in good running condition.

Availability is the proportion of time the equipment is available for use for its intended purpose, whether it is being utilized or not. The trend should be consistent and should not be lower than 98%. Calculation of availability can be reported on a daily, weekly, or monthly basis.

Available Time is given as the total time in a given period. 24 hours for one day, 168 hours for one week, 720 hours for 30 days, 744 hours for 31 days, or 8760 hours in a year.

Barcode is a machine-readable form of information on a surface that can be scanned by a barcode scanner. They are often known as UPC codes. The barcode is read by using a special scanner that reads the information directly. This information is transmitted into a database where it can be logged and tracked.

Basic Maintenance Discipline is the fundamental activity that should be performed on the equipment before proceeding with any other advanced or specialized disciplines. It also serves as the foundation of any maintenance management strategy. This includes training, KPI, Basic Equipment Condition, Autonomous Maintenance, and Preventive Maintenance.

Bathtub Curve is a conditional probability curve representing the age-reliability relationship of certain items or characterized by an early infant mortality region, a region of relatively constant reliability, and an identifiable wear-out age.

Bending Stress is the force an object encounters when subject to a load at a given point which causes the object to bend or warp. This type of stress usually occurs in objects when subjected to a tensile load.

Beta Filtration Rating is derived from the Multi-pass Test Method for Evaluating Filtration Performance of a Fine Filter Element (ISO 4572). The automatic particle counter measures the upstream particles and quantity per unit volume of fluid, the particle size, and the quantity downstream of the filter during the test.

Bill of Materials is a list of materials, items, spares, and consumables needed either for repair or Preventive Maintenance. Having a PM Bill of Materials will allow the people involved to identify easily what items and materials will be needed to conduct a repair or a scheduled routine Preventive Maintenance for a particular asset.

Blending is a refinery operation that blends different component streams into various grades of gasoline. For gasoline, the fuel is blended to achieve a higher octane rating standard, thereby creating different gasoline types. Most gasoline stations offer three octane levels, including regular, about 87, mid-grade, about 89, and premium, with an octane rating from 91 to 94.

Bottleneck is a constraint or congestion in a production operation that occurs when the products' inventory is built up quicker than the equipment can process. In manufacturing, the bottleneck equipment includes those assets that frequently fail in the process, or have a lower UPH (units per hour) than other equipment in the process.

Boundary lubrication is affiliated with metal-to-metal contact between two sliding surfaces as the asperities come in direct contact with one another. This mostly happens during start-up and shutdown.

Breakdown Occurrences refer to the frequency of failure or breakdown of the system or component. The trend should be the lower, the better. This can be reported on a weekly or monthly basis. Care should be taken on what to include and not to include as a breakdown.

Breakdown Loss sometimes referred to as equipment failure loss, is a loss when the machine stops since its function completely failed. The Japanese term for breakdown is kosho.

Brittle Fracture is a fracture that involves little or no permanent deformation of metals. Many non-metals lack ductility and are subject to brittle fractures. A brittle fracture occurs when a part is overloaded and breaks with no visible distortion and deformation.

C-level Executives are high-ranking executives of a company or any organization in charge of making company-wide decision-making. C stands for chief. This will include the Chief Executive Officer (CEO), Chief Operating Officer (COO), and Chief Information Officer (CIO).

Calibration is the process in which instrumentation and equipment used in industries are monitored and maintained to ensure they continue to give accurate and reliable results and ensure that deviations in measurements are corrected.

Capital Spares are the items within inventory that are purchased as spare parts for depreciable assets (e.g., capital equipment, backup engines, and redundancies). As such, these capital spares within the inventory can depreciate. In most cases, capital spares are not included as inventories since they are part of a company's property, plant, and equipment (PPE).

Carrying or Holding Costs is the accumulated cost of parts held in stock inside the storeroom. The holding or carrying cost begins to accumulate the day the items are put inside the storeroom. Usually, this will be around 10 % to 30% of the overall costs per year.

Glossary of Terms

Causal Factor in root cause investigation is any unplanned, unintended contributor to an incident or failure. It can also be stated as the contributing cause to the incident. It can also be said that the causal factor is just one of the reasons that contributed to the incident.

Chronic Losses are usually made up of a wide variety of causes and frequently occur over time. The word chronic means repeating or recurring, which can be seen in defects and failures. Chronic Losses are always a combination of different causes.

Circadian Rhythm which means around the day. It comes from the Latin words circa which means about and dies which means a day. These circadian variations are governed by a biological clock located in the human brain.

Classification is used in FMEA/FMECA to highlight those failure modes with a severity rating of 9 to 10 or failures with the highest impact or consequences on the equipment or system being analyzed in the FMEA analysis.

Cleaning Materials Inventories are items and materials needed to sustain and maintain the plant's cleanliness and facilities, comfort rooms, toilets, and office spaces, used by the janitorial services of the plant.

CMMS or Computerized Maintenance Management Software is a maintenance software used by the maintenance department to streamline and automate the maintenance process. CMMS should support these functions by capturing and automating administrative maintenance tasks and gathering relevant information to perform this process.

Combination of Tasks is the use of two or more types of maintenance to address a particular mode such as Preventive and Predictive Maintenance which can be done at different intervals and with different people. For example, if the failure mode is a bearing failure, operators may be performing lubrication, while the Predictive Maintenance group is performing Vibration Monitoring, while other operators are carrying out a daily reading of temperature using a thermal gun.

Complex Item refers to an item, equipment, or asset whose functional failure can result from numerous failure modes.

Compression is when we applied a downward force on a vertical cylinder, the object will have an equal upward force, which is equivalent to the normal force. Once we increase the downward force, then we also increase the upward force.

Conditional Probability of Failure refers to the probability that an item will fail during a particular age interval, given that it survives to enter that interval.

Condition-Based Maintenance or Predictive Maintenance checks the equipment's actual condition using sophisticated measuring non-destructive instruments with precision accuracy. Predictive Maintenance instruments are just a higher form of the human senses. Performing inspection through the use of human senses is also included in On-Condition tasks for RCM.

Consequences of Failure result from a given functional failure at the equipment level and for the operating organization classified in the RCM analysis: which may be a hidden failure consequence, safety consequences, environmental consequences, operational consequences, or non-operational consequences.

Consistency depends on the type and amount of thickener used and the viscosity of the base oil. Consistency is also known as the resistance to deform caused by an outside force. The measure of consistency is called penetration.

Containment is a short-term action being initiated to keep the defects or failures from further damage. Containment is just a temporary solution and will not fix the problem.

Contamination in oil is anything that should not be present in the oil besides its base and additives. Contaminants may be solids, liquids, or even gases in bubbles that cause foaming.

Consumable spares are regularly consumed and used parts and items such as fasteners, seals, belts, oil filters, grease, lubricants, gloves, WD-40, rags, face masks, cleaning materials, chemicals, and so on. These items are stored in the storeroom. They are used by maintenance regularly. They are considered fast-moving items in the storeroom.

Contract Maintenance is defined as the contract between 2 parties that creates the agreement that one party will maintain an asset owned by another party. This is also defined as the contract agreement signed between the industry and a third-party contractor who will perform a maintenance activity based on the terms of agreement on industries equipment, and assets. Usually, the scope of maintenance work will be dictated by the maintenance head, and performed by a third party and not by the plant's regular maintenance crew.

Corrective actions are actions taken to address the probable cause of identified non-conformances or incidents, manage their consequences, and prevent or reduce the likelihood of recurrence of failure. Identification and execution of corrective measures must be done for both the short term and the long term.

Corrective Maintenance is a maintenance task that has different meanings. In the majority of industries, this refers to repairing a failure after it happened. Other industries also refer to this as Predictive Maintenance tasks which are the activities that are done when the P-F interval is nearing its functional failure. For those industries implementing TPM or Total Productive Maintenance, corrective maintenance means performing an improvement on the equipment.

Corrosion is a wear process of materials caused by chemical, electrochemical, or other reactions. The main difference between corrosion and rust is that corrosion occurs due to the chemical reaction on metal surfaces, while rust only affects iron metals exposed to air or moisture.

Corrosion inhibitor is a chemical compound substance that decreases the metal or alloy's corrosion rate when it comes in direct contact with the oil when added to the lubricating oil. Corrosion inhibitors are additives to the fluids that surround the metal or related object. The corrosive inhibitor's nature depends on the material being protected, which are commonly metal objects, and on the corrosive agent that needs to be neutralized.

Corrosive wear is the deterioration of metal due to chemical or electrochemical reactions with its environment. Corrosion can occur when the substance is exposed to air or some chemicals while rust mainly occurs when a metal is exposed to air or moisture.

Glossary of Terms

Cortisol is a stress hormone that triggers your heart to work harder. Cortisol works with certain parts of our brain to control our mood, motivation, feelings, anxiety, and fear.

Cost avoidance focuses on taking actions that avoid incurring costs. For industries, these are measures to reduce their expenses. Any actions to avoid an increase in expenses are considered cost avoidance.

Cost-cutting refers to measures implemented by a company to reduce its expenses and improve profitability. Cost-cutting measures are typically implemented during times of financial distress for a company or during economic downturns.

Cost savings is the reduction from last year's spending on the same item. They will appear on the company's budget and in financial statements as a decrease in spending.

Cracking Process is used to maximize the effectiveness of heavier oils. Heavier oil contains large strings of hydrogen and carbon molecules. Using a catalyst, these long strings of molecules are broken into small chains that transform heavier oil into lighter fluids such as gasoline and diesel fuels.

Criticality is a measure of how important an asset is to your process. The more critical the asset, the more impact it will have when failure happens. Criticality is based on what could happen if the failure occurs.

Critical Failure involves the loss of function or damage that could directly affect safety and environmental consequences. It can also have operational consequences with devastating impacts or aftermath.

Culture means how people perceived and do these things around their plants. It refers to their common values and beliefs, while others refer to them as shared thoughts and feelings. According to Schein, culture is the pattern of basic assumptions that a given group had invented, discovered, or developed in learning to cope with its current problems of external adaptation and internal integration that had worked well enough to be considered valid which will be taught to new members as the correct way to perceive, think, and act concerning their day-to-day activities.

Dedicated Machines refer to manufacturing machines that produce only one type of product and there is no set-up or conversion involved.

Defect and Rework Loss is a type of equipment loss caused when defects on products are found in which the product has to be reworked or scrapped. Defects on the products occur for a variety of reasons such as machine, method, raw materials, human error, or environment.

Detection in FMEA/FMECA is an assessment of the design and machinery controls' ability to detect or capture that a failure mode is occurring independently. This is also termed as confidence.

Detergents are oil additives that keep the hot metal components free from deposits and neutralize acids, forming sludge and residues. The detergents used in lubricating oil can be organic soaps and salts of alkaline earth metals such as barium, calcium, and magnesium.

Design Speed Loss is a type of equipment loss on production caused by the difference between the design or theoretical speed and its actual operating speed. This is usually given in the PM manual document provided by the OEM.

Die, Tool and Jig Losses are losses that include the cost of the physical consumption of spare parts or refurbishments of items that are used on the line. Examples of this include the cost of spares, cost of replacement, and maintenance of tooling, dies, and jigs.

Diffusional Interception is a filtration mechanism where the particles that strike through random moving gas molecules impact the medium and are held by adsorptive forces. The probability of removal is further increased by the effect of Direct Interception, Inertial Impaction, and Diffusional Interception combined together.

Direct Interception is where the contaminants are trapped in the pores between the medium fibers. It can also catch contaminants that are smaller than the pore size through bridging. In direct interception, the contaminants come in physical contact and are attached to the filter media.

Dispersants are added to the lubricant to prevent the accumulation or particle attraction of sludge and dirt in the oil. The dispersants' function is to suspend the contaminant in the oil rather than for the contaminant to settle at the bottom of the crankcase and form deposits so that it can be removed by the oil filter easily when the oil flows and circulates inside the engine system.

Dissolved water is when the water is dispersed in the oil in a homogeneous molecular solution. The gas or moisture cannot be removed from a conventional' nominal filtration; however, several absolute oil filtration on the market can remove a certain amount of moisture in the oil.

Distillation Process is a refinery process where crude oil will undergo a process called distilling or the distillation process. The oil is heated in an atmospheric distillation vessel until the crude oil reaches its boiling point, which will turn oil into vapor.

Distribution Loss is a manpower loss that includes losses that occur due to incorrect or inefficient delivery of raw materials, packaging, or products to and from the factory or the production line. Example: Incorrect delivery of materials from supplier to store, late deliveries, excessive handling of deliveries (double handling).

Downtime is the period when equipment or asset is not operating. This is the time that the equipment is unavailable for use. The equipment's unavailability may result from equipment failure, malfunction, or non-machine-related downtime such as PM shutdown. Downtime is classified as machine-related (unplanned downtime) and non-machine-related (planned downtime). It refers to the amount of time spent on doing corrective maintenance on the equipment and assets.

Dropping Point is a test on grease that determines the temperature at which the grease drips off the testing unit in a non-decomposed condition. It indicates the grease's heat resistance or up to what temperature the grease can remain firm.

Glossary of Terms

Economic Consequences refer to the consequences of the failure mode, which is evident and does not affect safety, and the environment and has no operational consequences. The only consequence, in this case, is the direct cost of the repair.

Economic Order Quantity (EOQ) is one technique that can be used to optimize inventory levels by ordering the right quantity at a specific time interval to minimize inventory cost but still meet the users' demands. EOQ will answer how many orders need to be placed per year and how many items to place per order.

Electronic Data Interchange (EDI) can be best used by Purchasers in automating their transactions with their vendors and suppliers. It provides a direct link between the vendor and the buyer by automating the Purchase Request and Purchase Order system directly linked with an internal system.

Empowerment means power, control, authority, or dominion. The prefix "em" means to put on to or to cover with. Empowering is the passing of authority and responsibility to an individual. Empowerment occurs when the power goes to the operators, who then experience a sense of ownership and control over their jobs.

Emulsified Water can be in the form of microscopic droplets of water distributed in the form of an emulsion. Emulsified water is when the amount of dissolved water is greater than the saturation point. When the water is in an emulsified state in lubricating oil, its color turned into a white or milk state in layman's terms.

Energy Losses are losses in the input energy which cannot be used effectively for processing. Examples can include start-up losses and insufficient compressed air for pneumatic machines. Usually occurs on the facilities/utilities in industries.

Enterprise Asset Management (EAM) software provides a holistic view of an organization's physical assets and infrastructure throughout their entire life cycle, starting from the design, procurement, installation, commission, operations, and finally, its retirement or disposal.

Environmental Consequences mean that a failure mode has environmental consequences if it causes a loss of function or other damage, leading to the breach of any known environmental standards or regulations. This means that the consequences of the failure go beyond the industry and affect society, people, or any government laws. This is considered the worse consequence of all.

Equipment failure refers to an event in which the equipment cannot accomplish its intended function and objective. It may also mean that the equipment stopped working, is not performing as desired, or is not meeting its target expectations.

Ergonomics originated from the Greek word ergon, meaning work, and nomoi, meaning natural laws. It is the science of refining the design of products to optimize them for human use. It is also sometimes known as human factors engineering. Ergonomics is a science that deals with designing and arranging things so that people can use them easily and safely.

Erosive Wear is a type of wear due to mechanical interaction between the surface and a fluid, a multi-component fluid, impinging liquid, or solid particles. Erosive wear occurs by the impact of particles upon a surface with a resultant material loss due to fracture.

Evidence is a piece of information that supports a conclusion. Evidence in its broadest sense includes anything used to determine or demonstrate the truth of an assertion. It is the lifeblood of any Root Cause Failure Analysis investigation. The three types of evidence include people, paper, and physical evidence.

Evident failures are failures that will become evident to both the maintenance and operators when they occur independently.

External set-up includes the activities, which can be performed while the machine is still running or is running the last production lot before the actual conversion will take place.

Extreme Pressure additives (EP) an oil additive that is usually used in gearboxes to provide a layer of film on the metal's asperities so that when the two opposite asperities meet, instead of breaking up, they will simply slip or slide with each other. It is not recommended to use EP additives on yellow metals.

Fact-Finding Group is a group of people composed of the Principal Investigator and Evidence Gathering Team. Its mission is to find out the root cause of the failure, learn from it, and do something to prevent, mitigate, or eliminate its occurrence.

Fail-Safe System is a system whose function is replicated or duplicated so that the function will still be available to the equipment after the failure of one of its sources, making failure tolerable and allowed to happen.

Failure is the inability of the equipment to perform its required function. The failure of a component is viewed as terminating its life. In general, failure refers to the state or condition of not meeting a desirable or intended objective. In life, failure may be viewed as the opposite of success.

Failure Development Period is the time where the potential failure is spotted by the instrument until the time the equipment finally reached its functional failure or this is when the equipment is actually declared fail. This is also known as the P-F Interval.

Failure Finding Tasks are maintenance tasks designed for hidden or unrevealed failures such as protective devices and redundant functions. These tasks are being performed to reduce or mitigate the risks of multiple failures wherein both the protective device and protected function are both in a failed state. It is also termed as functionality inspection or Detective Maintenance.

Failure Effects describe, most likely, what happens when each failure mode occurs on its own. Writing the failure effect should compose of 20 to 60 words.

Failure Modes are the most likely, or probable cause of failure. These are not the root cause but are considered the possible or most likely cause of why the failure happened.

Failure Mode and Effects Analysis (FMEA) discipline were developed in the United States Military. The procedure for MIL-P-1629, titled Procedures for Performing a Failure Mode, Effects and Criticality Analysis, was written on November 9, 1949. It was used as a reliability evaluation technique to determine the effects of system and equipment failures.

Glossary of Terms

Failure Rate is the expected rate of failure or the number of failures in a specified period. It is expressed in failures per million or billions of hours. It is also the probability of a failure in a stated unit of time and is the reciprocal of MTBF.

False Brinelling is caused by the continuous vibration of the duty equipment, thereby causing fretting damage on the bearing for standby equipment. The best way to address false brinelling is to rotate the shaft manually around 360 to 720 degrees + 60 degrees so that the ball's position changes from its position on the outer raceway of the bearing.

Fatigue Wear is a type of wear that results from a continuous cyclic slip under repetitive load applications for many thousands or millions of load cycles. Fatigue is the phenomenon leading to fracture under repeated or fluctuating stress having a maximum value less than the material's tensile strength. Fatigue fractures are progressive, which can start as micro cracks and propagate into larger ones that cause complete destruction of the components.

Fault-Tree Analysis was developed by Boeing Aerospace in the 1950s for use in the development stages of the design process. This is a mathematical tool that yields probabilities. Its primary intent is to predict the probability of a specific failure.

Filtration can be defined as the process of removing contaminants from a fluid, liquid, or gas stream through some means of a porous medium. An oil filter's function is to remove contamination from a fluid, liquid, or gas medium through some porous medium to achieve a required fluid cleanliness level.

Finished Goods Inventory refers to the finished products of the plant available for customer purchase. These products will either be shipped to their clients, picked up, or ready to sell.

Fire Point is defined as the lowest temperatures at which vapor of the material will catch fire and continue burning even after the ignition source is removed. The fire point is usually higher than the flashpoint, typically plus 10 to 24 °C or 50 to 75 °F beyond the flashpoint. The vapors produced at the flashpoint are not sufficient to ignite the fuel.

Flashpoint is the temperature at which the oil gives off vapors that can be ignited with a flame over the oil. The lower the flashpoint, the greater the oil's tendency to suffer vaporization loss at high temperatures and burn.

Flow rate is how much volume of a liquid can pass a certain point in a given time, measured in gallons per minute or liters per second.

Foaming is the formation of air and bubbles in a petroleum product, lubricant, or fuel oil that can reduce the product's effectiveness. Foaming can cause sluggish operation, air binding of oil pumps, and overflow of tanks or sumps. It can also result in excessive turbulence, improper fluid levels, air leaks, cavitation, or contamination with water or other foreign material.

Four-ball EP (Extreme Pressure) is a grease test called the 4-Ball EP, Four-Ball Extreme Pressure, 4-Ball Weld, or Load Wear Index. The Four-Ball EP test (ASTM D2596) measures the grease's ability to prevent wear during sliding contact under extreme pressure caused by heavy loads.

Fracture refers to the separation of a solid body into two or more pieces imposed upon by stress, which is usually static and at temperatures relative to the material's melting temperature. The applied stress can be considered tensile, compressive, shear, or torsional.

Free Water can be any water or gas which is not dissolved in the lubricating oil. Free water is when the water and the oil separates. Since the water is denser and heavier than the oil, it will settle at the bottom which can easily be removed from the drain plug.

Friction is a force that is created whenever two surfaces move relatively across each other. As long as there is friction, heat is generated, and wear takes place. The area of contact will always be the point of wear. The main function of lubrication is to reduce friction.

Friction Modifier additives affect the frictional properties between two rubbing surfaces. These additives prevent scoring, reduce wear, and noise, and prevent micro-pitting in industrial gear lubricants. Friction modifiers are commonly used in gasoline engine oils.

FSNO Analysis stands for Fast-moving, Slow-moving, Non-moving items, and Obsolete items. This form of classification identifies the items frequently issued, less frequently issued for use, items that are not issued for a longer period, say, 2 years, and those that are no longer used.

Function is the normal or characteristic actions of an item, sometimes defined in terms of performance capabilities. This is also what the users want the equipment or asset to do.

Function Loss Failure or failure of the primary function is when a failure occurs where the machine or equipment will totally stop and will definitely halt or stop operations completely.

Functional Failure is defined as the inability of an item to meet its specific performance standard. This was when the asset had failed completely. This can also be called a function-loss breakdown.

Grease is defined as a solid to a semi-fluid product of dispersion of a thickening agent in a lubricant. It comes from the Latin word Crassus, which means fat. Grease is a thick, oily lubricant consisting of inedible lard, the fat of waste animal parts, mineral or petroleum-derived or synthetic oil, plus a thickening agent.

Grease Consistency depends on the type and amount of thickener used and the viscosity of the base oil. Consistency is also known as the resistance to deform caused by an outside force. The measure of consistency is called penetration.

Hard skills are technical knowledge gained from training needed to develop the skills to perform people's jobs correctly.

Hershey Number is the dynamic viscosity (η) multiplied by the speed (N) divided by the normal load (P) per length of the tribological contact. The load also means the average pressure.

Heterodyning is a process in ultrasonic analysis that translates these frequencies into the audible range. Heterodyning is the mixing of two waves, which produces both the sum and difference of their original waves, which allows the shifting of a high-frequency sound to the audible or sonic range. It converts the ultrasonic range to an audible or sonic range through this process which is audible to the human ear.

Glossary of Terms

Hidden Failures are failures that will not become evident to the operator or user of the equipment when the failure occurs independently. To consider a hidden failure, a secondary failure should occur. Hidden failures will only be applicable for protective devices and redundant components.

Hidden Failures Consequences refer to the risks of multiple failures due to an undetected earlier failure of a hidden function item. This mostly refers to the aftermath when protective devices and redundancies fail.

Horizontal Replication, also called fan-out, is the process of repeating or doing the proposed tasks derived from the RCM analysis to similar equipment with the same operating context or condition. This may also apply to modification, redesign, or improvement as long as the equipment replicated also possessed the same problems.

Human Cause is the second level of Root Cause Failure Analysis, which refers to human errors, omissions, or commissions resulting from the physical roots. Either someone did something wrong, or the person did the wrong thing.

Human Nature is a concept that denotes the characteristics of human beings such as the way they think, their feelings, and the way they feel and react naturally. Humans react based on their instinct.

Human Sensory Perception means humans are gifted with five senses which include smell, touch, hear, feel, and taste. For industries, forget the sense of taste. Some failure modes will give some sort of warning or symptoms that they are on the verge of failing, such as smell, excessive vibration, noise, heat, or anything that can be detected by the human senses.

Hydraulics is the transmission and control of forces and motions through the medium of fluids. It is also the science of transmitting force or motion through the medium of a confined fluid. In a hydraulic system, the power is transmitted by compressing a confined fluid. This transfer of energy takes place because a quantity of liquid is subject to pressure.

Hydrocracking is a flexible catalytic refining process that can upgrade a large variety of petroleum products. Hydrocracking is commonly applied to upgrade the heavier fractions obtained from crude oils' distillation, including their excess.

Hydrodynamic Lubrication (HL) is also termed or called full-film or fluid film lubrication. Hydrodynamic lubrication is where the film is thick enough to separate interacting surfaces to minimize friction.

Hypothesis is a term used in root cause failure analysis which means the probable or most likely causes that need to be verified. In RCM, they are termed failure modes, while in a crime scene, they are referred to as the suspects.

Industrial Internet of Things (IIoT) uses wireless smart sensors and actuators to enhance manufacturing and industrial processes by hooking them on the equipment and providing data directly to your computer.

Inertia Impaction is a process that helps remove contaminants smaller than the pore size of the filter medium. The contaminants are retained mechanically or through adsorption. This type of removing contaminants is more effective in gases than in fluids. This process is efficient for particles greater than 0.5 to 1 micron.

Infant Mortality Failures are failures, which occur at the beginning of life. Others refer to this as commissioning failures, start-up failures, or debugging failures that occur after conducting major Preventive Maintenance activities which include overhauls and replacements.

Infrared is a type of light in the electromagnetic spectrum that is not visible to our naked eyes. Our eyes can only see a small portion of the electromagnetic spectrum, the visible light spectrum, or the rainbow's colors in layman's terms. Infrared radiation lies between the visible and microwave portions of the entire electromagnetic spectrum.

Infrared Thermography is the science of actually allowing us to see the heat. An infrared inspection can detect heat that would normally be invisible to the naked eyes and represent them as an image, which we can see.

Inherent Reliability Level is the level of reliability of an item, equipment, or asset that is attainable by utilizing all the available maintenance tasks on RCM.

Initial Cleaning in Autonomous Maintenance Step 1 removes any form of unwanted abnormalities from the equipment. It consists of removing dirt, contaminants, grime, excess oil, grease, and other foreign objects that affect equipment parts and components that have accumulated over time in the equipment.

Initial task intervals also called proposed tasks are maintenance task intervals assigned before the service maintenance program, subject to adjustments based on the actual operating experiences.

Inspection Tasks refer to scheduled tasks requiring testing, measurement, visual inspection, or human senses for detecting failure evidence done by both operators and maintenance. For inspecting hidden failures, this will default to Failure Finding tasks, while for evident failure inspection, it will default to On-Condition Tasks.

Instinct can be defined as an inborn impulse or motivation to action typically performed in response to specific external stimuli. It can be considered as an intuition, feeling, impulse, or gut feeling. This is how we react to certain situations.

Insurance Spares is a spare part that will replace a failed part in a piece of equipment whose penalty cost for downtime is very high. These parts do not become obsolete until the equipment is retired from service. In most cases, these parts are big and classified as non-moving parts.

Intangible Measurements cannot be quantified, measured, or contribute to the plant's bottom-line results. They can also be termed as qualitative measurements.

Intermediate Discipline is a discipline on World-Class Maintenance that refers to the different reliability and maintenance strategies that can be adopted once basic equipment conditions have been established in the asset. This includes Lubrication Management, MRO Spare Parts Management, Life Cycle Management, Root Cause Failure Analysis, Reliability, and Maintenance Strategies.

Glossary of Terms

Internal Set-up are those activities that can be performed during conversion only when the machine had been totally shut down for conversion since a different product will be required to operate on the equipment.

Inventory Turnover is a ratio that measures the number of times an inventory is withdrawn or consumed in a given time by the end-user. Inventory turnover is sometimes called stock turns or stock turnover.

ISO 55000 is a standard described as the parent document of ISO 55001. It provides a general overview of asset management as a discipline and contains definitions of terms used in the ISO 55000 series of standards. This ISO standard aims to establish a clear and consistent understanding of the principles and requirements applied when developing and implementing an Asset Management System.

Karl Fisher Titration Test is an oil analysis instrument that measures the amount of dissolved moisture content in the oil. In this method, water reacts quantitatively with a Karl Fischer reagent. This reagent is a mixture of iodine, sulfur dioxide, pyridine, and methanol.

Kauro Ishikawa developed the Ishikawa or Fishbone Diagram in 1969. A fishbone is constructed by assigning the 4M's and 1E. 4M refers to man, machine, method, and materials, while 1E refers to the environment. The most common probable causes that relate to the problem are listed and grouped accordingly. The team brainstorms and focused on the most likely or probable causes of the failure.

Key Performance Indicators (KPIs) are measurements that are tracked and monitored regularly to indicate if an organization is on the right track to hit its goals and objectives.

Kinematic Viscosity is a viscosity test, which takes the time for the oil to travel through a glass orifice of a capillary tube under the force of gravity.

Kosho is the Japanese term for breakdown. It is a lost time due to equipment failure that may or may not cause downtime.

Knowledge-Based Mistakes are types of mistakes that occur when someone is, confronted with a situation that has not yet occurred before and had not been anticipated. In other words, there are no rules or procedures to follow. In situations like this, the person has to decide quickly about an appropriate course of action, and a mistake occurs due to the wrong decision.

Lagging Indicators are indicators and measurements that indicate the bottom-line results. Examples of this include productivity, repair, revenue, and maintenance costs. This will also depend on who is reading the KPIs, meaning that the lagging indicator for a maintenance manager will be the availability but not for the CEO of the plant.

Lapse occurs when someone misses out on a key step in a sequence of events or activities. For example, a mechanic leaves a tool behind after working on a machine or simply forgets to fit a key component while reassembling it. If a person will do steps 1, step 2, step 3, step 4, and step 5, and the person missed out on step 4 in the process, then a lapse occurred. As humans age, our memory will not be as sharp compared to when we were young, just like in our high school days. Lapse is when people tend to forget things as they age.

Latent Cause is the last level of Root Cause Failure Analysis, which refers to hidden or concealed causes that need to be exposed so that industries can learn from the things that go wrong. This is about looking at the man in the mirror and admitting that each of us is also part of the problem.

Lead time is the time the storekeeper noticed that the part is below the reordering point and makes a request for the part until the part is finally delivered and received in the storeroom.

Leading Indicators are the process of how the results were achieved. These indicators influence the final outcome of the results. These are also the indicators that lead to the lagging results.

Life Data Analysis is when a product, part, or component has operated successfully, or the time it operated before it failed, measured in hours, miles, cycles, strokes, minutes, or other measures.

Life Cycle Cost is the sum of the initial costs and the running costs of the equipment. It is the sum of the overall cost of equipment throughout its entire lifespan, including the costs incurred initially during the design stage up to the stage the equipment will be put out of service, disposed of, or decommissioned.

Line Organization Loss is a loss that results from a shortage of operators on the production floor where the operator needs to operate other machines that were originally planned. This often results in a shorter break time for the operators.

Lubricating oil, also called a lubricant or lube oil, is a class of oil used in equipment, engines, and machinery types to reduce the friction, heat, and wear between mechanical components to avoid metal-to-metal contact.

Lubrication Tasks refer to the scheduled tasks to assure the existence of completeness of lubrication films on the equipment. These tasks can either be scheduled routine tasks or condition-based maintenance tasks.

Machine-Related Downtime, also known as unplanned downtime, is when the equipment is not operating due to machine-related downtimes such as breakdowns, set-up, and conversion, cutting tool change, minor stoppages, and quality defects.

Maintenance are the activities that are done before an asset or equipment will fail. These are the activities of preserving or sustaining the assets. This means ensuring that the assets continue to do what the users want them to do.

Maintenance backlog is a time indicator that consists of delays in performing the scheduled maintenance works. This consists of pending scheduled planned activities allotted on the maintenance that has already passed its due date.

Maintenance cost is a universal and common measurement for all types of industries, unlike other KPIs. By definition, maintenance costs are any expenses incurred on the equipment, machines, people, or asset by the maintenance function to sustain and preserve it.

Maintenance Induced Failures are direct failures caused by the maintenance itself. Examples of maintenance-induced failures include infant mortality failure or doing intrusive maintenance, incorrect overhauling practices, and inducing early failures right after endorsing the equipment back to operators. In this case, the maintenance reassembled the equipment incorrectly.

Maintenance Management is the art or science of managing maintenance resources. It is also the manner of managing to keep our physical assets in an existing state or condition.

Maintenance Prevention according to JIPM, Maintenance Prevention is defined as the use of the latest maintenance data and technology when planning or building new equipment to promote greater reliability, maintainability, economy, operability, and safety while minimizing maintenance cost and deterioration.

Management Loss includes losses caused by the waiting time that is lost due to management problems and delays such as no available MRO spare parts, lack of manpower resources, lack of insufficient utilities, and work instructions.

Mastery is the last level of skill development where the person is capable of transferring both his skill and knowledge to another person.

Marking is the process in the semiconductor industry of identifying, traceability, and distinguishing marks on the integrated circuit package at the end of line process.

Material Hardness is the measure of the material's resistance to plastic deformation. The most common method to determine the hardness of the material is the Rockwell Hardness Test, which can test the hardness of metals and alloys. Usually, a compressive force will be applied.

Maxi-Event or large-scale RCFA is performed on maxi events or failures with a large impact or consequences in the industry. It is recommended that an outsider do this event or an independent third party person act as the principal investigator to avoid bias. There will be three persons assigned to the evidence-gathering team for the maxi event, which will require a stakeholder meeting.

Mean Time between Assists (MTBA) is the average time the equipment performs its intended function between assists. It is also the productive or the operating time divided by the number of assists, errors, or minor stoppages. Common unit used is in minutes.

Mean Time Between Failure (MTBF) is defined as the average time between failures and therefore is a measure of the trouble-free time. The common unit used is in hours. MTBF is derived from the US MIL-STD 217 for testing electronic parts. It is a reliability engineering term that means the average amount of operating time between the occurrences of breakdowns that requires repair divided by the frequency of failures. The frequency of reporting MTBF should be either on a weekly or monthly basis. MTBF trend should be the higher, the better.

Mean Time to Fail (MTTF) is the expected time to fail of a system, part, or component. It is a basic measure of reliability for non-repairable systems. It is the meantime expected until the first failure of a piece of equipment. This is also equal to the MTBF minus MTTR.

Mean Time to Repair (MTTR) is the average time required to repair a component. Other terms used are Mean Time to Restore, Mean Time to Recover, or Mean Time to React. It is also the average time required to perform corrective maintenance or repair on all removable items in a piece of equipment, product, or system.

Mean Units Before Assists (MUBA) indicates how many units were produced before an assist or minor stoppage is encountered on the equipment.

Measurement and Adjustment Loss are losses caused by frequent measurement and adjustment to prevent the recurrence of problems. Examples include excessive inspection integrated into the process as a result of poor quality and failure to find the root cause. Adjustment loss is experienced when adjusting equipment back to the standard after routine cleaning and periodic consumable changes.

Mercaptan was introduced by William Christopher Zeise in 1832. The Latin word mercurium captan means capturing mercury because the thiolate group bonds strongly with mercury compounds. These mercaptans are used as an odorant in sour gasoline into disulfides.

Micron is also known as a micrometer and is exhibited by the Greek symbol Mu or Mμ. A unit of one-micron length is equivalent to 39 millionths of an inch or 0.000039 or 0.0009906 millimeters. The human eye can see around 40 microns and beyond.

Midi-Event or Medium Scale RCFA is performed on midi or medium-scale events. A Principal Investigator leads the investigation with one or two persons assigned as the evidence-gathering group. This will require a stakeholder meeting.

Mini-Event or Small Scale RCFA event is usually performed on small scale failures where the Principal Investigator will be the ones collecting the evidence. This type of RCFA will not require a stakeholder meeting.

Molding is the process of modern molding used to encapsulate plastic material components to protect IC or passive devices at the end of line process in semiconductor industries.

Moisture Vapor Transfer Rate (MVTR), sometimes called Water Vapor Transmission Rate (WVTR), is a measure of water vapor passage through a substance. It is a measure of the permeability of vapor barriers. There are many industries where moisture control is critical such as semiconductor industries.

Moment represents the effectiveness of either a rotating, bending or twisting force upon an object. All materials have their respective hardness. The hardness of the material is the measure of the material's resistance to plastic deformation.

Motor oil, also called engine oil, is a lubricant used in internal combustion engines, power cars, motorcycles, diesel engines, engine-generators, mobile, mining equipment, etc. Inside the main engine are metal parts, which move relatively against each other, and the friction between these moving parts consumes more power by converting kinetic energy into heat.

MRO Spare Part is defined as a part of a machine ready to replace an identical part if it becomes faulty. It is also defined as those parts of the machine which are kept on standby to be substituted when a part of equipment fails, a repair is required, or the part simply becomes worn out and needs to be replaced

Glossary of Terms

MRO Wear and Tear Spares are spare parts that must be replaced every time the equipment undergoes Preventive Maintenance or Shutdown, and the equipment is disassembled and re-assembled for overhauls and parts replacement.

MSG-1 Document is a working paper prepared by the Boeing 747 Maintenance Steering Group published in July 1968 under the title: Handbook Maintenance Evaluation and Program Development (MSG1), which includes the first use of the decision diagram techniques to develop an initially scheduled maintenance program.

MSG-2 Document is a refinement of the decision diagram procedures in MSG1 published in March 1970 under the title MSG-2 Airline/Manufacturer Maintenance Program Planning Document, which is the immediate precursor of the RCM document. This document was used to develop a scheduled maintenance program for Lockheed 1011 and Douglas DC10, military aircrafts such as Lockheed S3 and P3, and McDonnell F4J.

MSG-3 Document refers to the Operator/Manufacturer Scheduled Maintenance Development developed by the Airlines for America (A4A), formerly the Air Transport Association or ATA. This document was developed in 1980, revised in 1988 and 1993, and to this day process is used for maintaining all types of civil aircraft.

Multipass Test is a controlled laboratory test where the effluent fluid is recirculated through the filter element while a new contaminant is continuously added. This test is used to determine the efficiency and the beta rating of the oil filter.

Multiple Failure is where the protected function had miserably failed because the protective device is said to be in a failed state.

Multi-purpose grease can be defined as grease combining the properties of two or more specialized greases that can be applied in more than one application. This permits the use of a single type of grease for a wide variety of applications.

Natural deterioration means that the part or item has reached its lifespan and eventually gradually wears out. This is similar to age-related failures.

Near Miss can be considered as minor accidents or close calls that have the potential for an injury, accident, property loss, or even death. A person was standing in front of a Wrecking Ball Crane. After a few minutes, the ball fell which nearly smashed the head of the person just a few feet away.

Needed Non-Productive Work includes those that cannot be totally eliminated in maintenance but are still needed. Examples of this include lunch breaks, attending training, mentoring operators about their equipment, conducting improvements, RCFA investigations, and the like. Although in this situation maintenance is not doing Planned Scheduled Work, these activities should also be part of the maintenance function

Net Operating Time is used to identify whether the equipment is operated at the stabilized speed within the unit time.

Noack Volatility test was named after Kurt Noack. It is measured by the principal European test called NOACK. It is the amount of oil lost (light molecules) over time at a given temperature and pressure. It directly impacts high engine temperature and oil effectiveness, especially on viscosity, emissions, and oil consumption. Today's oil has a NOACK volatility limit of 15 %. Oil's Volatility rate should be less than 15%.

Nominal Filtration refers to the average particle size of contaminants that will remain in the fluid after filtration. The efficiency of nominal filtration is 50%, and the beta rating is 2.

Non-Dedicated Machines are those machines in manufacturing that are designed to manufacture different products using the same machine. For example, in manufacturing, some machines are used to manufacture different products, hence, one measurement of importance will be the set-up or conversion time.

Non-Machine Related Downtime, also known as planned downtime, is when the equipment is not operating due to non-machine-related downtimes such as PM activities, operators' break time, meetings, or any other downtime which is not caused by the machine.

Non-Maintenance Induced Failures are failures on the equipment, which are not caused by the maintenance function, but by other factors such as how the equipment was operated, design errors, commissioning errors, or management cutting costs.

Non-Operational Consequences is one of the consequences of a failure mode, which does not directly affect safety, environmental, or operational consequences. The only consequences of this failure mode are the direct cost of repair and the chances of secondary damages to the equipment.

No-Scheduled Maintenance is a default task on RCM which means that there is no feasible scheduled maintenance. Failure is allowed to occur since the consequences are minimal and acceptable. Also similar to the terms run to fail, breakdown maintenance, reactive maintenance, or corrective maintenance. When the failure occurs, it will be subject to repair.

Not-Needed Non-Productive Work includes all waste on maintenance. These are those added activities and resources which are wasted as a result of poor or inadequate planning. This includes wasted time for the wrong part number supplied, lack of manpower, wrong tools, MRO spare part not available, or going back once more to the storeroom since you needed this and that.

Occurrence in FMEA/FMECA is a rating according to the likelihood that a particular failure mode will occur within a specific period. Occurrence is an assessment of the likelihood that a particular failure mode will occur. This is also termed as the probability of failure to occur.

Octane Rating measures the fuel's ability to resist engine knocking caused by the air-fuel mixture. The higher the octane, the greater resistance the fuel has to knock during the combustion process. The octane ratings are a measure of the fuel's stability.

Office Supplies Inventory are inventory supplies used by the different plant offices ranging from printer ink, pencil, ball pen, bond paper, folders, envelopes, scissors, paper clips, memo pads, and so on in industries.

Glossary of Terms

Oil Analysis is a maintenance management tool that allows users to monitor the oil condition for maximum equipment life and maximum lubricant drain interval. It tells us the actual condition of the oil in the equipment.

Oligomerization is a polymerization process in which a few, usually three to ten, of the basic building block of molecules are combined to form the finished product. Therefore, the product is formed with varying molecular weights and viscosities to meet a broad range of requirements.

On-Condition Tasks are maintenance tasks that entail checking the equipment's actual condition or using human senses, pressure, temperature inspections, gauges, and SPC charts.

One-Point Lesson is a learning tool for communicating standards, problems, and improvements in work processes and equipment. Workers and supervisors use one-point lessons to provide key information about everyday work and improvement opportunities.

Operational Consequences are a type of consequences where operations will be affected. The primary function of most equipment in any industry is connected to the need to earn revenue or support revenue-earning activities. A failure mode has operational consequences if it has a direct adverse effect on operational capability.

Operator Motion Loss includes losses generated due to unnecessary or excessive movement by the operator, as a result of poor layout, and work for the organization due to excessive walking, wasted motion, unnecessary reaching, or equipment being far apart where the operator will take a longer time to transport the product to the next process in the manufacturing.

Overall Equipment Effectiveness (OEE) is the primary measurement used for plants and industries initiating Total Productive Maintenance. It is calculated by multiplying the equipment availability by its performance rate and quality rate expressed in percentage.

Oxidation is the breakdown of oil due to the extreme heat in the engine or equipment. This is the reaction between the oil and oxygen, causing acidic gases, sludge, and varnishes to form in the crankcase.

Oxidation inhibitors are added to the oil to extend its operating life. These additives are said to be sacrificial (perhaps like a suicide bomber), consumed while performing their duty of delaying oil oxidation, thus protecting the base oil. They are present in almost every lubricating oil and grease. This inhibitor aims to prevent oxygen from reacting with the oil, thus slowing its aging rate.

P-F interval is the interval between the emanation of potential failure and its decomposition to its final stage, which is a functional failure. At this stage, the equipment has already failed. The P-F interval is actually a warning period or the lead-time to failure better known as the failure development period

P-M Analysis is a problem-solving tool designed to analyze chronic losses according to the inherent principles and natural laws that govern them. It was developed by Mitsugu Kaneda, Shirose Kunio, and Yoshifumi Kimura. P stands for Phenomena and Physical. A phenomenon is a deviation from a normal to an abnormal state. M stands for Mechanism and the 4Ms.

Pareto's 80/20 Rule. Dr. Joseph Juran, a Quality Management pioneer, developed the use of the Pareto Principle for problem-solving. He worked in the US from 1930 to 1940. According to Vilfredo Pareto's observation, 80% of the land in Italy was owned by 20% of its population, where 20 percent of something is always responsible for 80 percent of the results. He recognized a universal principle he called the vital few and trivial many and placed it into writing.

Partial Functional Failure is when the asset is still functioning, but it is below the performance standards that the user wants the asset to function.

Penetration is the measure of the grease's consistency and depends on whether the consistency had been altered by handling. ASTM D217, and D1403 test, measures the penetration of unworked and worked greases.

Performance Rate is sometimes referred to as the throughput or equipment Performance Efficiency. It reflects whether the equipment is running at its full capacity or speed for individual products. Performance Efficiency measures Speed Losses and Idling and Minor Stoppages.

Physical Cause is the first level of Root Cause Failure Analysis that refers to the physical cause of why the part or component failed. This is the technical explanation of why things broke or failed. This is said to be the metallurgical aspect as to why the failure occurred. It is also referred to as the Failure Analysis. The analysis will end on the component or part level.

Physiological Factors are a type of human error that refers to environmental stress affecting human performance. These stresses can include high or low temperatures, loud or irritating noise, excessive humidity, excessive or high vibration, exposure to toxic chemicals, radiation, or working too long without adequate break time, not to mention the day-to-day pressure from the boss.

Planned Maintenance is one of the TPM pillars, aiming to improve the current maintenance activities, whose goal is to zero out all unplanned breakdowns of the equipment or asset. They are also the mentors of Autonomous Maintenance.

Planning is defined as the total process set up to ensure that the right resources and materials arrive at the right place, at the right time, and doing the right job in the right way. Planning also has a macro meaning when reference is made to an array of scheduled shutdowns for the year, the business plan, the marketing plan, the budget, and others.

PLCC (Plastic Leaded Chip Carrier) is a plastic, square, surface mount chip IC package that contains leads on all four sides. The lead pins extend down and back under and into tiny indentations in the housing. PLCC is a type of integrated circuit package that can be used to enable us to be mounted on a printed circuit board directly soldered either to the board or within a socket.

PM Compliance is a maintenance indicator that will measure how many PM tasks were actually completed and closed divided by the total number of tasks to be accomplished on or before the due date.

Potential Failure is defined as an identifiable physical condition that indicates that a functional failure is about to occur or is in the process of occurring.

Glossary of Terms

Poka-Yoke also referred to as Mistake Proofing, is mostly used by manufacturing industries, especially to help an equipment operator avoid errors and mistakes during operations. Its purpose is to eliminate product defects by preventing, correcting, or drawing attention to human errors which were developed by Shigeo Shingo, as part of the Toyota Production System. It was originally termed as Baka-Yoke, which was called foolproofing or idiot-proofing. Later on, Shigeo Shingo changed it to a milder name Poka-Yoke.

Pour Point is an oil analysis test that refers to the temperature at which oil will solidify due to extreme cold temperatures making the oil no longer capable of flowing. This oil analysis test only applies to countries with winter or extremely cold seasons. This refers to the W rating for engine oils. Note that W stands for winter.

Pour point Depressants are critical substances added to the additive compound to prevent the wax formation in the base oil from forming large crystal networks that can inhibit the flow of lubricating oil at cold temperatures, usually during the winter season.

Precision Maintenance involves performing maintenance work in a consistent, precise, and industry-accepted way. If properly implemented, maintenance should yield exactly the same results no matter who is performing the tasks whether the person is the most experienced or least experienced in the craft.

Predictive means to declare or indicate something in advance, especially foretell based on observation, experience, or scientific reason. Predictive comes from the Latin word "pre," meaning before, and "diction," also from the Latin word "dicare," meaning to proclaim.

Predictive Maintenance is a type of maintenance performed on the equipment based on the equipment's actual condition with specialized diagnostic monitoring tools or Predictive Maintenance instruments to determine potential failures. Predictive Maintenance is a maintenance activity geared to indicating where a piece of equipment is on the critical wear curve and predicting its remaining useful life.

Preventive Maintenance is a type of maintenance performed on a fixed schedule or time-based interval. Preventive Maintenance is used when the parts have a wear rate and are directly related to their age.

Primary function explains why the asset was purchased. It has something to do with the volume, output, speed, and quality

Proactive-Corrective-Maintenance, which I can define as any activity done on the equipment such as overhauling, replacement, or other means resulting from Predictive Maintenance tasks before the failure is about to happen.

Proactive Maintenance is about understanding the root cause of why a part keeps on failing and performing corrective measures to eliminate the problem completely and avoiding the recurrence of failure. In Proactive Maintenance, maintenance is ahead of failure.

Productive Work includes those works that are scheduled by the planner such as routine Preventive Maintenance activities or any future works which are planned and scheduled.

Protective Device refers to those devices placed in the equipment and assets to protect something against failure. Samples of protective devices include led, beacons, sensors, alarms, emergency stops, lighting arresters, and so on. The protective device's function is to indicate that the protected function is still working and is functional.

Psychological Factors can be grouped according to those, which are unintended, and those errors that are intended. Unintended errors can be grouped into slips and lapses, while intended errors can be grouped according to mistakes and violations.

Pumpability is the ability of the grease to be pumped or pushed through a system, which is similar to the viscosity of the oil. This indicates how easy or hard the grease can flow through lines, nozzles, and grease dispensing units.

Purchasing Cost is also called the Ordering Cost. This is the sum of the fixed cost that is incurred each time an item is ordered. This includes the physical activities required to process an order. Usually, the purchasing cost can include the purchaser's salary, telephone costs, receiving clerk salary, account payable costs, internet costs, electricity, and other costs.

Quality Rate measures the loss incurred due to rejected, reworked, and return products. Quality Rate is the product of the following two factors, the number of items produced and the number of products scrapped, reworked, or rejected.

Radio Frequency Identification (RFID) refers to a technology where the digital data encoded in RFID tags or smart labels are captured by a reader through radio waves. RFID systems consist of three components: an RFID tag, or smart label, an RFID reader, and an antenna. For industries, this is usually used for MRO Spare parts.

Random Failures are failures that can occur at any given period. This means that the probability that an item will fail in any one period is the same as it is in any other period. This means that the conditional probability of failure is not constant. This is also referred to as chance failures and wear out is not identifiable which means that the item can fail at any given time or period.

Raw Materials Inventory: These are inventory parts and items used by industries to produce their finished products. If you work in the automotive industry, this will refer to the different parts needed to assemble a car.

RCFA Logic Tree diagram is a tool that uses deductive logic to guide the thought process used to draw correct conclusions. Therefore, a logic tree is a disciplined methodology that prompts the user to answer questions that will eventually identify the root cause of a failed event.

RCM Decision Diagram is an algorithm that allows the users to make a decision-making process to identify the most appropriate and feasible maintenance tasks to address a particular failure mode

RCM Default Tasks are RCM tasks available when it is not feasible to perform a Preventive or an On-Condition task to address a particular failure mode. Default tasks on RCM include Failure Finding Tasks for hidden failures, Run to Fail, Switching intervals for redundant functions, and redesign or modification.

Glossary of Terms

RCM Streamline Version is a shortcut RCM version for improving the effectiveness of current maintenance programs and strategies. It starts with the existing maintenance program used within the plant.

Reactive-Corrective-Maintenance is the activity of repairing or troubleshooting the equipment after a failure occurs. This will be done after the failure happens, similar to the original definition of corrective maintenance.

Reactive Maintenance is a type of maintenance strategy that simply means fixing it only when it fails. As the saying goes, when it ain't broke, don't fix it; when it fails, then we come and fix it. Other terms used to designate Reactive Maintenance are run to fail, run to destruction, corrective maintenance, fire-fighting mode, and stop the bleeding syndrome. RCM termed this as no-scheduled maintenance.

Recurrence Prevention includes a list of things and actions that need to be done to indicate that the failure has not repeated itself. This will usually be done in the case of the physical, human, and system cause only.

Redesign or Modification means that if the rest of the maintenance tasks are not feasible and worth addressing the failure mode, then the last option for maintenance will be to resort to redesign or modification, especially when the failure mode will have safety or environmental consequences. The goal here is to eliminate or reduce the consequences of failure from happening.

Redundancy or Standby means duplicating the system or component, which is not affected if failures and breakdowns occur. Failures are allowed or being tolerated through some forms of redundancy, standby, or when the asset has some form of duplicated function.

Redundant tasks are the same task done by different groups of people which can either be the PM Crew or a third party contractor. These are considered intrusive and should be deleted from the Preventive Maintenance lists since this will just end up in a waste of manpower and resources. What is important is to decide who is the best person or group to perform the task.

Reforming is a process designed to increase the amount of gasoline that can be produced from crude oil. Hydrocarbons in the naphtha stream have almost the same number of carbon atoms as gasoline, but their structure is generally more complex. Reforming rearranges naphtha hydrocarbons into gasoline molecules. The other products of reforming are light gases and a high-octane gasoline blending component called reformate.

Reliability is the probability that an item will operate without failure throughout a specified interval and that the item will perform its intended function under specified operational and environmental conditions under a given and specified time.

Reliability-Centered Maintenance is a process used to determine the most feasible maintenance requirements in its present operating context or state that it is being operated. This strategy is used to define the correct maintenance tasks for equipment or asset with the aid of an RCM decision diagram or algorithm.

Reordering point is the sum of the minimum stock, and in certain cases, we add the safety stock. Others called this the buffer stock. The Reorder Point is equal to the minimum stock plus the safety stock.

Repair are the activities that are done after an asset or equipment failed. These are the activities to be done to restore the equipment back to its operating state.

Replacement Asset Value (RAV) is also called the Estimated Asset Value (EAV). It is the cost of maintaining the asset which is measured against the value of the asset. It can be said that as the percentage of the cost to replace the asset. The lower the value of RAV, the more effective we are in maintaining and preserving the asset.

Return on Investment (ROI) tries to directly measure the amount of return for a given investment, relative to the investment's cost. In calculating the Return on Investment (ROI), the benefit or return gained from an investment is divided by the cost of the investment. The result is expressed as a percentage or a ratio.

Risk is something or someone that can cause a problem, harm, injury, damage, danger, hazard, or loss to the industry. Risk is the potential for uncontrolled loss that is of definite value to an organization. It is an intentional interaction with uncertainty.

Risks Priority Number or RPN in FMEA is the sum of severity multiplied by its detection and its occurrence. This is also the sum of the consequences multiplied by its criticality, confidence, and probability.

RNM Cost refers to the total repair and maintenance cost incurred on the equipment or system for a given period. The trend we want should be that the lower the RNM cost, the better. RNM costs should be tracked and reported either weekly or every month.

Roll Stability of Grease: ASTM D1831 is a grease test that assesses how stable the grease is when it is subject to operating conditions. This test allows us to determine if the grease will handle the intended load and for how long before it begins to fail.

Root Cause Failure Analysis is a basic investigative tool that allows us to understand why things went wrong by identifying the problem's basic source or origin. Root Cause Failure Analysis is the investigation process or probing of a problem, which is entirely based on the evidence unfolded on equipment-related failures. The level of root causes will include the physical, human, system, and latent cause of the problem.

Rooticians are people who are knowledgeable in conducting a Root Cause Analysis or Root Cause Failure Analysis investigation. They are the fact-finding group deployed to find the cause of a problem or incident.

Rotable Spare Parts are reusable spares or components that can be reconditioned and reused, such as motors and engines. These can be a component or inventory items that can be repeatedly restored or refurbished to be used once again on the equipment into a fully serviceable condition after it failed. The value of rotable spare parts depends on the remaining useful life of the production equipment it supports.

Ruled-Based Mistakes usually occur when people believe that they are following the correct course of action when doing specific tasks based on a specific rule or procedure, but the course of action is inappropriate.

RULER (Remaining Useful Life Evaluation Routine ASTM D-6971-04 3 and ASTM D-6810-02) is an oil analysis test that measures the remaining useful life of the lubricant. This test can be used to determine whether the oil needs to be changed or can be extended. The RULER oil analysis test will measure the remaining level of antioxidant additive in the lubricating oil.

Run to fail is a maintenance strategy, which tells us that it is time to repair the failure when a machine fails. The failure will happen first, and then maintenance reacts by repairing it. This is the very essence of reactive maintenance. Maintenance is done at a point when there is repair or actual breakdown of the equipment. This occurs when repair action is taken on a problem only when the problem results in a machine failure or breakdown.

SAE JA1011 is also known as the Evaluation Criteria for Reliability-Centered Maintenance (RCM) Processes. This document describes the criteria to indicate that any process or analysis to derive the tasks is compliant with the classical RCM process requirements. This also separates the classical RCM from the Streamlined RCM.

Safety Consequences: A failure consequence where a failure mode can cause injury or kill someone else. Failure modes that can cause accidents or even near-misses, as a result, will be included in this category.

Safety stock is an addition of items, which is applied mostly to critical parts for the storekeeper to increase their confidence in avoiding stock out. This is also called the buffer stock.

Safety Supplies Inventory refers to items regularly supplied to employees by the EHS, such as gloves, hard hats, goggles, safety shoes, ear protectors, overhaul outfits used as part of the plant's uniform in conducting their regular work routine provided to operators and maintenance people working in the plant.

Saponification is the process used to develop the grease thickeners where fatty acids are used to react with an alkali to form a chemical soap.

Schadenfreude means the pleasure we gained from another person's distress, especially if they perceived the person deserved it because they engaged themselves in a bad deed.

Scheduled Discard Tasks refer to removing an item or spare to be replaced at a specified time interval before the part wears out completely. This is included as a Preventive Maintenance task. Also termed as scheduled replacement of parts. The frequency of scheduled discard tasks is governed by the age at which the item or component shows a rapid increase in the conditional probability of failure.

Scheduled Restoration Tasks are Preventive Maintenance tasks that entail re-manufacturing a single component or overhauling an entire assembly on or before a specified age limit, regardless of its condition at the time, also termed as Scheduled Rework. The frequency of a scheduled restoration task is governed by the age at which the item or component shows a rapid increase in the conditional probability of failure.

Scheduled Maintenance Tasks, also called Preventive Maintenance tasks, are scheduled tasks that need to be done on the equipment or asset at scheduled intervals. In RCM, these refer to as scheduled discard and scheduled restoration tasks.
SDE Analysis classification of spares is based on the lead time to acquire the part. This classification is carried out based on the lead time required to procure the spare part. Scarce (S) are those items that are imported and those items which require more than 6 months of lead time. Difficult (D) are Items that require more than a fortnight but less than 6 months of lead time. Note that one fortnight means 2 weeks or 14 days. Easily Available (E) are easily available items, usually with less than a fortnight's lead time.

Seal Swell is one of the functions of the oil in which the oil must be compatible with the seals and must not cause the seal to crack, shrink, or degrade; rather, it must cause the seals to expand slightly to ensure proper sealing.

Secondary function refers to the other functions of the asset besides the primary function. Most assets are expected to fulfill more than one function besides their primary functions

Secondary Damage refers to the damage to the equipment that is caused by the primary failure mode. For example, a bearing seizes that it caused a fracture on the shaft or other parts due to excessive vibration.

Set-Up Loss, Conversion, or Changeover in several manufacturing industries, equipment is not dedicated, producing different product types with the same equipment. Hence, when it is time to change one product to another, it will require the time required to remove dies, jigs for one product, clean up, prepare dies and jigs for the next product, reassemble the equipment, adjust the equipment, perform trial runs and make further adjustments until the product of acceptable quality is now obtained from the equipment.

Severity in FMEA/FMECA refers to the impact of the failure mode and its effects. Severity considers the worst scenario that can happen after a failure or breakdown takes place. It is a rating corresponding to the seriousness of the effects of a potential failure mode. This is termed the consequence of the failure mode.

Shear force is the sum of the effect of shear stress over a surface that results in a shear strain. The important thing to consider is that the force acting on an object is parallel.

Shear Stress is also called tangential stress. Unlike both tensile and compressive stress, the force applied is not perpendicular but parallel to the area. Once a force is applied in parallel in opposite directions, it can cause the object to deform. Shear stress is a force acting parallel to a surface or to a planar cross-section of an object.

Shutdown Loss is the time when equipment is shut down for scheduled maintenance. However, shut down related work generally affects the operating time of the equipment. Shutdown-related work must be regarded as a planned downtime loss and reduction of shutdown work time which must be reduced.

Single Minute Exchange of Dies (SMED) is a process developed by Shigeo Shingo to reduce set-up time and changeover for manufacturing. The success of this system was illustrated in

1982 at Toyota when the die punch setup time in the cold-forging process was reduced over three months from one hour and forty minutes to three minutes.

Situation is a set of things happening and the conditions that exist at a particular time and place. Merriam Webster dictionary states that situation is how something is placed concerning its surroundings. The outcome is something that follows as a result or consequence of the situation. It is a result or effect of an action, situation, or event.

Skill is the product of personal motivation and thorough training. The end result is mastery, which is used to perform one's job correctly and efficiently, developed over time.

Soft skills are training that improves our personal well-being that shapes how we work and interact with others. An example of this training includes Supervisory Skills, Leadership, Team-Building, Work Ethics, Effective Communication, Teamwork, Time-Management, and the like.

Soot is a by-product of the combustion process and can escape the piston rings through blow-by. This may be caused either by low compression, or too much idling due to stop and start, such as traffic jams, or air-fuel mixture. Soot can range from 3 microns and above,

Sour Crude Oil is a crude oil that contains a high amount of sulfur, which is one of the main impurities in crude oil. It is very common to find crude oil containing sulfur impurities. When the crude oil's total sulfur content is more than 0.5%, the oil is called sour.

Spare Parts Management is the science of managing parts inside the storeroom when operations and maintenance required them in case of a breakdown that seeks replacement or Preventive Maintenance routine and replacement activities.

Speed Operating Rate is the ratio of the Theoretical Cycle Time divided by the Actual Cycle Time.

Squirrel Stores are maintenance secret hiding places for their own spare parts. Squirrel Stores are unofficial stores in which maintenance keeps their own spare parts for their own need and benefit.

Squirrelling is the act where maintenance keeps the spare parts themselves. Why does maintenance do this? It is because there are a lot of stock-outs, which means that the part reflects in the system, but physically it is out of stock.

Slip occurs when somebody does something incorrectly or does something in the wrong sequence. If a person will perform something in sequences such as step 1, step 2, step 3, step 4, and step 5, and what the person did was step 1, step 2, step 4, step 3, and step 5, the person committed a slip. This is when a human action takes place that is not intended. The occurrence of a lack of one's trains of thought is derived mostly from unconscious behavior.

Snake Oil is a quack remedy or panacea. Snake oil is a euphemism for deceptive marketing that promises its customers a silver bullet solution to all their problems. As the word implies, these are considered too good to be true products. Examples of these are lubricant additives commercialized on Home TV shopping performing tests on oil beyond your wildest imagination but only to be recognized by the Federal Trade Commission as a fraud.

Spalling is a fracture of the running surface and subsequent small, discrete material particles removal. It is the pitting or flaking away of bearing material. Spalling can occur in the inner ring, the outer ring of the bearing's rolling elements.

Spectroanalysis (ASTM D5185 and ASTM D4951) is the analysis of metal content and additive package. This test will check 19 elements and reports them in ppm or percentage. This oil analysis tests limit its test from 10 microns and below, and its disadvantage is detecting particles larger than 10 microns.

Sporadic Failures are failures that often indicate sudden large and abrupt deviations from the norm. This refers to catastrophic failures and is the opposite of chronic failures. Usually, their cause is easy to detect since there are just one or a couple of causes.

Stakeholders are a group of people in Root Cause Failure Analysis that will include any of the following; the person being accused, the person whose behavior needs to change, the person who has to spend money, the person accusing, or anyone that the Principal Investigator and evidence gathering team think is involved in the incident.

Start-Up Loss is a type of loss that occurs during starting up the equipment. Start-up loss means that the material loss is caused at the initial stage of product launching, namely the loss caused during the period from a start-up of production to the stabilized production stage.

Strategy is a careful plan or method for achieving a particular goal, usually over a long period. It also includes the carrying out of plans to achieve the goal. It is particularly a long-term plan for success. The strategy also means to maintain and build a competitive advantage over their competition.

Static friction is the force that keeps a motionless object from being pushed or pulled across a surface. The only force that is acting on this block will be the weight of the object.

Strain is the change of shape of the object after compressive or tensile stress is applied, which can also be equal to the Modulus of Elasticity (E), which is equal to the stress divided by the strain of the object. For tensile and compressive stress, the force applied is perpendicular or at 90 degrees to the area. It is also the quantity that describes the amount of deformation that can occur within a material body whenever a load is applied.

Strength in the metallurgical sense is the metal part's property to resist the part's stress imposed upon the part.

Stress is defined as the force per unit area, often considered as the force acting through a small area within a plane.

Stock-Out is when maintenance requests a part in the storeroom where there is an actual inventory in the system, but the actual or physical inventory is zero.

Storeroom's space utilization is equal to the actual space consumed by the storage and spares divided by the total space of the storeroom.

Sweetening is replacing a certain amount of oil so that new additives can be added to the existing used oil inside the equipment. Unlike changing oil, where the oil's total volume is replaced, around 25 to 30% of the used oil will be replaced with new fresh oil during the sweetening process so that new additives will be mixed with the existing oil.

Synthetic oil will fall on Group IV and V Base Stocks for lubricating oil. Synthetic oil is the result of a chemical reaction called synthesis, in which the end result is a uniformly shaped molecules that are more resistant to heat and impossible to achieve through the crude oil refining process compared to mineral-based oil, whose molecules are uneven in size.

TAN / TBN Crossover is where the value of both TAN and TBN becomes equal. When fresh oil enters the service, the TAN will be low, while TBN will be high. As the engine runs, TAN will increase, and TBN will fall. When the TAN value becomes equal to the TBN, the TAN/TBN crossover had been reached which means that it is time to change the oil.

Tangible Measurements or quantitative measurements are those measurements that are easy to quantify and have a direct impact on the bottom line performance of the plant.

Tasks Duplication means that the same task is performed by different people or groups. Decide on who is the best and most qualified to perform the task.

Tensile Stress is when you have an object in a vertical position such as a cylinder, and a downward force is applied to an object, there will always be an opposite force which will be equal to the downward force. Both the upward and downward forces will cause the object to be under tensile stress.

Theoretical Cycle Time also called the Ideal or Design Cycle Time is the Ideal production rate of the machine and is sometimes referred to as the Design Speed. The reciprocal of Theoretical cycle time is the units per hour or UPH.

Tool Algo is an instrument that monitors and records the number of strokes and is stored on the main computer so that tools, dies, punches, or whatever is monitored will be replaced after reaching its dictated stroke or cycles on the machine.

Tool Cutting Blade Change is a loss in the equipment or machine incurred on swapping or changing any consumable tooling item when it has become worn-out, ineffective, or severely damaged.

Torsional Stress is the twisting of an object caused by force acting on the object's longitudinal axis. The twisting effect is known as torsion. Torsional stress can lead to deformation. The twisting is called torque which is also called the twisting moment.

Total Acid Number (TAN) is a measure of the total acid concentration present in a lubricant. Occasionally, the decrease of an additive package may cause an initial increase in the TAN of the fresh oil.

Total Base Number (TBN) is a measure of alkaline concentration present in a lubricant. Engine oil is formulated with alkaline additives to combat the build-up of acids in a lubricant as it breaks down. Once alkaline additives are depleted, the lubricant no longer performs its function, and the engine is at risk of oxidation, corrosion, sludge, varnish, and other problems.

Total Functional Failure occurs when there is a total loss of function on the equipment. This can happen both in the primary and secondary functions of the equipment.

Total Productive Maintenance is a plant improvement methodology system that enables continuous and rapid improvement of the manufacturing process through employee involvement, employee empowerment, and closed-loop measurement of results. TPM is a production-driven improvement methodology designed to optimize equipment reliability and ensure efficient management of its assets. TPM aims to build up a corporate culture that thoroughly pursues production system efficiency, improvement, and Overall Equipment Effectiveness. This methodology originated in Japan.

Training gap is defined as the difference between the skills required to complete the job and the existing skillset of any particular team member. Focusing all our training based on the needs may or may not still contribute to the bottom line results.

Training Needs Analysis (TNA) is the process in which the company identifies the training and development needs of its employees so that they can do their job effectively, depending on the specific needs of that function in the organization.

Treating in oil refineries has several options for their treating processes, but the primary purpose of the majority of them is the elimination of unwanted sulfur compounds. A variety of intermediate and finished products, including gasoline, kerosene, jet fuel, and gases, are dried and sweetened. Sweetening is a major refinery treatment of gasoline that treats sulfur compounds to improve the color, odor, and oxidation stability.

Tribology was coined by Peter Jost in 1966, derived from the Greek word tribos, meaning rubbing, and translating the word literally on the rubbing science. The word tribology was not widely used until the Oxford English Dictionary defines tribology as the branch of science and technology concerned with interacting surfaces in relative motion and associated matters such as friction, wear, lubrication, and design bearings.

Trim is the cutting of the dambar that shortens the leads together. The form is the forming and bending of the leads into the correct shape and position. Singulation is the cutting of the tie bars attached in the individual units to the lead frame resulting in the individual separation of each unit from the lead frame at the end of line process in the semiconductor industry.

Turnaround is a scheduled event where an entire system or process of an industrial plant such as an oil and gas refinery, petrochemical plant, power plant, pulp and paper mill, and similar plants will be shut down. This is usually done at a longer interval.

Unplanned Breakdowns are unexpected failures occurring on the equipment in which the operator will generate a work request where the role of the maintenance or technician is to troubleshoot the equipment until it becomes operational once again.

Uptime is the time in which our equipment and asset are operating. It is also the time during which our equipment and asset are working without failure. Other terms to designate uptime includes operating time, productive time, running time, or time the equipment is utilized.

Utilization is the proportion of time that the equipment is utilized for its intended function and purpose. It is also the time the equipment is operating and running.

Varnish is an oil contaminant made of oxidized or carbonaceous adhesive material covering the engine's internal surfaces. As the oil is oxidized, these compounds form a consistent, hard, and shiny substance. Varnish occurs because of many factors, such as oxidization, friction, and heat. These are the products of the evaporation of the oil.

VED Analysis indicates that the classification is based on the criticality of the spare. Vital (V) means that a spare part will be termed vital if its non-availability will be a very high loss due to production downtime or a high cost will be involved if the part is procured on an emergency basis. Essential (E) will be a moderate loss incurred due to the non-availability of the part. Desirable (D) are those that are desirable if the production loss is not very significant due to its non-availability. Most of the parts that will fall under this category are equipment that has some redundancy making it less critical in the plant.

Velocity is the average speed at which the fluid can move from a given point. The velocity is measured in either mile per hour, feet per second, or in SI or International System of Units. This is an important factor in sizing the hydraulic lines.

Vertical set-up time will include the time the equipment was installed, and the commissioning time until the time it is now ready for the start of operation.

Vibration can be defined by these common terms, such as swinging back and forth, oscillating, unbalanced, and shaking. The vibration occurs when a machine or machine component moves from its neutral or normal position to a lower and upper extreme limit of travel. This happens because the force is applied to the rotating equipment and components. Vibration can also be a cyclic or pulsating motion of a machine or machine components from its point of rest.

Vibration analysis is one of the most common types of Predictive Maintenance techniques that analyzes these plotted vibration signals to diagnose abnormal vibration. This is the most used technique for analyzing the condition of rotating machinery.

Violation occurs when someone knowingly and deliberately commits an error. Violations fall under three categories: routine violation, exceptional violation, and acts of sabotage.

Viscosity is the measurement of a fluid's resistance to flow. Viscosity indicates the oil's capacity to lubricate by providing a thin film to separate metal-to-metal contact. An increase or decrease in viscosity can indicate many things in the oil and the equipment. Viscosity is not a perfect Oil Analysis test as we still need to perform other tests to diagnose the exact problem of why the viscosity of the oil changes.

Viscosity Index (VI) of a lubricant describes how the oil's viscosity will change depending on the temperature change. This means that if the temperatures increase, the viscosity will decrease, and vice versa. The viscosity index is a characteristic used to indicate variations in the viscosity of the lubricating oil concerning temperature changes.

Visual controls are any means or devices used to provide ease in inspection and expose early problems and deviations to the operators. It easily alerts operators of problems in the equipment. It is a business management technique that originated in Japan, where information is communicated using visual signals instead of texts or other written instructions. The design is deliberate in allowing quick recognition of the information being communicated to increase efficiency and clarity. This is also used to spot the anomaly or abnormality on the equipment more easily. Also termed visual management.

Volatility is the property that describes the degree and rate at which the oil will vaporize under given pressure and temperature. This means that the higher the volatility rate, the higher the oil lost due to evaporation.

Waddington Effect means the less invasive the Preventive Maintenance is, the better the maintenance's outcome can be as more maintenance can lead to early failures. This can be said to be the original concept of infant mortality failures.

Water Vapor Transmission Rate (WVTR) is a measure of water vapor passage through a substance. It is a measure of the permeability of vapor barriers.

Water Washout is a test on grease that refers to ASTM D 1264, which covers evaluating the resistance of lubricating grease to water washout from a bearing.

Wear is defined as damage to a solid surface caused by the removal or displacement of materials by any means of mechanical action of a contacting solid, liquid, or gas. It may cause significant surface damage, and the process is usually gradual. The damage is usually thought of as gradual deterioration, correlated with the equipment's operating age.

Wear debris can be defined as metal particles produced from the breakdown of surfaces within a machine. These particles can range from submicron size to chunks of metal as large as those seen by the naked eye.

Wear Metal Debris Analysis Tests is an oil analysis test instrument that indicates the number of metal contaminants present in the oil. By knowing this information, maintenance can understand what parts inside the equipment are on the verge of wearing out. Each of these elements will have its own specific limits, as specified in the oil analysis tables.

Wear-Out Failures or Age-Related Failures are parts that will eventually survive and reach a specific age before they fail. Age specified may be in running hours, time, number of strokes, calendar days, number of revolutions, number of stress applied, or any other form.

Weibull Distribution is often used to describe the lifetime of parts. It is used to analyze and predict failure rates and describe the failure of parts and equipment.

Wire-bond is a semiconductor machine responsible for adhering or welding a thin wire to a bare semiconductor die-chip pad and the other end of the wire to a conductive pad on a substrate. The wire is usually made of 99% pure gold and is usually very thin, often 0.001 to 0.0013 of an inch.

Work-in-process (WIP) Inventory comprises of all the materials, components, parts, assemblies, and subassemblies that are being processed in production or are still waiting to be processed within the system.

Work Order for Preventive Maintenance is a document containing all the information needed to perform and accomplish the maintenance task and provides the process for completing that task. PM Work Orders include the scope of the work to be done, who will perform the task, and other relevant details.

World Class is the ability to compete anywhere globally to meet and beat any competitor anywhere in the world in terms of product, price, quality, and on-time delivery. (Extracted from Terry Wireman's book on World Class Maintenance)

World Class Maintenance Management is the art and science of managing maintenance resources performed by the best-in-class industries worldwide.

Wrench Time is an indicator that is used to measure how much time the maintenance people actually spend on doing actual maintenance work on the equipment. Others refer to this as the tool time or when maintenance is actually holding a tool which of course is not only the wrench.

Yield Losses can be defined as the total loss between the input of raw material and the output of finished goods. This may also include the losses incurred due to reject products or those that had been scrapped.

RSA Maintenance Books Collection in Series

This is not only about a technical book on maintenance; it is a book that makes everyone in maintenance feel proud that they belong to the maintenance function. If you have been living through the day-to-day pressures of doing maintenance, then this is your story. *Rolly Angeles*

I have written these books inspired by the millions of maintenance mankind from different industries who want only the correct maintenance strategies for their equipment and assets to function the way we want. These books are written in series, explaining each discipline's maintenance in detail based on its original concept on World Class Maintenance, The 12 Disciplines.

- Volume 1: World Class Maintenance Management–The 12 Disciplines
- Volume 2: Maintenance–Roadmap to Reliability
- Volume 3: Reliability–A Shared Responsibility for Both Operators and Maintenance
- Volume 4: Cutting–Edge Maintenance Management Strategies
- Volume 5: Problems and Solutions on MRO Spare Parts and Storeroom
- Volume 6: Lubrication Tactics for Industries Made Simple
- Volume 7: Decoding Reliability-Centered Maintenance Process for Manufacturing Industries
- Volume 8: RSA Reliability and Maintenance Newsletter Vault Collection, Subscribers Edition
- Volume 9: Investigating Equipment Failures through Root Cause Failure Analysis
- Volume 10: Maintenance Indices–Meaningful Measures of Equipment Performance
- Volume 11: Implementing Preventive Maintenance for Industries the Right Way

Rolly Angeles Reliability and Maintenance Encyclopedia

Available at https://www.amazon.com/Rolly-Angeles/e/B07B3T1TXC/ref=ntt_dp_epwbk_0

RSA Reliability Website: https://rsaonlinebookstore.company.site/

Draf2Digital: https://books2read.com/ap/RWJydN/Rolly-Angeles

Index

Index

www.ingramcontent.com/pod-product-compliance
Lightning Source LLC
LaVergne TN
LVHW062145200926
848484LV00007B/1859
* 9 7 9 8 8 8 5 2 6 0 0 6 0 *